INTERNATIONAL TECHNOLOGICAL UNIVERSITY
This Book is Donated by:
PROF. WAI-KAI CHEN

Date:

CHAOS
AND
COMPLEXITY

Workshop on

CHAOS AND COMPLEXITY

Torino, October 5 – 11, 1987

Editors

R. Livi
S. Ruffo
Dipartimento di Fisica,
Universita di Firenze,
ITALY

S. Ciliberto
Instituto Nazionale di Ottica,
Università di Firenze,
ITALY

M. Buiatti
Dipartimento de Biologia,
Università di Firenze,
ITALY

INSTITUTE
FOR SCIENTIFIC INTERCHANGE

World Scientific
Singapore • New Jersey • Hong Kong

Published by

World Scientific Publishing Co. Pte. Ltd.
P. O. Box 128, Farrer Road, Singapore 9128

U. S. A. office: World Scientific Publishing Co., Inc.
687 Hartwell Street, Teaneck NJ 07666, USA

CHAOS AND COMPLEXITY

Copyright © 1988 by World Scientific Publishing Co Pte Ltd.

All rights reserved. This book, or parts thereof, may not be reproduced in any form or by any means, electronic or mechanical, including photocopying, recording or any information storage and retrieval system now known or to be invented, without written permission from the Publisher.

ISBN 9971-50-567-3
      9971-50-568-1 pbk

Printed in Singapore by General Printing Services Pte. Ltd.

PREFACE

The only "chance" for the movement is just this disorder, which allows people to speak freely and which can flow into a certain form of self-organization.

From an interview of J.P. Sartre to
D. Cohn-Bendit, May 20 1968.

Statistical and probabilistic concepts appeared in physics about one century ago and led to the formulation of such significant scientific subjects as kinetic theory and later statistical mechanics. Also the study of dynamical systems can be traced back so far, although the discovery of unusual dynamical behaviours was not given the relevance that we recognize now [1]. Since then progress in the understanding of complex phenomena has been confined to specialistic fields of research, sometimes at the frontiers between fundamental and applied sciences. For instance, the rediscovery of the relevance of chaotic motion is due also to the contribution of a meteorologist as E. Lorenz [2].

Only more recently the study of chaos and complexity has assumed a relevance by itself. Increasing areas of fundamental research based on these concepts have developed in different sciences as mathematics, physics, chemistry, biology and computer science. Such concepts as well as self-organization and optimization have become common keywords.

Although these new research fields can no more be framed in the well defined area of statistical mechanics, it is also evident that a clear definition of them in terms of common methodologies and widely applicable concepts does not yet exist.

This workshop was organized with the aim of stimulating a critical discussion about concepts and methods originated in various disciplines around the different qualitative and quantitative definitions of chaos and complexity. In particular, it was our impression that a close contact between the scientific community studying dynamical chaos and the one interested in the statistical behaviour of highly interconnected systems was lacking. This situation was, to us, in patent conflict with the discovery of the important role played by dynamics in the statistical behaviour of systems composed of many objects in mutual interaction. Many features of chaos in low-dimensional systems are widely understood both from the theoretical side and from the experimental point of view [3]. On the other hand, conservative and dissipative systems with a large number of degrees of freedom (possibly infinite) display interesting and *complex* patterns in their dynamical evolution, which still lacks a theoretical understanding. One would like at least to introduce meaningful quantities, which can measure the degree of chaos and complexity of experimentally observed patterns, in order to reduce the efforts of following the dynamical configurations as a whole. The aim is somewhat similar to the one of statistical mechanics, which justifies the introduction of state functions (pressure, temperature), which *summarize* the information contained in the phase-space of the system. But we want to go a step further and retain a bit more of information coming from dynamics, not simply suppose that the phase-space is uniformly filled in time, as substantially statistical mechanics assumes. The new dynamical features, which are taken into account at this higher level of complexity bear important consequences on the statistical behaviour of the system.

The philosophy is the following. Instead of posing quite generic questions as, *what is a complex system? which is the relation between a chaotic and a complex behaviour?* we prefer to follow a more constructive approach trying to understand how such concepts are *operatively* defined in relation with some specific problems. Typical questions that we prefer to pose are: *has the transition to chaos observed in low-dimensional systems something to do with phase transitions in statistical physics? what is the role of disorder in a large system of interconnected objects with respect to its dynamical behaviour? does the violation of ergodicity, present in many disordered and regular extended systems, imply that we must not trust any more the predictions of equilibrium statistical mechanics?* This questions are somewhat more limited in scope, but still very difficult to answer. However we think that answering such kind of questions should at least lead to the clarification of the significance of concepts (or better, *names*) used in different research fields, or even scientific disciplines.

Recognizing analogies among concepts used in different disciplines can be of fundamental importance for establishing cross-fertilization processes. Of course the first step which makes this possible is devising *abstract models*, which capture some relevant features that characterize a well defined level of complexity of a system. One should be always very honest and try to make always explicit the hypotheses which are beyond the abstraction of any mathematical model from physical reality.

We think, in fact, that the task of avoiding illicit and dangerous generalizations is in the hands of the scientist himself (thought also in terms of the community which validates his theories).

1. Theory of Dynamical Systems

At the turning of this century H. Poincaré [4] had already observed that a fully deterministic dynamics does not necessarily imply explicit predictions on the evolution of a dynamical system. This can be considered a milestone in the approach to the study of *dynamical chaos*. The content of the work by H. Poincaré and J. Hadamard is much more conceptually deep and subtle than we have resumed here. Anyway, chaos as an effect of instability of orbits in dynamical systems has remained for a long time a sort of *pure mathematical subject*. Only in the fifties Kolmogorov-Arnold-Moser theorem [5] and the numerical experiment on a chain of nonlinearly coupled oscillators by Fermi-Pasta-Ulam [6] have stressed again the fundamental relevance of dynamical chaos not only on a mathematical, but also on a physical ground. Later on, the works by E. Lorenz [2], M. Henon and C. Heiles [7] and B. V. Chirikov [8] have provided new insights on the origin of chaotic behaviours in dissipative as well as in conservative systems. The main conceptual improvement is the observation that dynamical chaos is not necessarily a consequence of the many degrees of freedom present in a system; on the other hand such a system can well display, under certain conditions, ordered and, at the same time, very complex behaviours.

Violation of ergodicity in hamiltonian systems (and its relation with the foundations of statistical mechanics), limit cycles, strange attractors, multifractal objects have become new central concepts in the realm of dynamical systems. In fact, nowadays the main interest in the study of these systems points towards such a characterization of chaotic behaviour. There is a widespread agreement to consider dynamical chaos as a very rich and complex phenomenon, which in general cannot be naively interpreted in terms of a stochastic process. Mathematically speaking, dynamical chaos corresponds to a typical undecidability problem in the sense of Gödel theorem, as **Agnes** and **Rasetti** elucidate in their contribution through the analysis of a class of dynamical systems. Although the evolution is governed by a deterministic dynamics, the exponential amplification of any uncertainty in determining the state of a system leads to an effective loss of memory. The trajectory has the tendency to fill densely some region in the phase space. Liapunov characteristic exponents are typical quantities, which quantify this divergence of nearby orbits and provide a first global description of the *chaoticity* of the dynamics. As **Isola** shows in his paper such an approach naturally introduces a statistical description of chaotic dynamics in terms of Kolmogorov-Sinai entropy, while a more refined analysis should involve the identification of the typical time scales for the loss of memory by the measure of time correlation functions. A relation between such quantities in some simple hyperbolic systems is discussed by **Badii, Politi, Heinzelmann** and **Meier**. The first two authors in a companion contribution study the spectrum of the fractal dimensions of the strange attractor characterizing the dynamics of the generalized *baker map*. This is a peculiar example of a formal connection between a statistical approach and the global properties of strange sets in the phase space. In this framework *fractal dimensions* have become the dynamical counterpart of scaling exponents in statistical mechanics. Typically, diffusive problems defined in terms of random walks or random surfaces on a lattice are described by such quantities, as **Maritan** discusses in his paper.

These examples suggest that this level of description of the complexity of a dynamical system has assumed the role of an *universal* language, which applies successfully to a large spectrum of physical problems.

One can observe an analogous situation for what Liapunov characteristic exponents are concerned. Numerical and analytical techniques for their determination are used in many problems, ranging from localization in Solid State Physics to random matrices dynamics. The contributions by **Lima** and **Vulpiani** provide an overview on some recent results in this field, e.g. the existence of a limit distribution for the spectrum of Liapunov exponents as the size of the system is increased and their power-law behaviour typical of intermittent processes. Most of the models under examination refer to symplectic dynamics, when a fully chaotic regime is present.

An opposite dynamical situation can be recovered in the study of hamiltonian systems, where non integrability does not necessarily imply ergodicity. **Benettin** shows that in many physically relevant cases hamiltonian dynamics is governed by a *freezing* of the high-frequency degrees of freedom, which forbids equipartition of energy and, as a consequence, a fully chaotic regime. This mechanism of violation of ergodicity is shown to have many consequences on the foundations of statistical mechanics and its classical interpretation. It is important to stress that Nekhoroshev theorem [9] has provided a rigorous description of this *freezing* mechanism in terms of exponentially long time scales, which characterize the approach to equipartition for high frequency modes. Similar results have been obtained by **Alabiso** and **Casartelli** in analyzing the relaxation properties in nonlinear chains of oscillators and in a model of a radiant cavity. Proper dynamical quantities have been introduced as stochasticity indicators. It is particularly interesting to observe that in the radiant cavity model the thermalization of higher modes takes place with an *increasing difficulty*, thus leading to an effective lack of ergodicity.

In their contribution **Coullet, Gil** and **Lega** provide an example of a dynamical description based on a generalization of Landau theory of equilibrium systems. One can stress again the cross-fertilization between dynamics and statistics; in this specific case the transition to a turbulent state in a fluid is described in terms of bifurcations and spatio-temporal patterns, whose chaotic behaviour is driven by the presence of topological defects. It is worth mentioning the importance of this approach in the description of many experimental situations, where spacial, temporal and space-temporal chaos are observed, as shown also in some contributions appearing in these proceedings.

A nice relation between dynamical maps and cellular automata evolution is presented by **Wolf**.

2. Experiments on Spatio-Temporal Chaos

Spatio-temporal chaos is, without any doubt, an example of *complex* dynamics that is not yet completely understood, both from an experimental and a theoretical point of view. In a physical system three cases are possible: i)the system is ordered in space, but chaotic in time; ii)the system does not present an interesting dynamics in time but is disordered in space; iii) the system presents chaos both in space and in time. Each of the three cases may have different levels of complexity. During the conference several experiments have been presented that analyse all of the above mentioned cases. Chaos in Quantum Optics is a quite good example of a system in which the spatial degrees of freedom are in practice frozen, but the dynamics is very rich. A particularly interesting case is the phase chaos in which the laser produces pulses of equal amplitude but at chaotic intervals (**Arecchi**). Instead **Gollub, Simonelli** and **Goldberg** have presented results obtained from an experiment on surface waves. This experiment is very interesting also for the apparatus, whose complete automatization has allowed to study with large precision the phase space of the system. More specifically, it is a good example of how it is possible to understand and to measure the behaviour of a physical system about which very little is known from a theoretical point of view. The other contributions by **Dubois** and **Bigazzi, Ciliberto** and **Rubio** deal with Rayleigh-Benard convection in a fluid heated from below. All of them have shown that low-dimensional chaos observed in small aspect ratio cells (the aspect ratio is the ratio between vertical and horizontal dimensions) is related with the freezing of the spatial degrees of freedom. However **Dubois** gave an interesting example of a very small square cell in which the time dependent behaviour arose, producing a completely unpredictable complex spatio-temporal dynamics. Also **Bigazzi, Ciliberto** and **Rubio** showed that small aspect ratio cells are not completely frozen, but they may produce relevant phenomena either of localization or of propagation inside the cell.

In all of these experiments the role of the noise is of course very important. The influence of noise on the measurements of a bifurcation has been clearly analysed by **Fronzoni** with analogic simulations. This method turns out to be very efficient to study such a problem.

In any case, the common feature of all the talks concerning the experiments was the need of finding mathematical methods apt to analyse the spatio-temporal dynamics more quantitatively than by a simple observation of the patterns. Some methods such as the spectral complexity (**Gollub et al.**), spectral entropy (**Bigazzi et al.**), reduction to a sort of symbolic dynamics (**Bagnoli et al.**) have been proposed, but much theoretical work is needed in order to completely characterize the *complexity* of the spatio-temporal data obtained from the experiments.

3. Networks

The study of equilibrium properties of lattice models is a classical subject of statistical mechanics. Thermodynamic properties, including phase transitions, have been reasonably understood through both approximate (mean field, renormalization group) and rigorous (exact results) methods. Until recently little attention was paid to the dynamics of spin models. The main interest relied in finding simulation dynamics (Monte-Carlo, Langevin), which could reproduce equilibrium results.

The study of disordered models, like spin glasses (which lead in a natural way to violation of ergodicity [10]) has been of fundamental importance in the rediscovery of the role of dynamics. Many of the contributions to this workshop consider spin dynamics as an interesting problem by itself.

The richness of patterns generated by different dynamical evolution, which was before attributed little relevance to statistical properties, become now a way to understand such diverse problems as differentiation and cell development in biology or retrieval of patterns in neurological network models of associative memory.

Finding a deterministic dynamics for simulating the Ising model is an important challenge, since this would reduce computer time needs significantly with respect to stochastic algorithms. However, dynamical correlation effects are present, for instance in the Q2R cellular automaton rule, which, at the present time, prevent us from trusting the derivation of equilibrium statistical properties. Such an effect is related to the scaling behaviour of periods analysed in the paper by **Stauffer**. An interesting transition is moreover present in the scaling of periods with system size (from exponential to power law) both in Kauffman models [11] (see also **De Arcangelis**) and neural networks. This reminds very much the study of Poincaré return times in systems of coupled oscillators, performed long time ago by P. Mazur and E. Montroll [12]. Possible relations with similar transition effects present in continuous-time Hamiltonian dynamics remain to be explored [13].

The short review paper by **Nadal** is a nice introduction to the subject of pattern retrieval and storing capacity of neural networks; the original part concerns short term memory models and the question of how a memory "forgets". **Opper** discusses exact results for storing capacities and learning times in Hopfield models [14], while **Kurten** proposes that the experimentally observed activity dependent synaptic efficiency bears some consequences on the ordered or chaotic behaviour of neural networks. Finally **Serra** studies the phase space of modified Hopfield models. Although it is not stated explicitly in anyone of these contributions, it seems to us that they are all trying to fill the gap between theoretical models and realistic situations in *human-like* brains. This is clearly a challenging path, which is stimulating an interesting and fruitful interaction among physicists, neurophysiologists and psychologists. However, one should also take into account the difficulty of *real* experiments (see also **Madar** in this volume), which is likely to be the bottleneck of this line of research. This is probably the reason why people are considering other biological networks, like the immunological network described in the talk by **Parisi**.

The powerful simulated annealing technique [15] is used in two different problems: training a boolean network how to perform additions (**Patarnello** and **Carnevali**), and optimizing random binary sequences (**Bernasconi**). The striking feature is the ability of the network to *generalize*, a question which was not previously faced in Hopfield-type models. **Bernasconi**'s paper is a thorough investigation of how to choose the coding schedule according to the specific thermodynamic behaviour of the considered models.

If one can extract a common moral from all the contributions to this section, we should say that they are all puzzled by the questions of how to extract information from the richness of dynamical patterns, which could be relevant to statistical behaviour in some asymptotic limit (thermodynamic or infinite time).

4. Cellular Automata

Cellular automata were introduced in 1948 by J. Von Neumann and S. Ulam [16]. They were *designed* as fully discrete dynamical systems evolving on a lattice according to some local synchronous updating rule. The main task of these new *computing tools* was to provide a description of the emergence of global organization and complexity present in many systems in terms of simple *microscopic* evolution rules.

Cellular automata have been thought just from the beginning as parallel universal computing machines, particularly useful in numerical simulations. In this sense they have opened new perspectives in the numerical description of mathematical, physical and biological models: rather than adapting an intrinsically discrete object as a digital computer to describe models defined in terms of continuum mathematics (which necessarily imply an imprecise floating-point arithmetics) one could *reverse* this attitude and try to construct fully discrete models, which can be exactly computable on an electronic digital device.

Probably the best known example of a cellular automaton is the so-called *Game of Life*, proposed by J.H. Conway in 1970 [17]. This is a typical example of a simple local evolution rule determining a universe where life is an effect of a delicate balance between *overcrowding* and *isolation*. A typical trend towards self-organization is observed, which is characterized by the *birth* of ordered patterns from initial random distributions of living cells in a certain landscape.

Anyway, we want to observe that Cellular Automata have remained for some time a sort of *exotic* field of research, mainly related to discrete mathematics of k-valued algebras. Only more recently the widespread attitude in the design of dedicated electronic machines has renewed the interest of scientists from different disciplines on this objects. In the last decade we have been faced with a rapid growth of this field of research, which has enlarged its range of applications to many problems like Ising model, metastability/instability transitions, fluid dynamics, pattern recognition, immunological networks, etc. (For a review on such applications see ref. [18]).

An up to date recognition on some of the most relevant developments in modeling complex systems by cellular automata is contained in the contribution by **Vichniac**, where many aspects related to fundamental and technical problems are discussed. One of the main achievements of cellular automata models is the simulation of fluid instabilities by a lattice gas dynamics. This represents an alternative method in solving Navier-Stokes equations [19] (with some limitations, as discussed also by some authors in this book). **D'Humieres, Lallemand, Boon, Dab** and **Noullez** discuss in their paper the 2-D model on a triangular lattice and provide a large amount of information about numerical simulations by a dedicated machine and theoretical results up to the extension of such a model for the treatment of diffusion phenomena. Limits and perspectives (mainly related to the construction of 3-D models) are reviewed in detail.

A generalization of the 2-D version of this lattice gas model is proposed by **Chopard** and **Droz**. Their purpose is the introduction of temperature (through the equipartition theorem) in the model in order to obtain a system ruled by thermo-hydrodynamic equations. It is worth mentioning that heat propagation in a fluid at rest is well described by such a model, although some physical problems arise when the fluid is in motion due to the lack of Galilean invariance (an unavoidable disease in lattice models).

The same authors have presented another contribution on the analysis of non equilibrium phase transitions in terms of cellular automata. In particular, they have proposed a model for describing adsorption processes on a catalytic surface; numerical simulations on a dedicated machine provide results for the scaling exponents which are in good agreement with the theoretical predictions coming from mean-field analysis.

Another interesting field in cellular automata theory refers to the definition of probabilistic evolution rules. Probabilistic cellular automata rules have already been considered by many authors [20] and in some interesting cases have been interpreted as dynamical systems in D dimensions *describing* D+1 dimensional statistical systems. More precisely, it has been shown that a probabilistic cellular automaton evolution pattern can be mapped into an equilibrium configuration of a corresponding spin problem on a lattice. This naturally leads to the characterization of different evolution patterns by the typical language of phase transitions. Such an approach has been proposed by H. Chaté and P. Manneville [21] for modeling spatio-temporal chaos in fluids.

Along this line of thought **Bagnoli**, **Ciliberto**, **Francescato**, **Livi** and **Ruffo** have presented the study of a cellular automaton probabilistic rule describing a *phase transition* from localized to delocalized turbulent flow for a fluid heated from below in an annular cell.

All the contributions to this section have shown the impressive improvements of the applications of cellular automata to modeling many relevant dynamical problems. Qualitative and quantitative predictions are now obtained also for what experimental situations are concerned. Nevertheless, as many authors have stressed, there are still many open problems, which, in our opinion, do not reduce simply to technical points.

5. Biology, Evolution Theory and Biological Networks

The history of biological theories, for a number of reasons, including the complexity of life itself, has followed a path which is at variance with those of other sciences like, for instance, physics. Ancient theories were descriptive and strongly influenced by philosophical and religious prevailing currents of thought as a direct consequence of the fact that they have a direct bearing on our conceptions of man. Attempts to develop "*independent*" theories and to give them a predictive value through the use of mathematical tools and concepts are much more recent and can be considered to have become productive only in this century. The *modern synthesis* of the darwinian theory of evolution with the mendelian model of inheritance and the subsequent development of population genetics theories along with the construction, on the basis of the results of molecular biology of a series of models ranging from the origin of life to the development of individual organisms, may be considered examples of such a tendency.

More recently, however, many of these models, based on the extrapolation to complex systems of single component rules, on underestimation of the relevance of non additive interactions, random fluctuations, hierarchical rules (constraints) etc., have shown themselves to be inadequate in many cases. The overall crisis of scientific paradigmas has thus attained biology and led to an open discussion which seems far from a satisfactory conclusion. The papers included in this volume are part of this discussion and range from critical analysis of existing models to new specific proposals.

Thus **Buiatti** summarizes recent evidences in the fields of development and evolution and critically compares it to mechanistic interpretations derived from the central dogma; **Kauffman** describes models based on the concept that evolution is a combinatorial process and on the assumption that selection acts in combination with self organization on families of rugged adaptive landscapes. **Peliti** discusses models of the emergence of biological order derived from statistical mechanics, ranging from that of Dyson based on the study of the behaviour of the population of monomers to that of Anderson, which takes into account replication and competition between quasi-species in Eigen's hypercycles, to that of Abbot derived again from the spin glass system. Finally **Parisi** presents a model of immunological memory and compares it with spin glass models and what is known on neural networks, while **Madar** reports on some experimental results obtained on excitability in cortical brain slices.

Unfortunately we cannot include in this volume the reports on the very interesting talks by P. Bergé, E. Bienenstock, F. Fogelman- Soulié, A. Georges, P. Grassberger, R. Nobili, R. Swendsen and G. Weisbuch..

We gratefully acknowledge the hospitality and the financial support of the Institute for Scientific Interchange in Torino, where this workshop was held from 5 to 10 October 1987. We also thank L. Peliti, A. Vulpiani and G. Weisbuch for having shared with us the organization of the workshop.

A particular thank goes to M. Rasetti for his enthusiasm, which convinced us to organize this workshop and to T. Bertoletti and C. Novella for their numerous advices and very qualified secretary work.

M. Buiatti, S. Ciliberto, R. Livi and S. Ruffo

Firenze, May 1988

References

[1] D. Ruelle, *Determinisme et predicibilité*, Pour la Science, Aout 1984, p.58.

[2] E.N. Lorenz, Journ. of Atm. Sciences **20**, 130 (1963).

[3] P. Bergé, Y. Pomeau, Ch. Vidal, *L' ordre dans le chaos*, Hermann, Paris, 1984.

[4] H. Poincaré, *Sciences et methode*, Flammarion, Paris 1908.

[5] A. N. Kolmogorov, Dokl. Akad. Nauk. SSSR **98**, 527 (1954); V. I. Arnold, Russ. Math. Surveys, **18**, 9 (1963); J. Moser, Nachr. Akad. Wiss. Göttingen Math.-Phys. Kl., 2 **1**, 15 (1962).

[6] E. Fermi, J. Pasta, S. Ulam, in *Collected Papers of E. Fermi* , University of Chicago, Chicago 1965, vol. II, p. 978.

[7] M. Henon , C. Heiles, Astron. Journ. **69**, 73 (1964).

[8] B. V. Chirikov, Journ. Nucl. Ener. **C1**, 253 (1960).

[9] N. N. Nekhoroshev, Russ. Math. Surveys **32**, 1 (1977).

[10] G. Parisi, Lectures given at the 1982 "Les Houches" Summer School *Recent Advances in Field Theory and Statistical Mechanics*, North Holland (1984).

[11] S. Kauffman, Journ. of Theor. Biol. **22**, 437 (1969).

[12] P. Mazur and E. Montroll, Journ. of Math. Phys. **1**, 1 (1960).

[13] S. Isola, R. Livi, S. Ruffo, Europhys. Lett. **3**, 407 (1987).

[14] J. J. Hopfield, Proc. Nat. Acad. Sci. USA **79**, 2554 (1982).

[15] S. Kirkpatrick, C. D. Gelatt Jr., M. P. Vecchi, Science **220**, 671 (1983).

[16] J. Von Neumann, *The General and Logical Theory of Automata*, in J. Von Neumann Collected Works, A. H. Taub ed.

[17] M. Gardner, Scientific American **223:4**, 120 (1970).

[18] Physica **D10**, 1-274 (1984).

[19] U. Frisch, B. Hasslacher, Y. Pomeau, Phys. Rev. Lett. **56**, 1505 (1986).

[20] W. Kinzel, Zeits. für Phys. **B58**, 229 (1985).

[21] H. Chaté, P. Manneville, Phys. Rev. Lett. **58**, 112 (1987).

CONTENTS

Preface   v

Theory of Dynamical Systems

C. Agnes, M. Rasetti, *Complexity, Undecidability and Chaos:
A Class of Dynamical Systems with Fractal Orbits*   3

S. Isola, *Understanding Complex Behaviour.
Some Remarks on Method and Interpretation*   26

A. Politi, R. Badii, K. Heinzelmann, P. F. Meier,
*Decay of Correlation Functions and Expansion Rates in
Dynamical Systems*   36

R. Badii, A. Politi, *Phase-Transitions in
Hyperbolic Dynamical Systems*   42

A. Maritan, *Hierarchical Random Networks*   49

R. Lima, *Lyapunov Exponents and Co.*   64

A. Vulpiani, *Lyapunov Exponents for Products of Random
Matrices in Condensed Matter and Dynamical Systems*   74

G. Benettin, *A Completely Classical Mechanism for the "Freezing" of
High-Frequency Degrees of Freedom*   84

C. Alabiso, M. Casartelli, *Complexity and Relaxation in Nonlinear Chains and in a Model of Radiant Cavity*   93

P. Coullet, L. Gil, J. Lega, *Transitions in Systems Far from Equilibrium: A Ginzburg-Landau Approach*   99

M. Wolf, *A Complex Behaviour of the Gas of Particles Moving According to the Equation* $x_{n+1} = (Ax_n + B) \bmod C$ 113

Experiments on Spatio-Temporal Chaos

F. T. Arecchi, *Shil'nikov Chaos in Lasers*   121

J. P. Gollub, F. Simonelli, J. D. Goldberg, *Advances in Experimental Nonlinear Dynamics: Space/Time Patterns*   142

M. Dubois, *Dynamic of Rayleigh-Benard Spatial Configurations in Confined Geometry*   152

M. A. Rubio, S. Ciliberto, P. Bigazzi, *Transition between Localized Oscillations and Traveling Waves in Thermal Convection*   157

L. Fronzoni, *Analogue Simulation of Stochastic Processes by Means of Minimum Components Electronic Devices*   171

Networks

D. Stauffer, *Periods, Damage Spreading and Multifractality in Cooperative Systems*   185

L. De Arcangelis, *Global and Local Periods in Kauffman Cellular Automata*   201

J. P. Nadal, *Neural Networks: A Path from Neurobiology to Psychology?*   208

M. Opper, *Statistical Mechanics of Learning in Neural Network Models*   219

K. E. Kürten, *"Training" Quasirandom Neural Networks*   224

R. Serra, G. Zanarini, F. Fasano, *Generalized Hopfield Learning Rules*   230

S. Patarnello, P. Carnevali, *Boolean Networks which Learn to Compute*   235

J. Bernasconi, *Optimization Problems and Statistical Mechanics*   245

Cellular Automata

G. Y. Vichniac, *Cellular Automata and Complex Systems*   263

D. d'Humières, P. Lallemand, J. P. Boon, D. Dab, A. Noullez, *Fluid Dynamics with Lattice Gases*   278

B. Chopard, M. Droz, *Cellular Automata Model for Thermo-Hydrodynamics*   302

M. Droz, B. Chopard, *Nonequilibrium Phase Transitions and Cellular Automata*   307

F. Bagnoli, S. Ciliberto, A. Francescato, R. Livi, S. Ruffo, *Cellular Automaton Model for a Fluid Experiment*   318

Biology, Evolution Theory and Biological Networks

M. Buiatti, *Information Flux and Constraints in Development and Evolution: A Critical View*   331

S. A. Kauffman, *Origins of Order in Evolution: Self-Organization and Selection* 349

L. Peliti, *Statistical Mechanical Models of the Emergence of Biological Order* 388

G. Parisi, *Immunological Memory in a Network Perspective* 394

I. Madar, *On Slow Rythmical Fluctuation of Neuronal Excitability in Cortical Brain Slices: An Experimental Study* 407

List of Participants 423

Theory of Dynamical Systems

COMPLEXITY, UNDECIDABILITY AND CHAOS :
A CLASS OF DYNAMICAL SYSTEMS WITH FRACTAL ORBITS

Corrado Agnes and Mario Rasetti

Dipartimento di Fisica del Politecnico, Torino, Italy

1. Introduction

There exist mathematical statements formally consistent within the logical scheme of an axiomatic theory which can neither be proved nor disproved [1]. This is one of the possible definitions of undecidability, a disturbing notion which shockingly entered the realm of mathematics and mathematical logic in 1931 with the pioneering work of **Kurt Gödel** and **Emil Post**.

In the years there were several different formulations and extensions of Gödel's incompleteness theorem, which played an essential role in leading to the conception and rigorous definition of such notions as those of algorithmic complexity[2], randomness[3], and code information content in computation science and information theory. In this latter context the theorem was related - mainly by the work of **Michael Rabin**[4] and of **Gregory Chaitin**[5] - to Turing's analysis of the computation process[6], and in particular to the so called halting problem, namely the question of deciding whether a given program in a universal computing machine (referred to as Turing machine) will ever halt.

It is just through the notions of algorithmic complexity, unsolvability of halting problems, undecidability of word problems that Gödel's theorem and the concept of unpredictability by undecidability have recently entered the context of physics, mainly in connection with the theory of deterministic chaos in dynamical systems.

Physical processes transform the initial state of a system, determining its evoluted states according to physical laws. Theoretical models of physical processes, on the other hand, are computational schemata whereby :

i) the state of the system is identified by a configuration W in some phase space Γ, a configuration consisting in the collection of all relevant data to identify in one-to-one way the physical state (its information theoretical meaning is straightforward) ;

ii) a set of algorithms modelling the physical laws, i. e. such that the data evolve by their action in such a way as to maintain the one-to-one correspondence between phase space configurations and physical states.

Such algorithms can be characterized in a very broad sense by two features: they are iterative and intrinsically non - linear. In computational language their action corresponds to a **Turing's** process P, and it is possible to encode the corresponding program by strings S_α of binary digits. We refer to the ensuing representation as code(P). If S_{in} is the string corresponding to a given initial physical state represented by W_0 in Γ, P the program that describes the evolution algorithm, and S_{fin} the string associated with the output state W_1, one says that the integer n - sum of the lenghts of S_{in} and of code(P) - is the complexity of the dynamical system when there is no number smaller than n for which the program started with S_{in} will eventually halt with S_{fin}.

The concept of complexity thus defined has bearings on the prediction capacity of the model as well as on its capability of giving reliable results in a controllable amount of time. Randomness in physical systems is thereby essentially related to the NP - completeness[7] (non- polynomial, i.e. exponential complexity) of the related problem, namely to algorithmic irreducibility.

Chaos on the other hand is an ubiquitous phenomenon, appearing - often unexpectedly - even in very simple dynamical systems (i.e. systems with a low number of degrees of freedom and characterized by relatively simple dynamical evolution laws). It can be ascribed to the high sensitivity of the system dynamics to the initial conditions, which implies an exponential amplification of even infinitesimal uncertainities in such initial conditions due to the non - linearity of the evolution law, and to an extended notion of ergodicity, whereby the whole phase space is accessible from any initial point.

More specifically chaos can be related to the existence of positive **Lyapunov** exponents[8] and can be characterized by the existence of strange attractors[9], which may be schematically thought of as the product of a **Cantor** set and an interval. All of this leads to a rigorous definition[10] based on the homeomorphism of chaotic dynamical systems with transitive topological mappings whose symbolic dynamics is isomorphic to that of **Smale** horse-shoe systems, invariant under **Bernoulli shifts**[11].

Recently these latter notions, together with the idea, due to **Marston Morse**[13] and beautifully revived by **Caroline Series** and **Rufus Bowen**[14], of coding orbits in terms of a group alphabet, i.e. of words in a finitely presented group, have been utilized by the authors[15] to prove rigorously that indeed, at least for a class of dynamical systems which are defined in terms of geodesic flows on Riemann surfaces of negative curvature, chaos is formally equivalent to undecidability, in the sense of **Gödel** and **Post**, of the word problem for a particular group : the mapping class group of the surface modulo its limiting set.

In the present note, stimulated by the latter examples we construct a new class of abstract dynamical systems, originated by undecidable recursive schemes of number theory, and show how they exhibit chaotic dynamical behaviour.

Once more a conceptual identification between deterministic chaos and undecidability in the sense of formal logic emerges.

The dynamical systems considered are not Hamiltonian and their orbits turn out to be fractals (i.e. self - similar discontinous varieties). The onset of chaos has in two cases the feature of intermittency, in the others those of a fully developed turbulence.

2. The Sequences

The following conjecture, which has been *undecidably* attributed to either **Stanislaw Ulam** or **Lothar Collatz**, and has bearings in number theory, is the origin of the first sequence considered. We shall refer to the latter as **Ulam** sequence. Starting from any natural number N_0, the elements of the sequence are generated by the recursive rule[16]:

$$S_u : \begin{cases} N_{n+1} = \frac{1}{2} N_n & \text{if } N_n \text{ is even} \\ N_{n+1} = 3 N_n + 1 & \text{if } N_n \text{ is odd} \end{cases} \qquad (2.1)$$

The conjecture (which has been checked for N_0 up to $7 \cdot 10^{11}$) [17] states that for any N_0 the sequence shall terminate after a finite, possibly arbitrarily large, number n_f of steps, at $N_{n_f} = 1$. More precisely, the sequence hits for $n = n_f$ a limit cycle of order three given by $\{1, 4, 2, 1\}$.

Proving the undecidability of the conjecture is straightforward, by mapping it onto a problem equivalent to the halting problem of a **Turing** machine, which is known to be undecidable[4].

To this end one writes first N_n as a binary number,

$$N_n = \sum_{l=0}^{L_n} a_l^{(n)} 2^l \quad ; \quad a_l^{(n)} \in \mathbb{Z}_2 \quad , \qquad (2.2)$$

where

$$L_n = \left[\log_2 N_n \right] \quad , \qquad (2.3)$$

[x] denoting the largest integer $\leq x$. Thus N_n is represented by a string of lenght $L_n + 1$ of 0's and 1's. Now the **Ulam** sequence (2.1), which can be written in

compact form as

$$S_\mathbf{u} : N_{n+1} = \frac{1}{2} N_n(5N_n + 3) - (5N_n + 2)[\frac{1}{2} N_n] \qquad (2.4)$$

gives raise to a recursive relation among strings. Indeed, upon inserting (2.2) in (2.4), one gets the following recurrent generating scheme for the binary string $\{a_0^{(n)}, a_1^{(n)}, \cdots, a_{L_n^0}^{(n)}\}$:

$$a_k^{(n+1)} = a_0^{(n)}\{a_k^{(n)} + 3\sum_{j=0}^{k-1} 2^{-(k-j)} a_j^{(n)} + 2^{-k}$$

$$- \sum_{j=1}^{k-1} 2^{-(k-j)} a_j^{(n+1)}\} + (1 - a_0^{(n)}) a_{k+1}^{(n)} \quad ; \quad \mod 2 \qquad (2.5)$$

Eq.(2.5) holds for $0 \leq k \leq L_{n+1}$, besides $L_{n+1} \leq L_n + 2$.

Notice that, if $a_0^{(n)} = 0$ (i.e. if N_n is even), the recursion (2.5) is but a **Bernoulli** shift. Moreover, one needs a global information on the first portion of the new string in order to be able to decide each new digit of it, as well as its lenght.

Thus no algorithmic procedure of lenght less than $\sim n^n$ can be devised to decide whether for a given n ,N_n equals 1 (or equivalently $a_0^{(n)} = 1, a_0^{(k)} = 0, \forall k : 1 \leq k \leq L_n + 1$).In other words the problem is NP-complete.

The input string $\{a_0^{(0)}, \cdots, a_0^{(L_0)}\}$ toghether with the instructions for the recursive scheme coded in binary form in (2.5), constitute the abstract (**Turing**) machine [18] mimicking, by the sequence of its internal states, the sequence of operations that take place in the process of computation.

The halting problem, namely the problem of distinguishing by a finite procedure which - if any - among the internal configurations of the machine lead to the program eventual halting, and which cause the program to compute forever, has been proved to be undecidable for any **Turing** machine by **Michael Rabin**[4].

The proof sketched above therefore holds for all the sequences which shall be considered in the sequel, for all of which the states are mapped into strings over a finite alphabet.

The next sequence considered, referred to as **Conway** sequence,has the following definition.Starting from any natural number M_0, one generates recursively the other elements by the rule $(n \geq 0)$:

$$S_c : \begin{cases} M_{n+1} = \frac{3}{2} M_n & \text{if } M_n \text{ is even} \\ M_{n+1} = \left[\frac{1}{4}(3M_n + 1)\right] & \text{if } M_n \text{ is odd} \end{cases} \quad (2.6)$$

Notice how (2.6) in fact is more subtle than **Ulam** sequence, in that it discriminates not only between even and odd integers, but between odd integers of the form $(4k+1)$ and $(4k+3)$ as well. Indeed (2.6) could be written as follows:

$$M_{n+1} = \begin{cases} 3k & \text{if } M_n = 2k \\ 3k+1 & \text{if } M_n = 4k+1 \\ 3k+2 & \text{if } M_n = 4k+3 \end{cases}$$

Representing, as in the previous case, M_n in binary form,

$$M_n = \sum_{\ell=0}^{L_n} c_\ell^{(n)} 2^\ell \quad ; \quad c_\ell^{(n)} \in \mathbf{Z}_2 \quad , \quad L_n = \left[\log_2 M_n\right] \quad (2.7)$$

formulas (2.6) can be cast in the form of recursion relations among the strings $\{c_0^{(n)}, c_1^{(n)}, \cdots, c_{L_n}^{(n)}\}$:

$$c_k^{(n+1)} = c_0^{(n)} \left\{ c_{k+2}^{(n)} + 3 \sum_{j=2}^{k+1} 2^{-(k-j+2)} c_j^{(n)} + 2^{-(k-c_1^{(n)})} \right\}$$

$$+ (1 - c_0^{(n)}) \left\{ c_{k+1}^{(n)} + 3 \sum_{j=1}^{k} 2^{-(k-j+1)} c_j^{(n)} \right\} - \sum_{j=0}^{k-1} 2^{-(k-j)} c_j^{(n+1)} \quad , \quad \text{mod } 2 \quad (2.8)$$

with $L_{n+1} \leq L_n - 1$ and $0 \leq k \leq L_{n+1}$.

The third sequence considered, denominated **Kay** sequence, is a generalization of **Ulam**'s sequence. Initiating with any natural number Z_0, letting p and q be two prime numbers, with $p > q \geq 2$ and setting $r = p - q$, the elements of the sequence are denoted by:

$$S_k : \begin{cases} Z_{n+1} = \frac{1}{p} Z_n & \text{if } (p \setminus Z_n) \\ Z_{n+1} = q Z_n + r & \text{if } (p \nmid Z_n) \end{cases} \quad (2.9)$$

where the symbol $(m \backslash \ell)$ means that integer ℓ is exactly divisible by integer m, and of course $(m \not{\backslash} \ell)$ its negation.

Now one expands Z_n as a p-adic number,

$$Z_n = \sum_{\ell=0}^{\Lambda_n} b_\ell^{(n)} p^\ell \quad , \quad b_\ell^{(n)} \in \mathbf{Z}_p \quad , \qquad (2.10)$$

namely one represents the number Z_n as a string $\{b_0^{(n)}, b_1^{(n)}, \cdots b_{L_n}^{(n)}\}$ of length $L_n = [\log_p Z_n]$, over the alphabet $\{0, 1, \cdots, p-1\}$. In terms of strings the recursive scheme (2.9) writes:

$$b_\ell^{(n+1)} = b_{\ell+1}^{(n)} \qquad \text{if } b_0^{(n)} = 0 \quad , \qquad (2.11)$$

whereas, if $b_0^{(n)} \neq 0$,

$$b_\ell^{(n+1)} = \{q b_\ell^{(n)} + r_{\ell-1}^{(n)}\} \quad \mod p \qquad (2.12)$$

for $2 \leq \ell \leq \Lambda_n$, with

$$r_\ell^{(n)} = \left[\frac{1}{p}(q b_\ell^{(n)} + r_{\ell-1}^{(n)})\right] \qquad (2.13)$$

and

$$b_0^{(n+1)} = \{q(b_0^{(n)} - 1)\} \quad \mod p$$

$$b_1^{(n+1)} = \{1 + q b_1^{(n)} + r_0^{(n)}\} \quad \mod p$$

$$b_{\Lambda_n+1} = \left[\frac{1}{p}(q b_{\Lambda_n-1}^{(n)} + r_{\Lambda_n-1}^{(n)})\right] \qquad (2.14)$$

where

$$r_0^{(n)} = \left[\frac{1}{p}q(b_0^{(n)} - 1)\right]$$

$$r_1^{(n)} = \left[\frac{1}{p}(1 + qb_1^{(n)} + r_0^{(n)})\right] \tag{2.15}$$

One can check that $L_{n+1} \leq L_n + 1$

Finally we define the sequence that we name after both **Conway** and **Kay**. Starting from any initial natural integer K_0 and letting once more q and p be primes, now such that $2 \leq p < q < p^2$, and $r = q - p$, the sequence is generated by

$$S_{\text{ck}} : \begin{cases} K_{n+1} = \frac{q}{p}K_n & \text{if } (p \setminus K_n) \\ K_{n+1} = \left[\frac{1}{p^2}(qK_n + r)\right] & \text{if } (p \not\setminus K_n) \end{cases} \tag{2.16}$$

Notice that (2.16) coincides with (2.6) for $q = 3$ $p = 2$.
Upon writing

$$K_n = \sum_{\ell=0}^{\ell_n} d_\ell^{(n)} p^\ell \quad ; \quad d_\ell^{(n)} \in \mathbf{Z}_p \quad , \quad \ell_n = \left[\log_p K_n\right] \quad , \tag{2.17}$$

the relations in $\mathbf{Z}_p$ corresponding to (2.16) are now, setting $\alpha = [q/p]$ and $\beta = q \bmod p$:
if $d_0^{(n)} = 0$,

$$d_\ell^{(n+1)} = \{\alpha d_\ell^{(n)} + \beta d_{\ell+1}^{(n)} + s_\ell^{(n)}\} \bmod p \tag{2.18}$$

for $1 \leq \ell \leq \ell_n - 1$, with

$$s_\ell^{(n)} = \left[\frac{1}{p}(\alpha d_{\ell-1}^{(n)} + \beta d_\ell^{(n)} + s_{\ell-1}^{(n)})\right] \tag{2.19}$$

for $2 \leq \ell \leq \ell_n - 1$, and

$$s_1^{(n)} = \left[\frac{1}{p}\beta d_1^{(n)}\right] \qquad (2.20)$$

and moreover

$$d_0^{(n+1)} = \{\beta d_1^{(n)}\} \quad \mod p$$

$$d_{\ell_n}^{(n+1)} = \{\alpha d_{\ell_n}^{(n)} + s_{\ell_n}^{(n)}\} \quad \mod p$$

$$d_{\ell_n+1}^{(n+1)} = \left[\frac{1}{p}(\alpha d_{\ell_n}^{(n)} + s_{\ell_n}^{(n)})\right] \qquad (2.21)$$

whereas,
if $d_0^{(n)} \neq 0$,

$$d_\ell^{(n+1)} = \{\alpha d_{\ell+1}^{(n)} + \beta d_{\ell+2}^{(n)} + t_\ell^{(n)}\} \quad \mod p \qquad (2.22)$$

for $0 \leq \ell \leq \ell_n - 2$, with

$$t_\ell^{(n)} = \left[\frac{1}{p}(\alpha d_\ell^{(n)} + \beta d_{\ell+1}^{(n)} + t_{\ell-1}^{(n)})\right] \qquad (2.23)$$

for $1 \leq \ell \leq \ell_n - 1$, and

$$t_0^{(n)} = \left[\frac{1}{p}\left((\alpha - 1) + \alpha d_0^{(n)} + \beta d_1^{(n)} + \frac{1}{p}\beta(1 + d_0^{(n)})\right)\right] \qquad (2.24)$$

Moreover,

$$d_{\ell_n-1}^{(n+1)} = \left\{\alpha d_{\ell_n}^{(n)} + t_{\ell_n-1}^{(n)}\right\} \mod p$$

$$d_{\ell_n}^{(n+1)} = \left[\frac{1}{p}\left(\alpha d_{\ell_n}^{(n)} + t_{\ell_n-1}^{(n)}\right)\right] \quad (2.25)$$

Notice that $\ell_{n+1} \leq \ell_n + 1$.

3. The Maps

We associate now an abstract one-dimensional dynamical system, defined as a map of the unit interval onto itself, for each of the sequences described in the previous section. More precisely each of the systems considered will turn out to map the subset $\mathbf{Q}_0$ of the rationals less than one into itself.

The definition proceeds as follows. To each natural integer N associate the string $S \equiv \{A_0, A_1, \cdots, A_\lambda\}$ of digits in $\mathbf{Z}_p$ corresponding to its expansion as adopted in the sequence considered ($p = 2$ for **Ulam** and **Conway**, p prime > 2 for **Kay**, and ≥ 2 for **Conway-Kay**):

$$N = \sum_{l=0}^{\lambda} A_l p^l \quad , \quad \lambda = [\log_p N] \quad .$$

Assign successively to S the rational number

$$Q = \sum_{l=0}^{\lambda} A_l p^{-(l+1)} \quad , \quad A_l \in \mathbf{Z}_p \quad . \quad (3.1)$$

This induces a one-to-one correspondence between N and Q.

The map $\mu_\alpha : \mathbf{Q}_0 \longrightarrow \mathbf{Q}_0$ is then constructed by connecting Q to the rational P whose corresponding integer is $N' = S_\alpha(N)$, where $\alpha = \mathbf{u}, \mathbf{c}, \mathbf{k}, \mathbf{ck}$ labels the sequences taken into consideration.

As an example, notice that if **Ulam** conjecture were true, the associated map should have a limit cycle of order three:

$$\{1/2 \to 1/8 \to 1/4 \to 1/2\} \quad .$$

Šarkovskiĭ's theorem [19] states that if a map has a periodic point of period k, it also has other periodic points whose periods are all the positive integers which follow k in the reordering of the field of natural numbers in which ascending odds come first, descending powers of two come last, and ascending odds times (ascending) powers of two in between. In particular, if a map has a periodic point of period three, it also has infinitely many other periodic points whose periods are all the natural integers; a condition which implies chaos [20]. It is interesting that, even though for the **Ulam** map **Šarkovskiĭ's** theorem cannot be assumed to hold, in that it requires that the map is at least piecewise smooth and monotonic - which is not the case, as will be shown in the sequel - nonetheless μ_u exhibits chaos.

Looking at the graphic representation of μ_u, Fig.1, one can guess a mechanism for the onset of chaos, which is indeed, unexpectedly, very closely correlated to the **Šarkovskiĭ's** ordering, that we write explicitly here for the sake of reference:

$$3, 5, 7, \cdots, 2\cdot 3, 2\cdot 5, \cdots, 2^2\cdot 3, 2^2\cdot 5, \cdots, 2^n\cdot 3, 2^n\cdot 5, \cdots, 2^m, 2^{m-1}, \cdots, 2^2, 2, 1.$$

If in S_u the input integer N_0 is a power of two, then the iteration of the sequence generating algorhitm makes it drop to one,touching all the integers to its right in the above ordering. This corresponds to strictly deterministic laminar regime. If on the other hand N_0 is of the form $2^n \cdot q$, q odd, than the iteration of S_u makes it jump leftward n times, touching all the blocks on its left at the homologous points $2^k \cdot q$, with $k = n-1, n-2, \cdots, 1, 0$, until it hits the odd q in the first block. Thus far the process is once more in a way laminar, but now the representative integer is reinjected from the odds into the even numbers, and where it lands, namely how the process will continue is an event which can be predicted only if the whole past history of the process has been followed in detail.

In other words the **Ulam** sequence acts on **Šarkovskiĭ's** ordered integers as a horse-shoe shift in a quite natural way. The features described imply that μ_u exhibits a chaotic dynamics which is similar to intermittency,even though the map representing it is nowhere differentiable [21]. One can check from Fig 1. and Fig. 2, as well as from (2.5) and (3.1) that

$$\mu_u = 2q \quad \text{for} \quad 0 < q < \frac{1}{2}, \tag{3.2}$$

identical, as expected, with a **Bernoulli** shift; whereas for $1/2 < q < 1$ the graph representing $\mu_u(q)$ is an everywhere singular function, all of whose points are isolated points, i.e. it is nowhere differentiable.

It is immediate to guess from the figures that $\mu_u(q)$ it is indeed a discrete fractal curve, in which a characteristic arrow-head shape is repeated by self-similarity infinitely many times, and whose **Hausdorff** dimension [22] is as

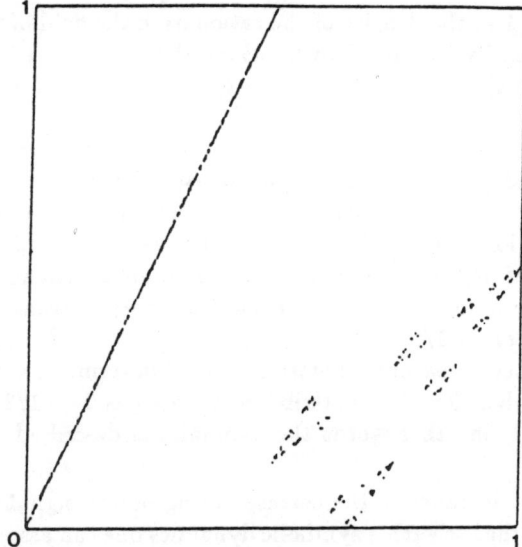

Fig. 1 – The full scale Ulam map.

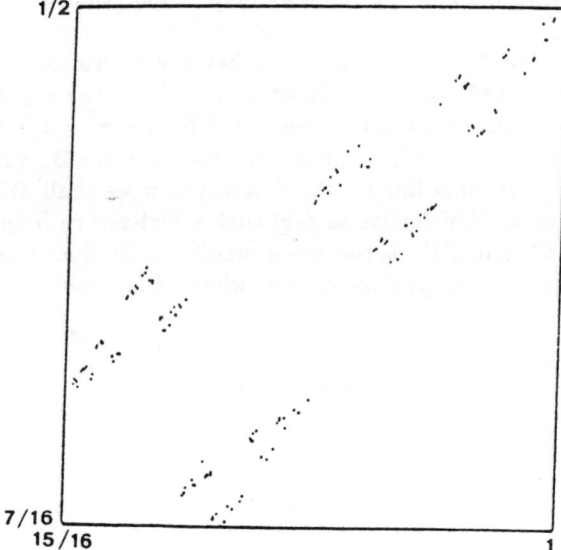

Fig. 2 – Fractality by self-similarity of the Ulam map.

infinitely close to 1 as the density of the rationals in the field of reals. The shape of the fractal is easily explained by checking that for each point the two lines of slopes 3/4 and 4/3 contain the images of the point itself under the squared action of μ_u.

The dynamics induced by μ_u has a typical intermittency feature: the part of the map beyond $q = 1/2$, whenever hit produces a random reinjection in the region $0 < q < 1/2$. On the other hand the representative point may sit in the latter for quite a long time before returning to the right hand side.

The iterates of μ_u versus n exhibit therefore an alternance of laminar and chaotic behaviour accordig to whether the trajectory is transiting in $0 - 1/2$ or in the fractal region $1/2 - 1$.

Such intermittency feature is better described in terms of a symbolic dynamics where the symbol (level) 0 is attributed to any $q \in 0 - 1/2$ and the symbol 1 to $q \in 1/2 - 1$ (in other words the dynamics is described only in terms of $a_n^{(0)}$).

Fig.3 shows an instance of the corresponding binary signal spectrum.

We show now that to such a symbolic dynamics one can associate a continous orbit, in an abstract configuration space, which may be knotted, and whose structure is connected once more to a word problem. The construction holds not only for the **Ulam** map but for the **Kay** map as well, thus we shall describe it in general.

We need first a few basic definitions. To begin with we shall consider the first-return (or **Poincaré**) map associated to μ_α: fixed the separation point $m \in \Im$ of the two basins of attraction ($m = 1/2$ for μ_u, $m = 1/p$ for μ_k), the image $\mu_k(q)$ of any $q \in \Im - \{m\}$, is reprojected onto the same $\Im$, which can then be thought of as a branching line for the abstract flow we shall define. For the sake of simplicity we simply denote as $f(q)$ such a first-return map.

We further recall that if Σ_A is the space (under the product topology) of all doubly-infinite sequences of symbols $n_i \in A$, where A is some alphabet

$$\mathbf{n} = \cdots n_{-1}, n_0, n_1 \cdots , \tag{3.3}$$

the shift map $\sigma : \Sigma_A \to \Sigma_A$ is defined by[11]

$$\{\sigma(\mathbf{n})\}_i = n_{i+1} , \quad i \in \mathbf{Z} . \tag{3.4}$$

Given an $n \times n$ matrix M, where $n = dim A$, whose enties are 0's and 1's, let $\Sigma(M)$ denote the subset of Σ_A consisting of all the sequences n such that for each i, the element $m_{n_i,n_{i+1}}$ of M is 1. Then the shift map σ leaves $\Sigma(M)$

invariant. We call the restriction of σ (which is the full **Bernoulli** A-shift) to $\Sigma(M)$ the subshift of finite type corresponding to M.

We can now specify the symbolic dynamics sketched above in a more precise way.

Let $q \in \mathfrak{I}$ and define the finite (or possibly infinite) sequence

$$\aleph = \aleph_0(q), \aleph_1(q), \aleph_2(q), \cdots \quad (3.5)$$

by

$$\aleph_0(q) = \begin{cases} x & \text{if } q \text{ is to the left of } m \\ 0 & \text{if } q = m \\ y & \text{if } q \text{ is to the right of } m \end{cases} \quad (3.6)$$

and by

$$\aleph_i(q) = \begin{cases} x & \text{if } f^{(i)}(q) < m \\ 0 & \text{if } f^{(i)}(q) = m \\ y & \text{if } f^{(i)}(q) > m \end{cases} . \quad (3.7)$$

$\aleph_i(q)$ is defined if and only if $f^{(i)}(q)$, the ith iterate of f is defined:

$$f^{(i)}(q) = \overbrace{f(f(\cdots f(q)\cdots))}^{i \text{ times}}$$

Notice that the sequences $\aleph(q)$ can be lexicographically ordered by setting $x \prec 0 \prec y$ (where the symbol $\prec$ means "precedes"). Let now Λ be a periodic orbit, if any exists, under μ_α of minimum period ν, and q_0 the leftmost point of Λ on the branch line $\mathfrak{I}$. Then $f^{(\nu)}(q_0) = q_0$, and $\aleph(q_0)$ is a periodic sequence of the form $\aleph_0(q_0), \aleph_1(q_0), \cdots, \aleph_{\nu-1}(q_0), \aleph_0(q_0), \cdots$.

Define $W = \aleph_0(q_0), \aleph_1(q_0), \cdots, \aleph_{\nu-1}(q_0)$. W cannot be periodic, because if it were we should have, say, $W = U^l$ with $\nu = l \cdot k$, and the orbits of q_0 and $f^{(k)}(q_0)$ would lie together on the same side of the point m after ν iterations of f, and hence after all possible iterations of f. This is clearly impossible because the orbit cannot remain forever on the right hand side of m (it is indeed reinjected to the left whenever it touches there) and on the left hand side the "derivative" of f, which can be defined by a straightforward limiting procedure, is > 1.

Thus the first time, say j, when the orbit of q_0 and that of $f^{(i)}(q_0)$, for some $i = 1, 2, \cdots, \nu - 1$ are on different sides of m, we have $f^{(j)}(q_0) < m < f^{(i+j)}(q_0)$, whence $W(q_0) \prec W(f^{(i)}(q_0))$, $i = 1, 2, \cdots, \nu - 1$.

There follows that the map $q \to \aleph(q)$ is a one-to-one order preserving correspondence between the points of the branch set and the lexicographical ordering of the set of all sequences defined as in (3.5)-(3.7) and such that $\aleph(q)$ terminates with $\aleph_j$ if $\aleph_j = 0$. Moreover the periodic orbits of μ_α correspond with the cyclic permutation classes of finite aperiodic words $\{W\}$ in the free monoid generated by x and y: we denote such classes "cyclic words".

By following a periodic orbit one straigtforwardly finds its "word" W and the corresponding subshift-generating matrix M : the question arises whether this process is reversible, namely if one can devise a procedure whereby given a finite set of distinct aperiodic cyclic words $W_1, W_2, \cdots, W_n$ (i.e. cyclic words such that **no W_i is a cyclic permutation of W_j for $i \neq j$**) one can reconstruct an equivalent orbit for μ_α.

By equivalent we mean "topologically equivalent", in the sense that one can find a manifold H, "holder of the orbit", supported on the branch line,such that the set $\{W_i\}$ essentially generates its fundamental group $\pi_1(H)$, and such that the orbit on H induced by the word structure intersects $\mathfrak{F}$ at the same points as the map μ_α.

We find that the answer is positive, and that one can actually design an algorithm whereby such orbits can be drawn on H and identified with a collection of knots, links and open self linked curves.

The algorithm runs as follows. Denote by $\tilde{W}_i$ the infinite periodic word $W_i \circ W_i \circ \cdots$ and let R be the operation which removes the first letter from a word. Define then recursively

$$R^{(0)} = I \quad , R^{(1)} = R \quad , R^{(j)} = R \circ R^{(j-1)} \quad .$$

There follows that , due to the assumed aperiodicity of W_j's and the fact that no W_j is a cyclic permutation of W_i, the set

$$\Omega = \left\{ R^{(j)} \circ \tilde{W}_i \mid j = 0, 1, \cdots \mid i = 1, 2, \cdots, n \right\}$$

has λ distinct words in it ($\omega_k \quad k = 1, 2, \cdots, \lambda$), where λ is the sum of the lengths of the words $W_1, W_2, \cdots, W_n$.

The words ω_k can be alphabetized according to the usual lexicographical ordering $x \prec y$, and one notices that the points $\aleph^{-1}(\omega_k)$ occur in $\mathfrak{F}$ in just this order. Thus - up to obvious topological equivalence - one can trace the orbits corresponding to W_i by starting at the point representative of $\tilde{W}_i$; connecting it to the right or left on H depending on whether $\tilde{W}_i$ begins or ends with x or

y, and then continue recursively, after describing a curve looping around one or the other of two "foci" L and R , boundary points of $\mathfrak{I}$, connecting then the point $R^{(i)} \circ \tilde{W}_i$ to the representative of $R^{(2)} \circ \tilde{W}_i$ in the same manner and so on.

Note finally that the trajectories hitting the branch line can be divided in two groups, a group of p left strands, circling around L, and a group of $q = \lambda - p$ right strands circling around R, such that strands in the same group never cross one-another, whereas strands in one group – say the left – always pass over those in the other. The pair of integers (p, q) and the permutation π induced by the first-return map f clearly define a braid B, closed into a link or a collection of links L_k.

The permutation π is completely determined by the collection of integers $\{\pi_1, \pi_2, \cdots, \pi_p\}$ where $\pi_i = \pi(i)$, since the remaining q follow from the natural ordering of the strings.

If π is cyclic, $W = x^{n_1} y^{m_1} x^{n_2} y^{m_2} \cdots x^{n_t} y^{m_t}$ and $\Lambda(W)$ is the unique periodic orbit of μ_α associated with W, π is recovered from W by the inequalities

$$f(q_0) < \cdots < f^{(n^1)}(q_0) > f^{(n_1+1)}(q_0) > \cdots > f^{(n_1+m_1)}(q_0) < \cdots$$
$$\cdots < f^{(n_1+m_1+n_2)}(q_0) > \cdots < \cdots > f^{(n_1+m_1+\cdots+n_t+m_t)}(q_0) \qquad (3.8)$$

if q_0 is the leftmost point of $\Lambda(W)$ on $\mathfrak{I}$, in that the same inequalities must be satisfied by the powers of π acting on 1:

$$\pi(1) < \cdots < \pi^{n_1}(1) > \pi^{n_1+1}(1) > \cdots > \pi^{n_1+m_1}(1) < \cdots$$
$$\cdots < \pi^{n_1+m_1+n_2}(1) > \cdots < \cdots > \pi^{\Sigma_{i=1}^t (n_i+m_i)}(1) \quad . \qquad (3.9)$$

If π is not cyclic, one has simply to repeat the whole argument for each of its cyclic factors.

It was conjectured by **Joan Birman** and **Robert Williams**[23], who studied by similar methods the **Lorenz** map, that if one denotes by τ the sum $\Sigma_j t_j$ where t_j are the numbers of syllables in W_i (a syllable being a maximal subword of the form $x^{n_j} y^{m_j}$), then the link L_k may be represented by a braid of τ strands.

It is easy to check that also τ can be computed from π: in fact t_j is the number of x symbols in W_j followed by y symbols, hence

$$\tau = \mathbf{cardinality} \ \{\pi_j \in \pi \,|\, \pi_j > p\} \qquad (3.10)$$

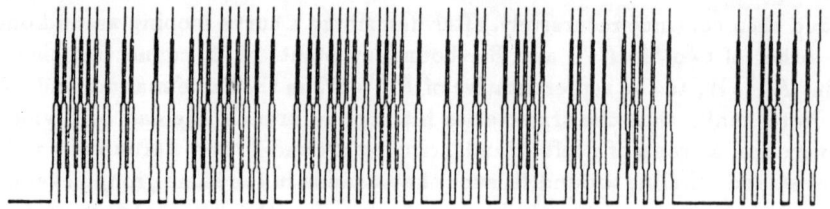

Fig. 3 – Intermittent binary signal spectrum for the **Ulam** map.

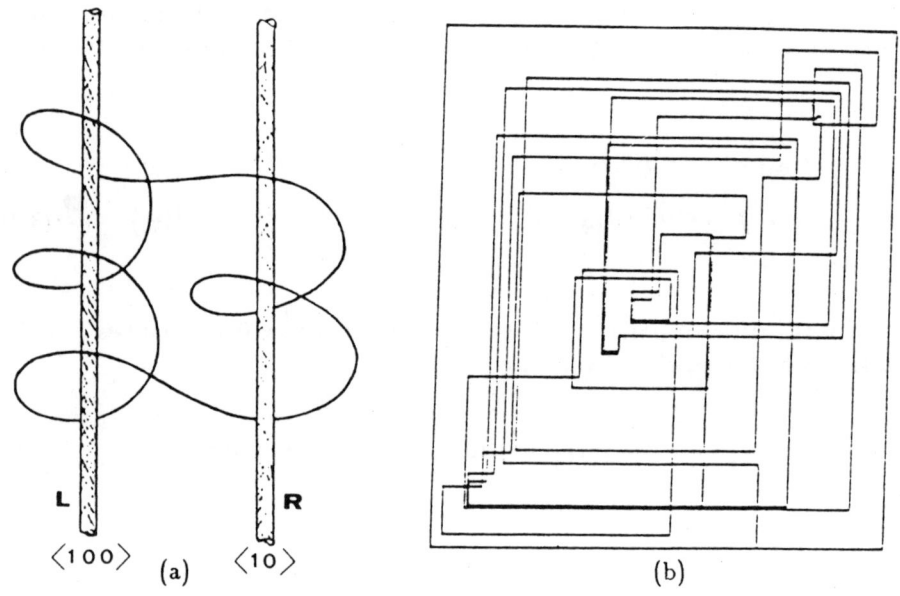

Fig. 4 – (a) : an **Aizawa**'s link for the **Ulam** map ;
(b) : a full topological link L_k for the **Kay** map.

It is worth pointing out how the construction sketched is much more effective than the customary one.

From the sequence represented in Fig.3 one simply gets a sequence of binary digits, where the appearance of 1's is isolated between repeated appearance of 0's :

$$1, \{0\}^{n_1}, 1, \{0\}^{n_2}, \cdots, 1, \{0\}^{n_l}, \cdots$$

One could, as done by **Yoji Aizawa**[24] , qualitatively ascribe a "focus" to each binary level, and construct a closed curve in 3-space, such that its winding number around the left focus equals$(n_{2l} + 1)$ and that around the right focus $(n_{2l+1} + 1)$.

The construction presented provides a much more exhaustive description of the topological properties of these new abstract orbits, and allows to associate with the fractal map μ_α a strange attractor in an extended space. Fig.4 compares the two procedures. Notice that the existence of such an attractor and its topological complexity are once more connected with the decidability of a word problem.

A behaviour similar to that of **Ulam**'s system is shown by **Kay**'s map, as Figs.5 and 6 illustrate.

Quite different is the behaviour of **Conway**'s and **Conway-Kay**'s maps, where the laminar part, in which the return map has positive derivative, is missing (see Figs. 7-10). Here we expect fully developed chaos, of turbulent nature. We conjecture the existence of knotted pathological orbits, represented by wild knots such as e. g. those connected with **Alexander**'s horned sphere or **Antoine**'s necklace fundamental group[25] (notice that **Antoine**'s necklace is homeomorphic with a **Cantor** set, as is the set of non-locally-flat points of **Alexander**'s sphere).

4. Conclusions

We constructed a class of very simple dynamical systems which exhibit chaos as the effect of undecidability (or NP-completeness) of an associate recursive algorithm. The latter is in turn interpretable as a code defining a **Turing** machine, whose complexity is intractable. It turns out that the models proposed indeed generate, by such simple assumption of undecidability, all known mechanisms of onset of chaos. In particular the intermittency phenomenon, the knotting and braiding of orbits, the fractality of **Poincaré** phase space portrait, full developed turbulence as untameness of knotted trajectories were shown (the latter in fact only conjectured).

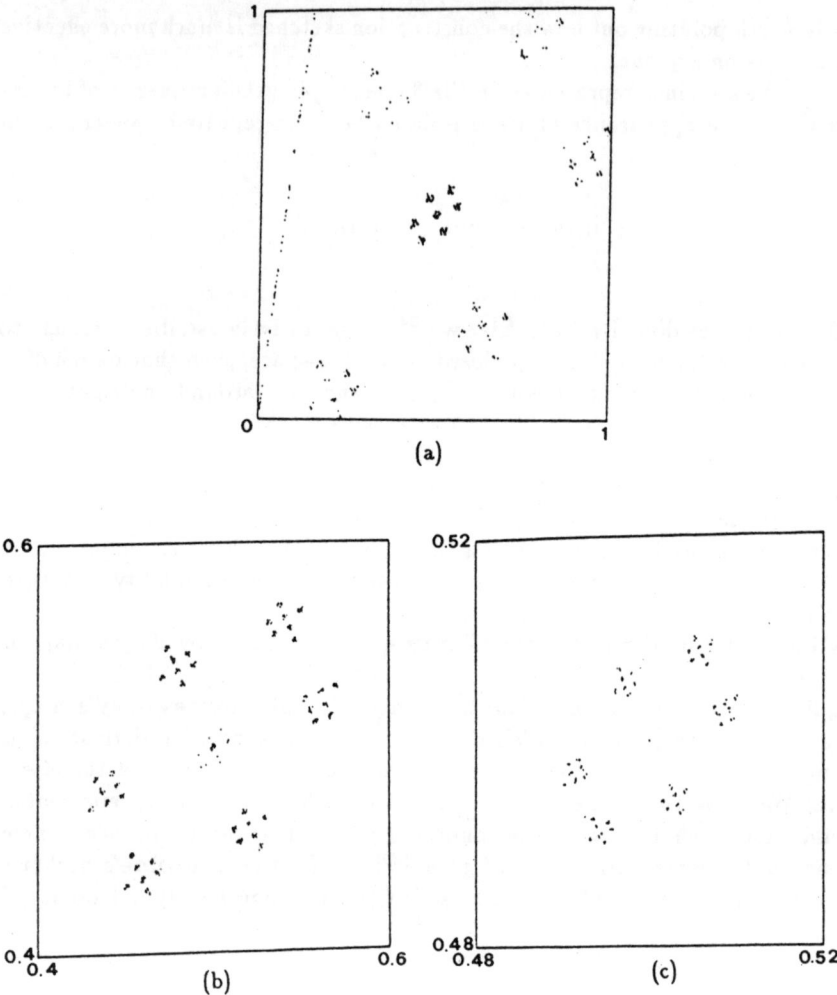

Fig. 5 – The **Kay** map : full scale [(a)] and fractality by self-similarity [(b) and (c)], (for $p = 7$ $q = 5$).

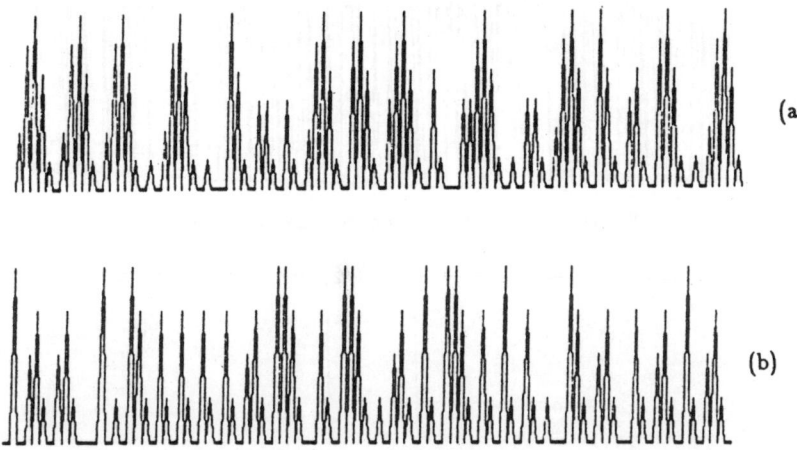

Fig. 6 – Intermittent binary signal spectrum for the **Kay** map [(a) : $p = 7\ q = 5$; (b) : $p = 5\ q = 3$].

Fig. 7 – The full scale **Conway** map.

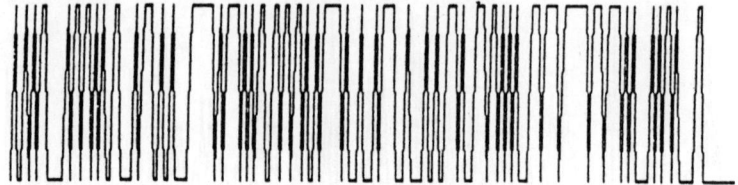

Fig. 8 – Binary signal spectrum for the Conway map.

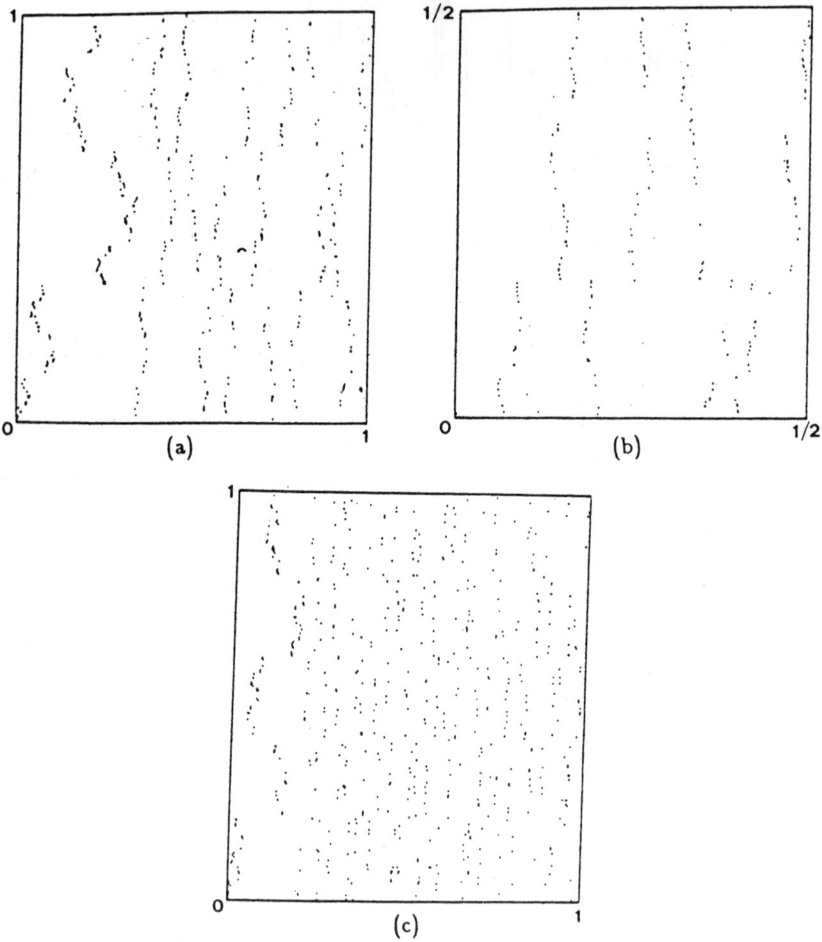

Fig. 9 – The full scale [(a) : $q = 5\ p = 3$; (c) : $q = 7\ p = 5$] and blow-up [(b), of (a)] of the **Conway-Kay** map.

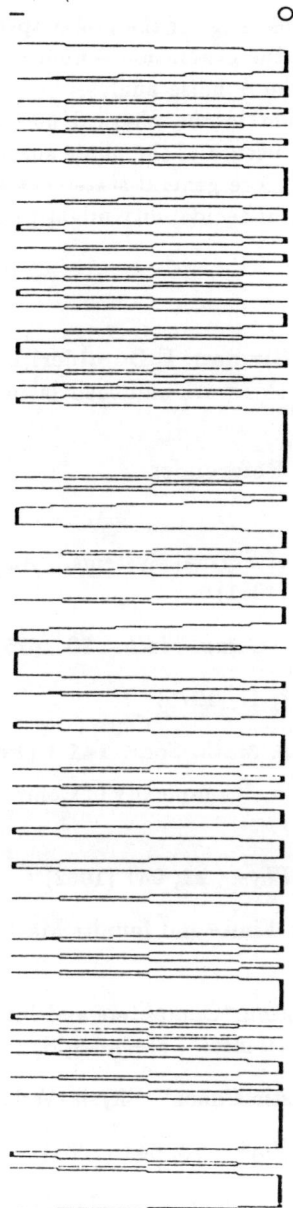

Fig. 10 – Binary signal spectrum for the Conway-Kay map.

We expect that further analysis, e. g. of the power spectrum of such events as depicted in Figs.3,6,8 and 10, the evaluation – numerical or analytical – of the **Lyapunov** exponents, and the $\frac{1}{f}$ noise analysis by renormalization group tecniques, may give further insight in the physical meaning of such models. Moreover, since once again the structure of the phase space is given in terms of word-problem, we believe that a more general statement of the equivalence of deterministic chaos and Gödelian undecidability might be close at hand.

Aknowledgements

The authors wish to thank **J.Birman, E.Bombieri, L.Kauffman, R.Livi** and **S.Ruffo** for their interest in this work and their inspiring comments.

References

1. R. M. Smullyan, "Theory of Formal Systems," *Ann. Math. Stud.* **47**, Princeton University Press (1961)

2. M. O. Rabin, *Comm. Assoc. Comp. Mach.* **20**, 625 (1977)

3. G. Chaitin, *Sci. Amer.* **232**, 47 (1975)

4. M. O. Rabin, *Transact. Ann. Math. Soc.*, **141** 1 (1969)

   M. J. Fisher and M. O. Rabin, "Complexity of Computation," *SIAM-AMS Proc.* **7**, 149 (1974)

5. G. Chaitin, *Intl. J. Theor. Phys.* **21**, 941 (1982)

6. M. Minsky, "Computation : Finite and Infinite Machines," Prentice Hall, Englewoods Cliffs (1967)

7. M. R. Garey and D. S. Johnson, "Computers and Intractability : A Guide to the Theory of NP–Completeness", Freeman, San Francisco (1978)

8. H. G. Schuster, "Deterministic Chaos", Physik-Verlag, Weinheim, (1984)

9. J. P. Eckmann, *Rev. Mod. Phys.* **53**, 643 (1981)

10. C. Agnes and M. Rasetti, Chaos and Undecidability: a Group Theoretical View, in *Advances in Statistical Mechanics*, A. I. Solomon ed. , World Scientific Publ., Singapore (1988)

11 S. Smale, *Bull. Ann. Math. Soc.* **73**, 747 (1967)

12 D. S. Ornstein and B. Weiss, *Israel J. Math.* **14**, 184 (1973)

13 M. Morse, "Symbolic Dynamics"; (mimeographed notes by R. Oldenburger of lectures given in 1937-1938); The Institute for Advanced Study, Princeton (1966)

14 R. Bowen, *I. H. E. S. Publ. Math.* **50**, 269 (1979)

   C. Series, *Acta Math.* **146**, 103 (1981)

15 C. Agnes and M. Rasetti, Word Problem, Undecidability and Chaos, in *Chaotic Behaviour in Non Linear Systems*; A. Vulpiani et al. eds. ; World Scientific Publ. , Singapore (1987)

16 C. Agnes and M. Rasetti, Undecidability and Chaos : a Finite State Example, in *Chaos in Education*, G. Marx ed. ; Publ. National Center for Educational Technology, Vészprèm (1987)

17 P. Halmos, private communication.

18 A. M. Turing, *Proc. London Math. Soc., Ser. 2*, **43**, 544E (1936)

19 A. N. V. Šarkovskĭi, *Ukr. Mat. Z.* **16**, 61 and **17**, 104 (1964)

20 Y. Oono,*Progr. Theor. Phys.* **59L**, 1028 (1978)

   Y. Oono and M. Osikawa, *Progr. Theor. Phys.* **64**, 54 (1980)

21 P. Manneville and Y. Pomeau, *Physica* **1D**, 219 (1980)

22 P. Grassberger, *J. Stat. Phys.* **19**, 25 (1981)

23 J. S. Birman and R. F. Williams, *Topology* **22**, 47 (1983)

24 Y. Aizawa, *Progr. Theor. Phys.* **70**, 1249 (1983)

25 D. Rolfsen, "Knots and Links", Publish or Perish, Berkeley (1976)

Understanding complex behaviour.
Some remarks on method and interpretation.

by Stefano Isola

Dipartimento di Fisica, Univ. di Firenze, I-50125, Firenze, Italy

> Ce qui est nouveau dans la
> "raison nouvelle", ce n'est pas
> tant la raison elle-même, que la
> façon dont nous l'utilisons pour
> "expliquer" le rèel.
> Henri Atlan[1]

Over the last years, in many fields of scientific investigation, a certain number of tacit beliefs on the behavior of natural systems have been brought out for examination. The "discovery of complexity" has created a new set of ideas on how very simple systems may give rise to very complex behaviours - and in many cases the "laws of complexity" have been found to hold universally, caring not at all for the details of system's constituents. Hence, we heard very often speaking about the birth of a "new science" which would produce a sort of cut across scientific disciplines offering a universal way of seeing structured patterns where formerly only noise and stochastic fluctuations had been observed.

The purpose of this paper is not to discuss the likelihood of this thesis but rather to utilize an example of "complex object" in order to stress the general problems (problems of method, but also of interpretation) that it generates. The opinion of the author is that, though the adjective "complex" has appeared in several scientific disciplines (often with very different meanings), this does not constitute a relevant problem in itself, at least from the point of view of scientific rationality. More precisely, there is no reason for which the discovery of some new physical/mathematical objects qualified as complex, together with new procedures to describe them, should involve a different rationality, providing a more comprehensive, "complex", vision of the world.

If the notion of complexity deserves some interest it is just for it allows to stress the very originality of modern science, that is the new interplay between abstraction and concreteness in explaining our observations. In fact, on one hand it is quite clear that whenever some observed facts have to be explained they have to be formerly introduced in some conceptual (i.e. abstract) frame, where they can be suitably handled by theoretical tools. Nothing new in this. But on the other hand it is undeniable that, approaching "complex" systems, the *a priori* selection of the "relevant properties" plays an essential role. *Something becomes observable only when it has become logically possible in an abstract context.* In this sense we can speak of a new relation between theoretical activity and empirical reality, where the former plays its role in selecting meaningful models and the latter is thought, and finally perceived, as a model *among many possible* models.

The case of chaotic behaviour

Now, let us see how the argument sketched above applies to a particular case of complex behaviour, that is the *chaotic* behaviour. Nearly two decades after Ruelle's claim that turbulence in fluids might have some strong relation with that infinitely tangled abstraction called "strange attractors"[2], the fact that many nonlinear dissipative dynamical systems do not approach stationary or periodic states asymptotically has become a fairly accepted notion. The nonlinearities present in the problem can in fact produce an extreme sensitivity with respect to the initial state of the system, so that even a small error on its knowledge will be exponentially amplified as the time goes on. A system with this property (i.e. *sensitive dependence on initial condition*) is *chaotic*: despite its deterministic nature it undergoes a sort of lack of memory about its past history so that precise predictions on the future becomes impossible. If a chaotic time evolution approaches an attractor asymptotically, then this attractor will be *strange*.

Strange attractors provide a good example of something which is becomed *observable* after its "logical possibility" has been encountered in abstract contexts; in this case the possibility being that a sequence of Hopf bifurcations leads, "likely", to a dynamical regime with sensitivity to initial conditions.

Then, the recognition of strange attractors has provided numerical explorers of a clear program to carry out: wherever a natural process seems to be behaving randomly, the presence of low-dimensional deterministic chaos can be tested. Indeed, it has been found that many systems coming from different domains have time evolutions that, although irregular and chaotic, nevertheless make them sharing common qualitative properties about their asymptotic behaviour. So, it seems that one of the most promising directions of research should be the systematic applications of ideas and techniques of non linear dynamics to all kinds of natural systems, including those of biology, economics and social sciences. For instance, at present there is who assembles huge amounts of stock market data and begins searching strange attractors there (see for example [3]). Others argued that earth's weather might lie on a strange attractor, but this raised a controversy [4], [5].

However, we are not interested in promoting metaphysical statements about the ubiquity of deterministic chaos. The problem here is to point out the methodological and conceptual novelties that arise whenever we study a system which exhibits it. Let us make this more precise. Usually, the question of understanding a set of observations which constitute what we consider a physical system will be answered by making it possible abstracting and formalizing these observation. This is considered as an explanation. A fairly general procedure for the mathematical study of a physical system starts with explication of the *space of states* of that system (for instance a compact manifold M), which will be finite dimensional if the system has a finite number of degrees of freedom. Then, the mathematical object which describes how the system evolves in time will be a *dynamical system* or a first order ordinary differential equation on the space of states.

Now, one side of the problem is that the formal representation of a chaotic time evolution with a differential equation on a suitable space does not eliminate at all its unpredictability and then the uncertainty about its behavior. On the other hand, starting from an experimental chaotic time series of the form $\{x(t_i)\}_{1\leq i \leq N}$ (which represents the time evolution of an observable of the system monitored at fixed time intervals), it is generally possible to reveal if this series could be generated by a deterministic non linear evolution equation, but not to specify how this equation looks like. Namely we can extract the deterministic nature of the time evolution (along with other information, see later) but any precise identification is generally impossible. This double-faced situation is the core of the problem stated above. It concerns the meaning of *understanding* a system whose observation cannot provide any identification with a precise mathematical object, like a differential equation, and, anyhow, even a precise identification is practically useless (it looses its *pertinence* in the sense of [6]) because it cannot provide any prediction of the observations. Then, the crucial point here is to reformulate what has to be considered as useful. In other words, what are the "good questions" for such a system. This is actually the turning point which lets the notion of complexity to make sense.

The statistical analysis of time series for nonlinear systems

Now follows a brief discussion on the current shape of the *statistical* approach to chaotic time evolutions in order to give some insight into general ideas and techniques (for reviews on the subject see [7], [8]). Let us give firstly some definitions.

Consider a time evolution

$$\vec{x}(t) = f^t \vec{x}(0), \qquad \vec{x} \in M$$

where M is a d-dimensional manifold (for example the Euclidean space E^d; note that for real experiments d is often very high if not infinite) and f^t is a map on it (which is unknown in experimental situations!). An *attractor* is a subset of M toward which almost all sufficiently close trajectories converge asymptotically, covering it densely as the time goes on. An *observable* is any continuous function $\phi : M \to R$.

Then, an experiment consists in picking an initial point $\vec{x}(0)$ on M and plotting $\phi(\vec{x}(t))$ as a function of the time t. The challenge is to extract physical sense from such a graph.

Usually one does not have access to time evolution of the d-dimensional vector $\vec{x}$. Instead one follows only one component of it. Therefore the typical object one deals with in physical experiments as well as in computer simulations is a *time series* of the form $\{x(t_i), i = 1, \ldots, N; x(t_i) \in R\}$. A general method which may provide a characterization of the attractor for a time series obtained by monitoring a scalar signal for a finite time $t_N = T$, is to *embed* the latter in a space of finite dimension d_e (the *embedding dimension*), and to reconstruct a d_e-dimensional orbit. This can be done for example by vectors of the form

$$\vec{x}_e(t_i) = (x(t_i), x(t_i + \tau), \ldots, x(t_i + (d_e - 1)\tau))$$

with τ some fixed interval. By this technique one can try to reveal if the time evolution has a deterministic nature by plotting the reconstructed d_e-dimensional orbit and looking for a d_e-dimensional projection of a "well structured" attractor. Once obtained a good representation of the attractor one can cut it along a *Poincaré section* and, given a point x on this section, construct the *first return map*, namely the map which will bring x to Px, again in the Poincaré section. Whenever the map P can be constructed, one has essentially succeeded in extracting a sort of deterministic model for the time evolution.

However, this procedure of geometric interpretation of a chaotic signal is feasible only when precise geometrical information about the shape of the attractor are accessible. In other words it is restricted to those cases where there is a low-dimensional fairly "simple" attractor. In order to extract useful information from a time evolution which is extremely complicated or lives on a relatively high-dimensional attractor a statistical approach is suitable. In this context one focuses its attention on the invariant probability measure ρ generated by the asymptotic density of points $f^t x$ in the space M, rather then on the attractor itself. In physical application this density is usually well defined, namely the time average

$$\langle \phi \rangle = \lim_{T \to \infty} \int_0^T \phi(\vec{x}(t)) dt$$

does exist for suitable initial point defining a probability measure ρ which is invariant under time evolution and ergodic. Clearly in the case of a time series (i.e. discrete time) the integral must be replaced by a sum, this will be implied in the following. Thus we can write

$$\langle \phi \rangle = \rho(\phi) = \int \rho(dx) \phi(x)$$

The measure ρ describes how frequently various part of the attractor are visited by the time evolution and the existence of such a measure makes possible to utilize *ergodic theory* as an abstract background to recognize *invariant properties* of the system. The general approach here is the following: some ergodic quantities, namely some functions of the measure ρ representing invariant physical properties of the dynamics, are firstly introduced. Then, a sort of "decodification" of a deterministic time evolution can be carried out by setting good techniques to measure such properties. As a matter of fact, these two "steps" come together very rarely. Due to the difficulty of handling chaotic behaviour, often no one knows how to attach numbers to some interesting quantities for quite a long time.

Before to discuss some relevant properties which can be measured in a statistical approach to chaos, it should be pointed out that our perspective is already quite drastically changed. In fact, we have seen that the sensitive dependence on initial conditions introduces some uncertainty at each iteration of the map f (or at each outcome $x(t_i)$ of the experiment), so that any small error after a certain time will make the state $\vec{x}(t)$ of the system completely indeterminate (it could be everywhere on the attractor). However, as far as *statistical* properties are concerned this must not trouble us because they do not depend on the details of the time evolution. In particular, small perturbations play an important role in selecting the right *physical measure* ρ among many possible probability measures. This is the measure which has just the attractor as support and is stable under small stochastic perturbations (see [7]).

Therefore the small errors always present in experimental situations (external noise) as well as in numerical simulations (round-off errors), though "fatal" in the classical perspective, become here "beneficial" in order to get a reliable description of the statistical behaviour of the system.

Now, let us see briefly which are the most relevant properties one shall try to extract from a chaotic time evolution.

• Ever since the notion of strange attractor has been introduced it has been clear that the *characteristic exponents* (sometimes called *Liapunov exponents*) might be employed in studying chaotic motions.

If we consider a d-dimensional ball of infinitesimal radius ϵ and we let it evolve according to the transformation f, we find it will be deformed into an ellipsoid with principal axes ϵ_i, $(i = 1, \ldots, d)$. Then, the characteristic exponents λ_i are determined by:

$$\epsilon_i(t) \sim \epsilon \exp(\lambda_i t)$$

If the asymptotic measure ρ has at least a positive λ_i then there is sensitive dependence on initial condition (note that the sum of the λ_i's, describing the global contraction of the volume, has to be negative). So, the most relevant property of chaotic behaviour can be extracted by measuring characteristic exponents. This can be done from the tangent maps $D_x f^t$. For instance, in the simplest case of a one-dimensional map of the form $x(t+1) = f(x(t))$ with $t \in Z$, the characteristic exponent λ is given by:

$$\lambda = \lim_{T \to \infty} \frac{1}{T} \sum_{t=0}^{T-1} \log \mid D_x f^t \mid = \int \log \mid D_x f \mid \rho(dx)$$

where $D_x f = \frac{df}{dx}$ and $f^t = \overbrace{f \circ f \circ \cdots \circ f}^{t \text{ times}}$.

Since in computer experiments the derivative is directly calculable whereas in physical experiments it has to be estimated by an analysis of the experimental data, the methods are somewhat different in the two cases and several techniques have been proposed (see [7]).

• Due to sensitivity on initial conditions, any uncertainty about seemingly insignificant digits in the sequence of numbers which defines an initial condition, spreads with time towards the significant digits producing a change in the *information* we have about the state of the system. This change can be thought as a creation of information if we consider that two initial conditions that are different but indistinguishable (within a certain precision), evolve into distinguishable states after a finite time. The *entropy* $h(\rho)$, or *Kolmogorov-Sinai invariant*, measures the asymptotic rate of creation of information during the time evolution. For the precise definition of $h(\rho)$ we refer to the literature on the subject (for instance [9]); let us say that in some cases (generally when ρ is not singular with respect to the Lebesgue measure) it is related to the *positive* characteristic exponents by the Pesin equality:

$$h(\rho) = \sum_{\lambda_i > 0} \lambda_i$$

In more general cases a corresponding inequality holds. Moreover some generalized entropies can be introduced and some of them turn out to be more easily accessible in experimental situations [10], [12].

• Now, a strange attractor typically arises when the time evolution stretches a volume element in some directions (those corresponding to positive λ_i) and contracts it in some others (with negative λ_i). But in order to remain confined to a bounded region the volume element gets folded over itself at the same time, so that it acquires progressively a complicated multisheeted structure. This fact makes the attractor a *fractal* object in the sense of Mandelbrot [11], namely it has locally a Cantor-set like structure in some directions. In other words, if we cover the attractor by d-dimensional hypercubes of side length l and consider the limit $l \to 0$, then the minimal number of cubes needed for the covering grows like

$$n(l) \simeq_{l \to 0} l^{-D}$$

and the exponent D, the *capacity* of the attractor, is not an integer number. Of course, D is a purely geometric measure. However one can introduce other related quantities depending of the frequency with which a typical trajectory visits the various part of the attractor, namely some functions of the measure ρ. An important one is the *information dimension* (for a precise definition see for example [7]), but there exists a complete hierarchy of generalized dimensions providing information on the local structure of the attractor as a fractal object (see [12]). Generally speaking, let us say that measuring *information dimension* (or related quantities) provides a reliable way to extract the possible deterministic nature of a chaotic time evolution [10]. Moreover it allows to compare different strange attractors to see which one is "larger".

- To conclude our list (without claims of completeness), it is worth mentioning *correlation functions* and *power spectra*.

Given two observables $A, B : M \to R$, the correlation function $C_{AB}(t)$ is defined as follows:

$$C_{AB} = \rho[(A \circ f^t)B] - \rho(A)\rho(B)$$

In the simplest case where the observables are the "position variable" x one gets the *autocorrelation function* of the x-coordinate. Picking an initial point $x(0)$ and another point $x(t)$ on the time evolution, this function tells us how much the deviations of these two points from their common mean value know about each other, on the average over all initial points. Therefore the correlation function yields another measure of the irregularity of the motion. A time evolution for which $C_{AB}(t) \to_{t \to \infty} 0$, for all choices of A, B, is called *mixing*. Then, the problem is to know *how*, for a mixing system, the correlation function decays to zero. This is important because it gives the rate at which the time evolution becomes independent of where the system began. However, it seems that at present there are no successful techniques to measure the decay rate directly from the correlation function. An hopeful direction is rather to study the properties of its Fourier transform:

$$S_{AB}(\omega) = \int_{-\infty}^{+\infty} dt \exp(i\omega t) C_{AB}(t)$$

Note that $S_{AA}(\omega)$ is the *power spectrum* of the signal $A(x(t))$. Usually a chaotic time evolution exhibits a continuous power spectrum and for this property it has been largely utilized as an indicator to distinguish between regular and chaotic motions. Nevertheless the function S_{AB} turns out to be useful also in order to investigate mixing properties, namely to distinguish between different chaotic regimes. In fact the decay rate of the correlation function is strictly related to the analytic properties of its Fourier transform S_{AB}. More precisely the location of poles of S_{AB} in the complex frequency plane (interpreted as a *resonances*) may provide an appropriate description of the behavior of $C_{AB}(t)$ (see [13], [14]).

Some concluding remarks

The picture presented above on the actual state of the statistical approach to chaotic time series is fairly incomplete. Some others quantities to be measured have been recently introduced and others properties have been investigated. However, the purpose was just to give a sight over the current ideas on this subject, in order to get an example of a possible direction to approach complex behaviour. But may be it gives also some insight into the meaning of the adjective "complex", as something else with respect to "complicated". Indeed, the picture sketched above suggests that speaking about "complexity" is really suitable in those situations where, in order to extract a physical sense from a set of observations, it is necessary to replace the questions inherited from the classical approach with some others "good questions". As a consequence, *understanding* a given phenomenon will acquire a different meaning.

This is particularly evident in the peculiar use of probability within the context of statistical analysis of chaotic time evolutions. Here the probabilistic approach is not an expedient to get an approximate description of a hidden reality which could be well described only by a Laplace demon. Rather, it is a fruitful procedure which is able to give a physical sense to chaotic behavior by the formulation of some "good questions". To use a geometric image, think of a grid such that, once superimposed to the set of observations, allows to perceive an intelligible sense. Moreover this sense has all the features of an abstract model which can be analysed by mathematical tools. Thus it constitutes an explanation.

Hence, the reality we have to consider as meaningful is just what the procedure brings forth, and, in this sense, it constitutes basically an "objective" reality. Actually, as we have seen in the previous section, the physical reality of a dynamical behavior lying on a strange attractor is determined operationally by measuring a series of invariant properties and therefore is a sort of *asymptotic* reality which appears progressively. Nevertheless *there is no longer reason to consider it as an approximation of some more profound reality.*

By the way, this is a very concrete and general statement. Its consequences permeate the modern scientific practice whenever it produces a new possible scenario where a piece of reality, if adequately introduced in it, can be found to be intelligible.

References

1) H.Atlan, *A tort et a raison. Intercritique de la science e du mythe*, Ed. du Seuil, Paris, 1986
2) D.Ruelle and F.Takens, 1971, Comm.Math.Phys., **21**, 21
3) J.A.Sheinkman and B.LeBaron, 1987, Preprint, Chicago
4) C.Nicolis and G.Nicolis, 1984, Nature, vol.311, 529
5) P.Grassberger, *Are there really climatic attractors?*, 1986, Preprint, Wuppertal B86-6
6) F.Bailly and I.Stengers, *Complexité. Effet de mode ou probléme?*, in: I.Stengers (ed.), *D'une science à l'autre. Des concepts nomades*, Ed. du Seuil, Paris, 1987
7) J.-P.Eckmann and D.Ruelle, 1985, Rev.Mod.Phys. **53**, 643
8) D.Ruelle, *Chaotic evolutions and strange attractors*, Notes prepared by S.Isola from the "Lezioni Lincee" (Rome, 1987), Cambridge University Press, Cambridge UK, 1988
9) P.Billingsley, *Ergodic theory and Information*, Wiley, New York, 1965
10) P.Grassberger and I.Procaccia, 1983, Physica **9D**, 189
11) B.Mandelbrot, *The fractal geometry of Nature*, Freeman, San Francisco, 1982
12) P.Grassberger, 1983, Phys.Lett. **97A**, 227; G.Paladin and A.Vulpiani, 1987, Phys.Rep.
13) D.Ruelle, 1985, Phys.Rev.Lett. **56**, 405; 1986, J.Stat.Phys. **44**, 281
14) S.Isola, 1988, Comm.Math.Phys. **115**

Decay of Correlation Functions and Expansion Rates in Dynamical Systems

A. Politi *, R. Badii *, K. Heinzelmann *, P.F. Meier *

° *Istituto Nazionale di Ottica, I-50125, Firenze, Italy*
* *Institut für Theoretische Physik, Universität Zürich, CH-8001, Switzerland*

Abstract

Correlation functions of simple hyperbolic dynamical systems are computed exactly and the decay rates are related to generalized Lyapunov exponents.

Dissipative dynamical systems are usually characterized by evaluating metric entropies, Lyapunov exponents and fractal dimensions, while more standard statistical tools like correlation functions have only been discussed for simple systems [1], or in connection with particular phenomena. In fact, very little is known about time correlation functions, even for strictly hyperbolic systems [2]. In two and more dimensions, the difficulties arise essentially from the contribution of all Lyapunov numbers (multipliers) to the time-decay. Recently, it has been shown that the decay rate of correlation functions can be related, in some cases, with generalized Lyapunov exponents [3]. In this work, we review the main results of Ref.[3] on the behavior of correlation functions of the form

$$C_{AB}(\tau) \equiv \langle A[\mathbf{x}(t)]B[\mathbf{x}(t+\tau)]\rangle - \langle A\rangle\langle B\rangle \equiv$$
$$\lim_{T\to\infty} \frac{1}{T} \int_0^T A[\mathbf{x}(t)]B[\mathbf{x}(t+\tau)]dt - \langle A\rangle\langle B\rangle ,$$
(1)

where A and B are functions of the position $\mathbf{x}(t)$. Time t can be either continuous or discrete: in the latter case, the integral in Eq.(1) is replaced by a sum. The averages can be rewritten as

$$\langle A\rangle = \lim_{T\to\infty} \frac{1}{T} \int_0^T A[\mathbf{x}(t)]dt \equiv \int A(\mathbf{x})\rho(\mathbf{x})d^E x ,$$
(2)

where an invariant probability measure $\rho(\mathbf{x})$ (ergodic with respect to the dynamical system) has been defined and E is the phase-space dimension.

We first show that the decay of correlation functions in one-dimensional maps with piecewise constant invariant measure ρ, is connected with dynamical quantities like the generalized Lyapunov exponents $\Lambda_k(q)$ ($k = 1, \ldots, E$) [4,5]. More precisely, the correlation function is shown to coincide (up to multiplicative factors) with the average inverse multiplier $\langle \mu_1^{-1}(\tau) \rangle$, computed over a number τ of iterations. Accordingly, the decay rate γ is equal to $\Lambda_1(2)$ in all cases in which μ_1 has constant sign along the trajectory. More in general, γ is larger than $\Lambda_1(2)$, indicating a faster decay.

Then, we give an analytic solution of the $2-d$ generalized baker transformation [6] in terms of symbol sequences. The resulting exact expansions for the two variables allow the evaluation of the correlation functions. Three different exponentially decaying terms have been identified. Besides the contribution already found in the $1-d$ case, the average second multiplier $\langle \mu_2(\tau) \rangle$ and the average ration $\langle \mu_2(\tau)/\mu_1(\tau) \rangle$ are present. For some dynamical systems, the new terms may yield the relevant contribution.

Let us start discussing the relations between the decay of the correlation function $C(\tau)$ and dynamical quantities like generalized Lyapunov exponents, for one-dimensional maps $y_{n+1} = F(y_n)$ of the interval $[0,1]$. Equation (1) can be rewritten as

$$C_{yy}(\tau) = \int y_0 F^\tau(y_0) \rho(y_0) dy_0 - \langle y \rangle^2 , \qquad (3)$$

where $F^\tau(y_0)$ denotes the τ-th iterate of the initial condition y_0. Next, let us introduce a partition of the interval $[0,1]$ in the largest intervals on which the function F^τ is still invertible. It can be easily shown that, apart from irrelevant differences, such a partition coincides with the τ-th refinement of the standard generating partition. Further restriction to hyperbolic systems allows to approximate the width of the j-th sub-interval with $\Delta_j = \mu_1(\tau)^{-1}$, since the fluctuations of the multiplier with the initial condition y_0 vanish in the limit $\tau \to \infty$. Accordingly, we can rewrite Eq.(3) as

$$C_{yy}(\tau) = \sum_{j=1}^{N} \int_{\Delta_j} y_0 F^\tau(y_0) \rho(y_0) dy_0 - \langle y \rangle^2 = \qquad (4)$$

$$\sum_{j=1}^{N}\{\int_{\Delta_j} u\,F^\tau(\bar{y}_j+u)\rho(\bar{y}_j+u)du + \bar{y}_j\int_{\Delta_j} F^\tau(\bar{y}_j+u)\rho(\bar{y}_j+u)du\} - \langle y\rangle^2,$$

where the change of variables $y_0 = \bar{y}_j + u$ has been made, with $\bar{y}_j$ denoting the center of mass of the j-th interval. The first integral in Eq.(4) can be easily estimated by expanding $F^\tau(\bar{y}_j+u)$ around $\bar{y}_j$. The zero-th order term vanishes by definition of $\bar{y}_j$, while the first order one yields the leading contribution $\langle \mu_1(\tau)^{-1}\rangle$, where the u-dependence has been neglected in the invariant density. It is easy to show that higher order terms vanish faster. The computation of the second integral in Eq.(4) is, in general, much more complicated. Here, we analyze the simple cases in which the τ-th iterate of the invariant measure, restricted to the j-th subinterval, coincides with the invariant measure itself. This class of systems includes piece-wise linear maps, for which the interval $[0,1]$ can be divided in a finite number of segments, preimages of the whole interval. In that case, independently of j, the second integral always yields $\langle y \rangle$ multiplied by the mass (or width) Δ_j of the interval. Therefore, after the evaluation of the sum, this term is seen to be equal to $\langle y\rangle^2$ and cancels exactly the last term of C_{yy}. Accordingly, the order of magnitude of the correlation function is given by the first term in the r.h.s. of Eq.(4):

$$C(\tau) \sim \langle \mu_1(\tau)^{-1}\rangle. \qquad (5)$$

To elucidate better the meaning of relation (5), we recall the definition of generalized Lyapunov exponents [4,5], starting from the expansion rate over a finite number τ of iterates:

$$\lambda_1(x_0;\tau) = \frac{1}{\tau}\ln\prod_{i=0}^{\tau-1}|F'(x_i)| = \frac{1}{\tau}\ln|\mu_1(\tau)|. \qquad (6)$$

The associated generalized Lyapunov exponent $\Lambda_1(q)$ is then defined as

$$\Lambda_1(q) = \lim_{\tau\to\infty}\frac{1}{\tau(1-q)}\ln\left\langle e^{\tau(1-q)\lambda_1(\tau)}\right\rangle. \qquad (7)$$

From Eqs.(5)-(7), in case of an everywhere positive (or negative) multiplier, we have

$$|C_{yy}(\tau)| \propto e^{-\tau\Lambda_1(2)}. \qquad (8)$$

When the multiplier changes sign during the time evolution, we are led to define the decay rate as

$$\gamma = -\limsup_{\tau\to\infty} \frac{1}{\tau} \ln |C_{ww}(\tau)| \qquad (9)$$

which is, in general, larger than $\Lambda_1(2)$. For the symmetric tent map, the average multiplier is 0 and, therefore, $\gamma = \infty$ as for a delta-correlated process [3].

We now investigate how the results found for $1-d$ maps are modified in higher dimensional cases. For simplicity, we consider $2-d$ generalized baker transformations [6] of the form

$$(x_{n+1}, y_{n+1}) = \begin{cases} (\alpha x_n, y_n/p) & \text{if } y_n \leq p, \\ (\beta x_n + 1 - \beta, (y_n - p)/q) & \text{if } y_n > p, \end{cases} \qquad (10)$$

where $\alpha + \beta < 1$. The asymptotic attractor is the product of a continuum (along the y-direction) by a Cantor set (along the x-direction). Since the baker map is hyperbolic, one can construct Markov partitions [7] and identify any point $\mathbf{x} \equiv (x,y)$ with the associated symbol sequence. The generating partition is obtained by cutting the square horizontally at $y = p$. Accordingly, we define the binary symbols $a_n = [1 + \text{sgn}(y_n - p)]/2$. Taking into account the symmetries of Eq.(10) [3], we obtain the "bit" expansions

$$x_n = \frac{1-\beta}{\beta-\alpha}\Big[-\alpha + (1-\alpha)\sum_{i=1}^{\infty} \alpha^i \prod_{j=1}^{i} \sigma^{a_{n-j}}\Big] \qquad (11)$$

and

$$y_n = \frac{p}{q-p}\Big[-p + q\sum_{i=1}^{\infty} p^i \prod_{j=1}^{i} \pi^{a_{n-1+j}}\Big], \qquad (12)$$

where the abbreviations $\sigma = \beta/\alpha$ and $\pi = q/p$ have been introduced.

Usually, in experiments on chaotic systems, a single scalar time series w_n is available for a reconstruction of the attractor, through embedding techniques. In the case of the baker map, neither x nor y can be used for this purpose: indeed the variable y is decoupled from x, while the evolution of the variable x cannot be inferred from the knowledge of its own past history alone. Therefore, it is convenient to consider the linear combination

$$w_n = x_n + y_n \tag{13}$$

as an appropriate embedding variable for this system. The autocorrelation function $C_{ww} = \langle w_n w_{n+\tau} \rangle - \langle w \rangle^2$ can then be evaluated by forming averages over the symbols a_n, using the expansions (11,12). Since the symbol sequence is a random process satisfying

$$\langle a_n \rangle = q \text{ , and } \langle a_m a_n \rangle = \begin{cases} q & \text{if } n = m, \\ q^2 & \text{if } n \neq m, \end{cases} \tag{14}$$

all expectation values can be explicitly calculated. Noticing that the correlation $C_{xy}(\tau) = \langle x_0 y_\tau \rangle - \langle x \rangle \langle y \rangle$ is identically zero, because the variable y does not depend on the variable x at all, one finally obtains [3]:

$$C_{ww}(\tau) = \kappa_1 \langle \mu_1^{-1} \rangle^\tau + \kappa_2 \langle \mu_2 \rangle^\tau + \kappa_3 \langle \mu_2/\mu_1 \rangle^\tau, \tag{15}$$

where we have introduced the average values of the one-step multipliers:

$$\begin{aligned} \langle \mu_1^{-1} \rangle &= p^2 + q^2, \\ \langle \mu_2 \rangle &= \alpha p + \beta q, \\ \langle \mu_2/\mu_1 \rangle &= \alpha p^2 + \beta q^2. \end{aligned} \tag{16}$$

Notice that successive values of μ_1, μ_2 are completely uncorrelated from each other: therefore, $\langle \mu_k(1) \rangle^\tau = \langle \mu_k(\tau) \rangle$. From Eq.(15) we see that not only the correlation function $C_{yy}(\tau)$ of the Bernoulli shift [1] decays exponentially, but all correlation functions of the $2-d$ baker map exhibit this behavior. By comparing Eq.(15) with the $1-d$ result (5), we see that two new terms contribute to the correlation function. If the y-variable in Eq.(10) evolves according to the Bernoulli shift, the dominating long-time term of C_{ww} is still the one which derives from the expanding multiplier. In fact, from the inequalities

$$\frac{\mu_2}{\mu_1} < \mu_2 < \frac{1}{\mu_1} < 1, \tag{17}$$

between single-step quantities ($\mu_1, \mu_2 > 0$), it is clear that $\langle \mu_1 \rangle^{-\tau}$ is the slowest term. A similar derivation of $C_{ww}(\tau)$ can be performed by letting y evolve according to the tent map and allowing the contracting multiplier to assume negative values as well [3]. The final result confirms the presence of three distinct contributions to $C_{ww}(\tau)$, proportional to $\langle \mu_1^{-1} \rangle$, $\langle \mu_2 \rangle$, $\langle \mu_2/\mu_1 \rangle$, in this more general map also. As a consequence, either $|\langle \mu_2 \rangle|$ or $|\langle \mu_2/\mu_1 \rangle|$ can yield, in suitable ranges of parameter values, the leading contribution to the correlation function, rather than $|\langle \mu_1^{-1} \rangle|$. Therefore, the sign of the multipliers has such an effect on correlation functions that it is not possible to infer the time behaviour from a simple unique prescription.

On the basis of these arguments, we expect $|\langle \mu_1^{-1}(\tau) \rangle|$ to provide an upper bound to the decay of the correlation function. Preliminary simulations confirm this conjecture showing, however, a finite difference between the two rates.

The analysis of simple two-dimensional maps revealed the existence of three terms in the correlation function. The problem of the possible occurrence of other combinations of the multipliers in generic systems remains open.

References

[1] a) H. Mori, B.C. So and T. Ose, Prog. Theor. Phys. **66**, 1266 (1981);
    b) H. Fujisaka and T. Yamada, Z. Naturforsch. **33a**, 1455 (1978).

[2] D. Ruelle, J. Stat. Phys. **44**, 281 (1986).

[3] R. Badii, K. Heinzelmann, P.F. Meier and A. Politi, Phys. Rev. A, February (1988).

[4] H. Fujisaka, Prog. Theor. Phys. **70**, 1264 (1983).

[4] P. Grassberger and I. Procaccia, Physica **13D**, 34 (1984).

[6] J.D. Farmer, E. Ott and J.A. Yorke, Physica **7D**, 153 (1983).

[7] V.M. Alekseev and M.V. Yakobson, Physics Reports **75**, 287 (1981).

Phase-Transitions in Hyperbolic Dynamical Systems

R. Badii* and A. Politi•

Institut für Theoretische Physik, Universität Zürich, CH-8001, Switzerland

•*Istituto Nazionale di Ottica, I-50125, Firenze, Italy*

Abstract

We study a three-dimensional hyperbolic map in the framework of the thermodynamic formalism for strange attractors, showing that the occurrence of phase-transitions is not restricted to non-hyperbolic systems.

The analogy between nonlinear dynamical systems and one-dimensional hamiltonian spin models, originally developed in Refs.[1] for $1-d$ expanding maps and Julia sets, has been recently utilized to improve the understanding of chaotic phenomena [2,3]. In particular, it has been shown that the dimension function $D(q)$ can exhibit non-analytic behaviour which is interpreted as a phase-transition in the associated spin model. These singularities can only appear in non-self-similar measures and are due to long-time correlations in the symbolic dynamics. Although this phenomenon is prevalently observed in non-hyperbolic systems [4,5,6], it can also occur in the hyperbolic case [7,8]. In this paper, we review the theory of thermodynamic ensembles for fractal measures and apply it to the analysis of a filtered hyperbolic dynamical system: the generalized baker map. Indeed, the effect of filtering on experimental signals has been found to be responsible for the breaking of self-similarity of strange attractors [9].

In order to characterize a fractal measure, it is useful to introduce the pointwise dimension $\alpha(\mathbf{x})$ as the ratio

$$\alpha(\mathbf{x}) = \lim_{\varepsilon \to 0} \frac{\ln P(\varepsilon, \mathbf{x})}{\ln \varepsilon}, \quad (1)$$

where $P(\varepsilon, \mathbf{x})$ is the mass contained in a ball of size ε centered at $\mathbf{x}$, and to study its probability distribution. Alternatively, a global estimate of dimension can be obtained through a suitable average over the pointwise dimensions $\alpha(\mathbf{x})$.

By covering the set with balls of size ε_j and mass P_j, the dimension function $D(q)$ is defined as [6]

$$\lim_{\varepsilon \to 0} \left\langle \frac{P^{q-1}}{\varepsilon^{\tau(q)}} \right\rangle \sim O(1) , \qquad (2)$$

where $\varepsilon = max_j\{\varepsilon_j\}$ and

$$\tau(q) = (q-1)D(q) . \qquad (3)$$

In the following, we shall restrict ourselves to constant-mass coverings ($P_j = p$, independently of the position), thus obtaining

$$\left\langle \varepsilon(p)^{-\tau(q)} \right\rangle \sim p^{1-q}, \text{ for } p \to 0. \qquad (4)$$

In order to establish the connection with statistical mechanics, let us assume that a generating partition composed of M elements has been detected, so that each ball of the covering can be represented by a finite sequence S_n of symbols: $S_n = \{\ldots, s_{i-1}, s_i, s_{i+1}, \ldots\}$, with $s_i \in [0, M-1]$. Furthermore, time is considered as a one-dimensional lattice with sites labelled by the index $i \in \{\ldots, -2, -1, 0, 1, 2, \ldots\}$. Accordingly, the sequence S_n is interpreted as a chain of n spins σ_i, which are defined as $\sigma_i = s_i - (M-1)/2 \in [-(M-1)/2, (M-1)/2]$. Therefore, each ball of the covering is associated to a $1-d$ configuration of spins and represents the *microstate*. Then, Eq.(2) can be rewritten as follows [8]:

$$Z_{can}(\mathsf{V},\mathsf{T}) \equiv e^{-\mathsf{F}(\mathsf{V},\mathsf{T})/k_B \mathsf{T}} = \sum_{\{S_n\}} e^{-(\tau/\alpha(S_n))\ln p} \sim e^{[1-q(\tau)]\ln p} , \qquad (5)$$

where $q(\tau)$ is the inverse function of $\tau(q)$ (Eq.(3)). Z_{can} is the canonical partition function, F the free energy, V the volume and T the temperature. The sum is over all sequences (i.e., balls) S_n. Energy, temperature and volume are defined as

$$\begin{aligned} \mathsf{E} &= \ln \varepsilon(S_n) < 0 , \\ \beta &\equiv 1/k_B \mathsf{T} = \tau , \\ \mathsf{V} &= -\ln p \geq 0 . \end{aligned} \qquad (6)$$

The free energy is then $\mathsf{F}(\mathsf{V},\mathsf{T}) = \ln p/D(q) = -\mathsf{V}/D(q)$.

The microcanonical ensemble consists of all balls with size $\varepsilon(S_n) = p^{1/\alpha(S_n)}$ in the range $[\varepsilon, \varepsilon + d\varepsilon]$. The microcanonical partition function is, then, defined in terms of the probability density $\hat{P}(\alpha; p)$ to observe a pointwise dimension α in a ball of mass p, chosen at random with respect to the natural measure [6]:

$$Z_{mic}(\mathsf{E},\mathsf{V}) \equiv e^{S(\mathsf{E},\mathsf{V})/k_B} \propto \hat{P}(\alpha; p) \sim \sqrt{|\ln p|}\, p^{1-f(\alpha)/\alpha} . \tag{7}$$

Taking the limit $p \to 0$, we obtain the entropy

$$\mathsf{S}(\mathsf{E},\mathsf{V}) = -k_B\big[f(\alpha)/\alpha - 1\big]\ln p , \tag{8}$$

where the function $f(\alpha)$ is called spectrum of dimensions [2,3]. It is concave in α, and its maximum coincides with the Hausdorff dimension of the fractal set. It can be shown that

$$\tau(q) = \inf_{\alpha}\{q\alpha - f(\alpha)\} , \tag{9}$$

i.e., the function $f(\alpha)$ is the Legendre transform of $\tau(q)$ [3]. This equation, according to the definitions (6) and (8), is equivalent to

$$\mathsf{F}(\mathsf{T}) = \mathsf{E}(\mathsf{T}) - \mathsf{T}\mathsf{S}(\mathsf{E}) \tag{10}$$

and all usual thermodynamic relations are recovered:

$$\left(\frac{\partial \mathsf{S}}{\partial \mathsf{E}}\right)_\mathsf{V} = \frac{1}{\mathsf{T}}, \left(\frac{\partial \mathsf{S}}{\partial \mathsf{V}}\right)_\mathsf{E} = \frac{\mathsf{P}}{\mathsf{T}}, \left(\frac{\partial \mathsf{E}}{\partial \mathsf{S}}\right)_\mathsf{V} = \mathsf{T} \text{ and } \left(\frac{\partial \mathsf{E}}{\partial \mathsf{V}}\right)_\mathsf{S} = -\mathsf{P}, \tag{11}$$

where $\mathsf{P} = 1/D(q)$ is the pressure.

We now apply this formalism to a simple model of a filtered chaotic signal [9]:

$$\begin{aligned}
(x_{n+1}, y_{n+1}) &= (r_1 \cdot x_n, y_n/p_1) & \text{if } y_n \leq p_1 , \\
(x_{n+1}, y_{n+1}) &= (r_2 \cdot x_n + 1 - r_2, (y_n - p_1)/p_2) & \text{if } y_n > p_1 , \\
z_{n+1} &= \gamma z_n + x_{n+1} + y_{n+1} ,
\end{aligned} \tag{12}$$

where $p_1 + p_2 = 1$, $r_1 + r_2 < 1$ and $0 \leq \gamma < 1$. The first two equations represent the generalized baker transformation [10], while the third one describes the effect of the filter, characterized by the cut-off frequency $\eta = -\ln\gamma$. By introducing the symbolic representation

$$s_{-n} \equiv \frac{1 + \text{sgn}(y_n - p_1)}{2}, \tag{13}$$

it is possible to give the following "bit" expansions for the x- and y- variables:

$$x_n = \frac{1 - r_2}{r_2 - r_1}\left[-r_1 + (1 - r_1)\sum_{i=1}^{\infty} r_1^i \prod_{j=1}^{i} \theta^{s_{-n+j}}\right],$$

$$y_n = \frac{p_1}{p_2 - p_1}\left[-p_1 + p_2 \sum_{i=1}^{\infty} p_1^i \prod_{j=1}^{i} \pi^{s_{-n-j+1}}\right], \tag{14}$$

where the abbreviations $\theta = r_2/r_1$ and $\pi = p_2/p_1$ have been introduced. The formal solution of the filter equation yields the expression for z_n as

$$z_n = \sum_{k=0}^{\infty} w_{n-k}\gamma^k = \sum_{k=0}^{\infty}(x_{n-k} + y_{n-k})\gamma^k. \tag{15}$$

Since the attractor is continuous along the y-direction, we take a section of it at $y_n = 0$, by setting $s_{-i} = 0$ for $i \geq n$. A projection of the set onto the x-axis yields the two-scales (r_1 and r_2) Cantor set while the scaling of the distances along the z-direction is more complex. Therefore, we evaluate the segment lengths

$$\varepsilon(S_n) \equiv z(S_n \cup \{1,1,\ldots\}) - z(S_n \cup \{0,0,\ldots\}) \tag{16}$$

between the two extrema of a generic segment at the n-th level of resolution: the two z-values are characterized by the same symbolic subsequence S_n, for the first n steps, and by the semi-infinite sequences $\{0,0,\ldots\}$ (lower extremum) and $\{1,1,\ldots\}$ (upper extremum), in the asymptotic limit. We shift time by n iterations and change its sign to obtain, after lengthy but straightforward calculations [8],

$$\varepsilon(S_n) = \gamma^n \left\{ a + \sum_{l=1}^{n}\left(\frac{r_1}{\gamma}\right)^l \prod_{m=0}^{l-1}\theta^{s_{n-m}} + b\sum_{l=1}^{n} p_1^l \prod_{m=0}^{l-1}\pi^{s_{n-m}} \right\}, \tag{17}$$

where a and b are constants. If the filter bandwidth is large ($\gamma < r_1 < r_2$), the first sum diverges for large n. Hence, $\varepsilon(S_n)$ asymptotically behaves as the segment-lengths in the x-direction and the set is self-similar. The pointwise dimension

(1) belongs to the interval $[\alpha_{min} = -\ln 2/\ln r_1, \alpha_{max} = -\ln 2/\ln r_2]$ (assuming, for simplicity, $p_1 = p_2 = 1/2$). When γ is increased above r_1, the asymptotic behaviour of $\varepsilon(S_n)$ changes: in fact, depending on the sequence S_n, either it exhibits the same scaling as before (powers of r_1 and r_2), or it decreases like γ^n and the self-similarity is broken. This phenomenon is illustrated in Fig. 1: the curve $f(\alpha)$ vs.

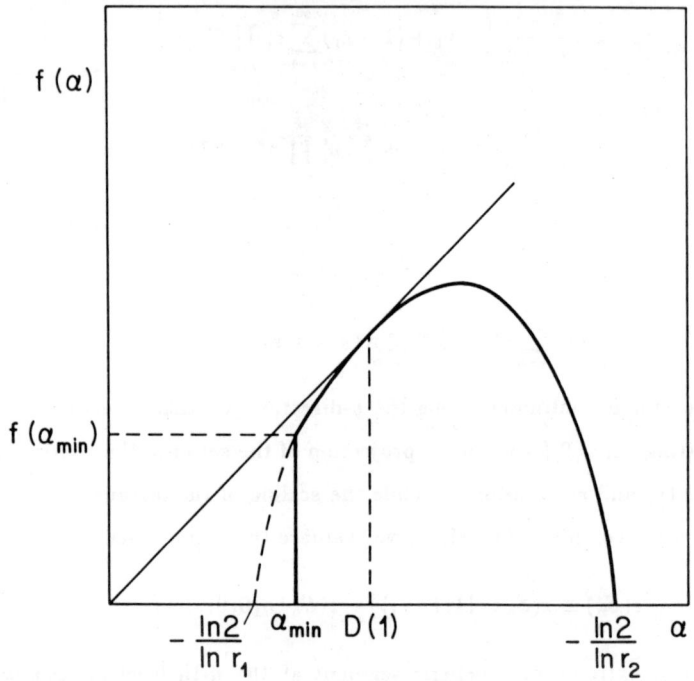

Figure 1: Schematic plot of the $f(\alpha)$-curve (solid line) for the filtered baker map with $r_1 < \gamma < \sqrt{r_1 r_2}$. Dashed line: same curve in absence of filter.

α (solid line) coincides with that in absence of filter (dashed line), for α larger than $\alpha_{min}(\gamma) = -\ln 2/\ln \gamma$, while displaying a cut-off at $\alpha = \alpha_{min}(\gamma)$. From Eq.(9), we then obtain

$$D(q) \sim \frac{q\alpha_{min} - f(\alpha_{min})}{q - 1}, \qquad (18)$$

for q larger than the critical value $q_c = \lim_{\alpha \searrow \alpha_{min}} f'(\alpha)$. The situation is illustrated in Fig. 2. The appearance of a corner point in the "free energy" $1/D(q)$ can be interpreted as the evidence for a first-order phase transition. According to Eq.(6), the hamiltonian H can be defined as the logarithm of the generic segment-length (17). By using the symmetrized spins $\sigma_i = \pm\frac{1}{2}$, it is possible to rearrange the terms

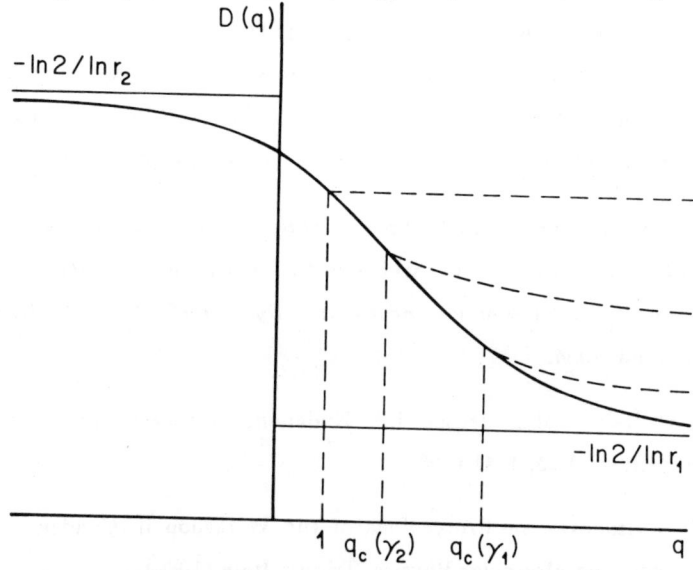

Figure 2: Dashed lines: dimension function $D(q)$ vs. q, for some values of the parameter $\gamma \in (0, \sqrt{r_1 r_2})$. Solid line: $D(q)$ in absence of filter.

in the hamiltonian obtaining, finally,

$$H = h_0 + \sum_{i=1}^{n} J_i^{(1)} \sigma_i + \frac{1}{2} \sum_{i \neq j=1}^{n} J_{ij}^{(2)} \sigma_i \sigma_j + \frac{1}{6} \sum_{i \neq j \neq k=1}^{n} J_{ijk}^{(3)} \sigma_i \sigma_j \sigma_k + \ldots + J^{(n)} \sigma_1 \cdot \ldots \cdot \sigma_n, \quad (19)$$

where h_0 is a constant. The magnitude of the interaction coefficients J does not depend on the distance. Furthermore, the interactions are between all orders of nearest-neighbours ($1^{st}, 2^{nd}, \ldots, n^{th}$), and involve up to n spins: these long-range interactions (i.e., long-time correlations) are responsible for the appearance of the phase-transition illustrated in Figs. 1 and 2 and are related to the occurrence of tangencies between the eigendirections in the contracting subspace $x - z$ [8]. This behaviour is, clearly, possible only for three- or higher-dimensional (hyperbolic) maps. In the non-hyperbolic case, two dimensions are sufficient (Hénon map), due to the homoclinic tangencies.

References

[1] Ya. G. Sinai, Russ.Math.Surv. **27**, 21 (1972);
R. Bowen, Lect.Not.Math., **470**, Springer, Berlin (1975);
D. Ruelle, *"Thermodynamic Formalism"*, vol. 5 of Encyclopedia of Mathematics and its Applications, Addison-Wesley, Reading, MA (1978).

[2] G. Parisi, appendix in U. Frisch, *'Fully Developed Turbulence and Intermittency'*, in Proc. of Int. School on *"Turbulence and Predictability in Geophysical Fluid Dynamics and Climate Dynamics"*, M. Ghil editor, North-Holland, 1984.

[3] T.C. Halsey, M.H. Jensen, L.P. Kadanoff, I. Procaccia and B. Shraiman, Phys.Rev. **A33**, 1141 (1986).

[4] P. Cvitanovič, in Proceedings of the Workshop in Condensed Matter, Atomic and Molecular Physics, Trieste, Italy (1986).

[5] D. Katzen and I. Procaccia, Phys.Rev.Lett. **58**, 1169 (1987).

[6] P. Grassberger, R. Badii and A. Politi, *Scaling Laws for Hyperbolic and Non-Hyperbolic Attractors*, J.Stat.Phys., to appear.

[7] R. Badii and A. Politi, Physica Scripta **35**, 243 (1987).

[8] R. Badii, *Conservation Laws and Thermodynamic Formalism for Dissipative Dynamical Systems*, Riv. N. Cim., to appear.

[9] R. Badii and A. Politi, in *"Dimensions and Entropies in Chaotic Systems"*, G. Mayer-Kress editor, Springer, Berlin (1986).

[10] J.D. Farmer, E. Ott and J.A. Yorke, Physica **7D**, 153 (1983).

Hierarchical Random Networks*

A. MARITAN

Dipartimento di Fisica dell'Università
and INFN - Sezione di Padova,
Via Marzolo 8, 35131 Padova, Italy

1. INTRODUCTION

Random surfaces (RS) are without any doubt one of the subjects of statistical mechanics where an increasing number of people are working [1-7]. This is due both because it is a fascinating problem on its own and because it has many physical applications ranging from high energy physics to condensed matter. It is also clear that, at least at the present, the most interesting applications are in the latter field. Suitable models of RS could describe two-dimensional networks of molecules[8] (e.g. biomembranes), (like self-avoiding walks model linear polymers) microemulsions, red blood cells and interfaces in general, surfaces of materials etc.

Here we will focus on some particular model and will discuss briefly some other.

A system which has been extensively studied recently is based on random embedding of a planar triangular network of size $L \times L$ (see fig. 1) in the d-dimensional euclidean space [6].

DFPD 2/88 February 1988

---

*Attilio Stella should be considered as a coauthor of the present report

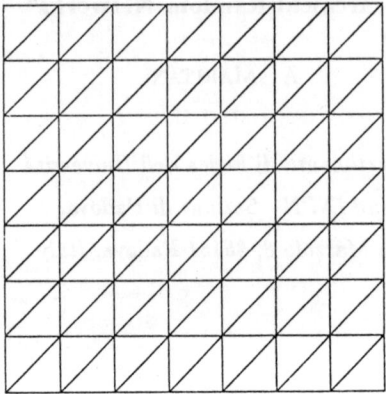

Fig. 1. Triangular Network with L=7

If i, j,... denote sites of the network and $\vec{x}_i, \vec{x}_j$,... their respective images in $\mathbf{R}^d$, the Boltzmann weight associated to a configuration $\{\vec{x}_i\}$ will be taken proportional to

$$exp(-H) = exp\{-\sum_{<ij>} v_{ij}(|\vec{x}_i - \vec{x}_j|) - \sum_{i,j} V(|\vec{x}_i - \vec{x}_j|)\}, \quad (1.1)$$

where the first sum is over nearest-neighbour sites of the network while the last one is over all couples of sites. V is an hard core potential which allows only configurations $\{\vec{x}_i\}$ such that $|\vec{x}_i - \vec{x}_j| \geq a$ for any i and j ($i \neq j$) and a is the size of the "atoms" we think the network is formed of.

The model (1.1) has been studied in ref. 6 with all $v_{ij} = v$ both with $V = 0$, i.e. without excluded volume effect, and with $V \neq 0$. In the former case a Migdal-Kadanoff approximation and numerical simulation indicate that the radius of gyration, ξ, of the RS, behaves asymptotically as

$$\xi \equiv [\frac{1}{2L^4}\sum_{i,j} < (\vec{x}_i - \vec{x}_j)^2 >]^{\frac{1}{2}} \sim \sqrt{\ln L} \quad (1.2)$$

and that the coarse grained hamiltonian should be the harmonic one i.e.

$$H_{effective} = K \sum_{<i,j>} (\vec{x}_i - \vec{x}_j)^2. \qquad (1.3)$$

The result (1.2) has been shown to be rather general at least for $d \to \infty$ (see ref. 4 and Bander and Itzykson in ref. 5) and it implies a fractal dimension of free RS

$$D = \lim_{L \to \infty} ln L^2 / ln \xi = \infty \qquad (1.4)$$

The inclusion of interaction makes, however, them non-trivial in all dimensions.

Indeed a Flory argument suggests that the fractal dimension of the self-avoiding RS (i.e. $V \neq 0$) is

$$D = \frac{d+2}{2} \qquad (1.5)$$

for any d[6,9]. In particular this prediction was tested by Monte Carlo (MC) simulation in d=3 up to L=16 [6] and the agreement is within the numerical uncertainty.

In force of the observations leading to (1.3) in the following we restrict ourselves to harmonic potentials v_{ij}, between nearest neighbours "atoms" and furthermore we allow the elastic constant K to depend on the position, i.e.

$$H = \sum_{ij} K_{ij} (\vec{x}_i - \vec{x}_j)^2 + \sum_{i,j} V(|\vec{x}_i - \vec{x}_j|). \qquad (1.6)$$

In section 3 we will be able to predict the exact asymptotic behaviour of ξ for the free RS when K_{ij} have a hierarchical distribution while an extension of the Flory argument will give us the possible asymptotic behaviours of the self-avoiding RS [10].

2. HIERARCHICAL RANDOM CHAIN (HRC).

When possible, it is always useful to make, comparison between RS and random chain (RC). Therefore before going to the very complicated case of RS we investigate for a while what happens to a linear network of L sites randomly embedded in R^d according to the statistical weight exp(-H) with H given in eq. (1.6) and

$$K_{ij} = \delta_{i,j-1} K(i), \quad K(2^n(2m+1)) = R^n, \tag{2.1}$$

$n = 0,..,N, m = 0,..,2^{N-n-1} - 1$ and $L = 2^N$ [10].

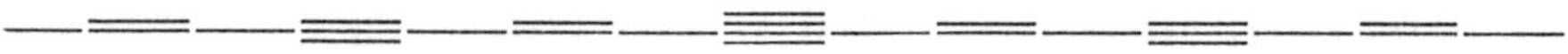

Fig. 2. Linear network involved in the definition of HRC where the bond multiplicity n gives the coupling constant strength according to (2.1).

This will be called hierarchical distribution in analogy with related problems [11–13].

In order to calculate the probability distribution for the free RC

$$P_{ij}(\vec{r}) = <\delta(\vec{x}_i - \vec{x}_j - \vec{r})> = Z^{-1} \int \prod_{k=1}^{L} d^d x_k exp(-H) \delta(\vec{x}_i - \vec{x}_j - \vec{r}) \tag{2.2}$$

($Z = \int \Pi_k d^d x_k exp(-H)$) we need the following result valid for an arbitrary network [14].

$$\int \Pi_{k \neq i,j} d^d x_k exp\{-\frac{1}{2}\sum_{\ell,m}(\vec{x}_\ell - \vec{x}_m)^2\}$$
$$= const \; exp\{(\vec{x}_i - \vec{x}_j)^2/\Omega_{ij}\} \tag{2.3}$$

where Ω_{ij} is the resistance between sites i and j of the network where conductances $K_{\ell m}$ are placed between the sites ℓ and m.

The proof is rather simple and it is sufficient to give it for d=1. The exponent in the integrand of eq. (2.3) can be rewritten as

$$\frac{1}{2}\sum_{\ell m}(x_\ell - x_m)^2 = \frac{1}{2}\sum_{\ell m} K_{\ell m}(V_\ell - V_m)^2 + \\ + \sum_{\ell m}(\sum_n K_{n\ell}\delta_{\ell m} - K_{\ell m})(x_\ell - V_\ell)(x_m - V_m) \quad (2.4)$$

where the configuration $\{V_\ell\}$ satisfies the Kirchhoff's law

$$\sum_m K_{\ell m}(V_\ell - V_m) = 0, \quad \ell \neq i,j \quad (2.5)$$

with $V_i = x_i$ and $V_j = x_j$ fixed external potentials. The dissipated power of the conductances network must satisfy the following equation

$$H\{V\} = \frac{1}{2}\sum_{\ell m} K_{\ell m}(V_\ell - V_m)^2 = (x_i - x_j)^2/\Omega_{ij} \quad (2.6)$$

where Ω_{ij} is just the total resistance between sites i and j. Eq. (2.3) follows immediately from (2.4,6).

Using the Fourier representation of the δ-function in eq. (2.2) one gets easily the interesting result

$$P_{ij}(\vec{r}) = (\frac{1}{\pi\Omega_{ij}})^{\frac{d}{2}} exp\{-r^2/\Omega_{ij}\} \quad (2.7)$$

which holds for a generic network.

For the HRC with $i = 1$ and $j = L = 2^N$ the eq. (2.7) gives the end-to-end distribution probability with

$$\Omega_{1L} = \begin{cases} \frac{R}{1-2R} R^{-N} & R < \frac{1}{2} \\ \frac{R}{2R-1} 2^N & R \geq \frac{1}{2} \end{cases} \sim L^\varsigma \quad (2.8)$$

where $\varsigma = Max\{1, |lnR|/ln2\}$ is known as the resistance exponent. It is belived that scaling laws of the type (2.8) always hold for self-similar structures like the one considered here.

From (2.7,8) the mean square distance between the chain extremes is

$$\xi^2 = <(\vec{x}_1 - \vec{x}_L)^2> = \frac{d}{2}\Omega_{1L} \sim L^\varsigma \qquad (2.9)$$

implying a fractal dimension $D = \lim_{L\to\infty} lnL/ln\xi = 2/\varsigma$ for the free HRC.

It is interesting to notice that if one chooses an independent random distribution of conductance for each couple of nearest neighbour sites of the type

$$\rho(K) = \theta(K_o - K)\frac{1-\alpha}{K_o}(K/K_o)^{-\alpha} \qquad (2.10)$$

with $0 < \alpha < 1$ then the (quenched) average of (2.7) for the case $i=1$ and $j=L$ becomes

$$\overline{P}_{1L}(\vec{r}) \equiv \int \prod_{i=1}^{L-1}[dK_{i,i+1}\rho(K_{i,i+1})]P_{1L}(\vec{r}) = r^{-d}f(L/r^{2(1-\alpha)}) \qquad (2.11a)$$

$$f(x) \propto \int_o^\infty dy y^{d-1} \int_o^\pi d\theta \, sin^{d-2}\theta exp\{iycos\theta - cxy^{2(1-\alpha)}\} \qquad (2.11b)$$

and $c = \Gamma(\alpha)4^{\alpha-1}$.

Eq. (2.11a) implies that the fractal dimension in this case is

$$D = 2(1-\alpha) \qquad (2.12)$$

For the hierarchical distribution (2.1) the fraction of conductances in the interval $(K, K+\Delta K)$ goes like $K^{-\alpha}\Delta K$ in the $K \to o$ limit with $\alpha = ln(2R)/lnR$ and $R < \frac{1}{2}$. Surprisingly enough the result (2.12) with this value of α coincides with the one obtained for the HRC ! $^{(10)}$ (related problems have the same peculiarity $^{(12)}$).

One could show rather easily that for the random distribution of conductances (2.10) the resulting configurations for the RC have the same asymptotic behaviour as the trajectories of the Levy flights $^{(15)}$ where the probability density for a step $\vec{\ell}$ behaves as

$$p(\vec{\ell}) \propto |\vec{\ell}|^{-d-\mu} \tag{2.13}$$

for large $|\vec{\ell}|$ and $\mu = 2(1-\alpha)$.

Recently self-avoiding chains with step distribution of the type (2.13) have been studied by several authors $^{(16-18)}$. In principle one should distinguish between two types of self- avoidance constraint: path avoiding and node avoiding chain. In the former case one requires that the polygonal trajectory does not intersect itself while in the latter case one requires only the existence of an excluded volume effect for the vertices of the polygonal.

At least in 1-d these seem to belong to different universality classes $^{(18)}$.

In particular the path avoiding chain in d=1 with the hierarchical distribution (2.1) is trivial and one gets

$$<|x_1 - x_L|> \propto \begin{cases} 2^N & R > \frac{1}{4} \\ R^{N/2} & R \leq \frac{1}{4} \end{cases} \tag{2.14}$$

i.e. a fractal dimension

$$D = Min\{1, \frac{\ell n 4}{|\ell n R|}\} \tag{2.15}$$

Thus for $R < \frac{1}{4}$ D is the same as the non-interacting chain i.e. the path-avoiding constraint is irrelevant. This result is just the same as of the random case (2.10) with $\alpha = \ell n(2R)/\ell n R$ and the same of ref. 18 with the distribution (2.13).

More difficult is the case of the node avoiding chain, with both hierarchical and random distribution. For the node avoiding Lévy flight there exist (MC) simulations in d=1 [17,18] and 2 [16] and ϵ-expansion around the upper critical dimension $d_c = 2\mu$[16].

A Flory approximation can be devised for the node-avoiding chain based on a energy-entropy argument. According to (2.7) with i=1 and j=L a mean field entropy, S, for a chain with a fixed end-to-end distance (= ξ), can be estimated proportional to ξ^2/L^ς where $\varsigma = 2/\mu$ and L^ς is proportional to the average of Ω_{1L}

The (repulsive) energy, U, due to two body interaction can be taken proportional to L^2/ξ^d. The minimization of the free energy $F(\xi) = U - TS$ gives $\xi \sim L^D$ with

$$D = \begin{cases} \frac{d+2}{2+2/\mu} & d < 2\mu \\ \mu & d \geq 2\mu \end{cases} \qquad (2.16)$$

Another proposal for the free energy [16,17] based on an entropy $\sim \xi^\mu/L$ gives $D = (d+\mu)/3$ for $d < 2\mu$.

Even if MC results are not so accurate they seem to give more support for (2.16) than for the last proposal. Indeed $(d+\mu)/3$ is belived to underestimate the true value of D[18] while $(d+2)/(2+2/\mu) > (d+\mu)/3$.

3. HIERARCHIRAL RANDOM SURFACES (HRS) [10].

The generalization to RS of the distribution (2.1) is straightforward and it is shown in figure 3 where the same conventions as in the previous figure are adopted.

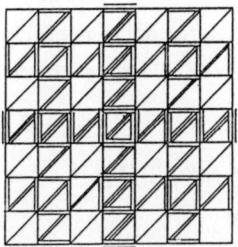

Fig. 3 with L=8.

The radius of gyration of the free RS will have, according to (2.7), the asymptotic behaviour

$$\xi \equiv [\frac{1}{2L^4} \sum_{i,j} < (\vec{x}_i - \vec{x}_j)^2 >]^{\frac{1}{2}}$$
$$\sim [\frac{1}{2L^4} \sum_{i,j} \Omega_{ij}]^{\frac{1}{2}} \sim L^{s/2} \qquad (3.1)$$

In order to calculate ξ for the hierarchical distribution we have to use a relation existing between the resistance in the network with conductances K_{ij} and the diffusion problem of a particle on the same network with transition rate probability between sites i and j proportional to K_{ij}. If the mean square distance of the diffusing particle behaves at long time as

$$< r^2(t) > \sim t^{2/d_w} \qquad (3.2)$$

then

$$\varsigma = d_w - d_F \qquad (3.3)$$

where d_F is the fractal dimension of the network (=2 in the present case). Eq. (3.3) is known as Einstein relation [19].

Then we have shifted the problem to the calculation of the fractal dimension d_w of a random walk in the structure shown in fig. 3. If we eliminate all the diagonal bonds in that figure then the diffusion problem can be solved exactly by a renormalization group technique of the type introduced in ref. 13.

The result is

$$dw = \begin{cases} ln(2/R)/ln2 & R < \frac{1}{2} \\ 2 & R \geq \frac{1}{2}. \end{cases} \qquad (3.4)$$

The original problem is however a little bit more complicated. Nevertheless we believe that the result (3.4) still holds also for this case. Indeed a scaling argument in support of it goes along the following line [20].

If all transitions rates were equal the diffusion process would be characterized by a fractal dimension $d_w^{(o)}$ (=2 in this case) i.e. after a time t the particle visits a region of linear size $L \sim t^{1/d_w^{(o)}}$. In the hierarchical network of fig. 3 a region Λ of linear size L is characterized by a total residence time $\tau(L) = \sum_{i,j \in \Lambda} K_{ij}^{-1}$ which scales like

$$\tau(L) \sim L^{d_\tau}, \quad d_\tau = Max\{2, ln(2/R)/ln2\} \qquad (3.5)$$

for large L. In terms of a temporal scale renormalized according to the average residence time per site the diffusion must be normal i.e.

$$L \sim (t/L^{d_\tau - d_F})^{1/d_w^{(o)}} \qquad (3.6)$$

Implying [20]

$$d_w = d_r + d_w^{(o)} - d_F. \qquad (3.7)$$

Eq. (3.7) with $d_w^{(o)} = d_F = 2$ together with (3.5) gives (3.4) again.

From the Einstein relation (3.3) and eq. (3.1) the fractal dimension for the HRS turns out to be

$$D = \frac{4}{\varsigma} = \begin{cases} 4\ell n 2/|\ell n(2R)| & R < \frac{1}{2} \\ \infty & R \geq \frac{1}{2} \end{cases} \qquad (3.8)$$

Thus for $R = \frac{1}{2}$ we have a sort of transition induced by the dynamical transition in the related diffusion problem [12]. For $R > \frac{1}{2}$ the asymptotic behaviour of HRS is the same as the one for ordinary RS considered in ref. 6 while for $R < \frac{1}{2}$ the fractal dimension $D < \infty$ and a finite upper critical dimension, $d_c = 2D = 8/\varsigma$ results for self-avoiding HRS.

A Flory argument for the excluded volume problem based on a free energy $F(\xi) = U - TS$ with $S \propto \xi^2/L^\varsigma$ and $U \propto L^4/\xi^d$ gives, after minimization with respect to the radius of gyration ξ, a fractal dimension

$$D = \begin{cases} (d+2)/(2+\varsigma/2) & d < 8/\varsigma \\ 4/\varsigma & d > 8/\varsigma \end{cases} \qquad (3.9)$$

which becomes (1.5) for $\varsigma = 0$.

It would be extremely interesting to know what happens if one chooses a random distribution of "conductances" consistent with the hierarchical one.

Already for free RS it is difficult to believe that, the resulting fractal dimension remains unchanged as it occured for the linear chain. This and other related issues are presently under investigation.

We would like to remark that relations like the one for free HRS, $D = 4/\varsigma$ and (3.9) for the self-avoiding case are similar to the ones obtaines in ref. 9 for a random gaussian embedding of self similar network (polymeric fractals) with H given in eq. (1.3). In this case non-trivial behaviour comes from the basic network structure while in our case it comes from the hamiltoniam (1.6). It is this difference that allowed us to give a concrete example of non-trivial RS.

4. CONCLUSION AND SOME PERSPECTIVES.

In this last section we would like to discuss briefly some general model of RS on cubic lattice.

Let us consider an emsemble $\mathcal{E}$ of closed self-avoiding RS made up of elementary plaquettes* whose partition function is defined as

$$Z = \sum_{\mathcal{E}} exp\{\beta(\mu_s S + \mu_c C + \mu_\chi \chi + \mu_N N + \mu_v V)\}, \qquad (4)$$

where

S=number of plaquettes in the ensemble $\mathcal{E}$,

C=number of couple of plaquettes at right angle,

χ=(number of sites in $\mathcal{E}$) -(number of links in $\mathcal{E}$) + S=Euler characteristic of $\mathcal{E}$

N=number of closed surfaces,

V=total volume enclosed by the surfaces,

and μ_s, μ_c etc. are their respective chemical potentials.

Such a model should be of some interest for the microemulsion problem. A microemulsion is a system of oil + water + surfactant [21] in equilibrium where the surfactant molecules saturate the oil/water interface [22]. To this purpose $-\mu_s$ is the bare surface tension, $-\mu_c$ is an energy associated to the mean curvature of the surface while $-\mu_\chi$ is associated to its Gaussian curvature, μ_N controls the number of connected components of the ensemble and μ_v is the relative chemical potential of the water and oil.

---

*Contact point are allowed but each link is shared at most by two plaquettes

Some particular case of (4.1) has been already considered in the recent literature. The $\mu_\chi \to +\infty$, $\mu_N \to -\infty$, limit, i.e. a single surface without handles, and $\mu_c = \mu_v = 0$ has been considered in ref. 1 and 2. Renormalization group analysis and a Flory argument indicate for them an upper critical dimension dc=8. [1].

However MC simulations of ref. 2 give contradictory predictions. In the limit of large dimensionalities rather general model of RS are dominated by configurations which are ramified structures of tubes [23] making RS very similar to branched polymers. The MC simulation in the first paper of ref. 2 seems to indicate that this situation persists in d=3 for self-avoiding RS.

A renormalization group calculation for this model with $\mu_c < 0$ would predict a "low temperature" phase where the surface is flat with fractal dimension D=2 [24]. On the basis of these results one could think therefore that there exists a critical regime for this model with two phases: the high rigidity and the branched polymer one separated by a tricritical point where the surface should be a real two dimensional object (not a ramified structure) with a non trivial D. It is curious that the above mentioned phases should have both D=2 (indeed the fractal dimension of branched polymer in d=3 is 2 [25].

The $\mu_c = \mu_N = \mu_v = 0$ case has been considered in ref. 26. This model and the one where it has been taken into account the inclusion of interactions for the contact points of different closed surfaces and for the links shared by more than two plaquettes (in the case of a relaxed self-avoiding constraints) shows a rather rich phase diagram.

The two dimensional analog of (4.1), i.e. a loop gas, in a square lattice, with a chemical potential for the total perimeter, for the total number of corners and for the number of contacts sites, is known to be equivalent to an eight-vertex model [27]. The resulting phase diagram is extremely rich and also a phase boundary with non-universal critical behaviour is present.

A generalization of this model for RS has been proposed by Karowski in ref. 2.

REFERENCES

1. Maritan A. and Stella A.L., Phys. Rev. Lett. 53, 123 (1984) and Nucl. Phys. B280, 561 (1987).
2. Glauss U. and Einstein T.L., J. Phys. A (1986); Karowski M., ibid A19, 3375 (1986).
3. Cates M.E., Phys. Lett. 161B, 363 (1985).
4. Gross D.J., Phys. Lett. 138B, 185 (1984): Duplantier B., ibid 141B, 239 (1984).
5. Boulatov D.V., Kazakov V.A., Kostov J.K. and Migdal A.A., Nucl. Phys. B275, 641 (1986); David F., ibid B257, 543 (1985); Bandes M. and Itzykson C., ibid. B257, 531 (1985); Ambjorn J., Durhuus B., Fröhlich J. and Orland P., ibid B270, 457 (1986).
6. Kantor Y., Kardar M., and Nelson D.R. Phys. Rev. Lett. 57, 791 (1986) and Phys. Rev. A35 3056, (1987).
7. Nelson D.R. and Peliti L., J. Physique 48, 1085 (1987).
8. See ref. 6,7 and references therein.
9. Cates M.E., Phys. Rev. Lett. 53, 926 (1984).
10. Maritan A. and Stella A.L., Phys. Rev. Lett. 59, 300 (1987).
11. Huberman B.A. and Kerszberg M., J. Phys. A18, L331 (1985).
12. Teitel S. and Domany E., Phys. Rev. Lett. 55, 2176 (1985) and 56, 1755 (1986).
13. Maritan A. and Stella A.L., J. Phys. A19, L269 (1986) and Phys. Rev. Lett. 56 1754 (1986).
14. Stephen M.J., Phys. Rev. B17, 4444 (1978); Coniglio A. in "Magnetic Phase Transitions", ed. by Ausloos M. and Elliot R.J., Springer Series in Solid-State Sciences, vol. 48 (Berlin 1983).
15. See e.g. Montroll E.W. and West B.J. in "Fluctuation Phenomena" ed. Montroll E.W. and Lebowitz J.L. (North-Holland 1979).
16. Halley J.W. and Nakanishi H., Phys. Rev. Lett. 55, 551 (1985).
17. Grassberger P., J. Phys. A18, L833 (1985).
18. Lee S.B., Nakanishi H. and Derrida B., Phys. Rev. A36, 5059 (1987).

19. Gefen Y., Aharony and Alexander S., Phys. Rev. Lett. 50, 77 (1983); for a derivation see also Cates M.E. J. Physique 45 1059 (1985).
20. See also Machta J., J. Phys. A19, L531 (1985) and Tremblay M.S. Phys. Lett. A116, 329 (1986).
21. See e.g. Jouffroy J., Levinson P. and de Gunnes P.G., J. Physique 43, 1241 (1982) and references therein.
22. Ober R. and Taupin C., J. Phys. Chem.84, 2418 (1980).
23. Drouffe J.M., Parisi G. and Sourlas N., Nucl. Phys. B161, 397 (1979).
24. Maritan A. and Stella A.L., unpublished results.
25. Parisi G. and Sourlas N., Phys. Rev. Lett. 46, 871 (1981).
26. Karowski M. and Thun H.J., Phys. Rev. Lett. 54 2556 (1985). See also Karowski M. in ref. 2.
27. Rys F.S., Phys. Rev. Lett. 51, 849 (1983).

LYAPUNOV EXPONENTS & CO.

R. LIMA
Centre de Physique Théorique,C.N.R.S.
Luminy Case 907
F-13288 Marseille Cedex 9,France

1. LYAPUNOV EXPONENTS

In many domains of physics it is important to have an insight into the asymptotic properties, in time or space, of the solutions of the equations entering into the problem.

Lyapunov exponents are one of the tools which have been used, since a long time, in order to express these types of properties.

The aim of this note is to describe some new results in this field. Even if the localizaton properties for the solutions of the Schrödinger equation are not in the scope of this note, we refer to [1] and references therein for a related treatment of the subject, including some material used during the conference.In this case, physical applications appear in solid state physics.

In particular, section 3 is directly inspired by Thouless formulae relating integrated densities of states and Lyapunov exponent in solid state physics models, which can be considered as a linear version of our result.

Instead, the non linearity is an essential tool for the understanding of almost all physical problems where dynamical instabilities are present. This is certainly the case if we try to describe a plasma to be confined in a Tokamak, using a strong toroidal magnetic field together with a poloidal field.In this case the stochasticity of the motion was first attributed to collision effects.But in case of high temperatures as for the Tokamak plasma, where collisions are rare events, the stochasticity of the particle orbits can, in principle, be understood as an effect of non linear fluctuations of the magnetic field, resulting for instance from short wavelength plasma microinstabilities.In this case, neighboring orbits diverge from each other exponentiallly, and the correspondent rate is exactly the Lyapunov exponent, which is then interpreted as the inverse of a correlation lenght.

Section 2 contains results obtained in collaboration with S. Ruffo [2].

Section 3 is partially in [3], which is a collaboration with A. Lambert and R. Vilela Mendes.

Section 4 is a small part of [4], a work done with M. Rahibe.

2. SCALING DIAMONDS.

The Poincaré map of an hamiltonian system of k degrees of freedom is a sympletic map of $\mathbb{R}^{2k}$. The corresponding tangent map is given by matrices of the following form, A being k x k matrices:

$$A = \begin{pmatrix} 1 & 1 \\ A & 1+A \end{pmatrix} \qquad (1)$$

For an integrable hamiltonian, expressed in action angle variables, A=0 at any step and therefore Lyapunov exponents, defined as the logarithms of the eigenvalues of the matrix:

$$\Lambda = \lim_{m \to \infty} (M_m^* M_m)^{1/2m} \qquad (2)$$

where

$$M_m = A_m \cdot A_{m-1} \ldots \cdot A_1 \qquad (3)$$

are all zero.

For near integrable systems KAM tori still exist and only Arnold diffusion is possible (if k > 2); far from integrabily, orbits became chaotic and a random approximation of the Poincaré map using a stochastic distribution of the matrices A is commonly accepted and numerically justified.

In this context it is an important question to know how Lyapunov exponents scale with $\| A \|$ since these exponents are related to the rate of exponential growth of the volume forms in phase space. In particular the greatest exponent λ_{max} characterizes, roughly speaking, the mean exponential rate of divergence of nearby trajectories.

In [5] G. Parisi and A. Vulpiani derive an estimation of the scaling of λ_{max} as a power law of $\| A \|$. These authors obtain the exponent 2/3 or 1/2 according to the case when the mean value of the elements of the A's is zero or different from zero.

We introduce a model reproducing and, up to a certain extend, explaining the scaling behaviour of all the Lyapunov exponents.

For 2 x 2 matrices we define a diamond to be a symmetric matrix with identical elements on the diagonal. A diamond flush is a 2q x 2q matrix which is a tensor product of q diamonds.

Now, any diamond A_m can be diagonalised by unitary matrices of the following universal (independent of A_m) form :

$$u = \frac{1}{\sqrt{2}} \begin{pmatrix} 1 & 1 \\ 1 & -1 \end{pmatrix} \quad (4)$$

The corresponding eigenvalues are the sum s_m and difference r_m of the two numbers appearing as elements of the diamond matrix A_m and therefore the mean value of the second eigenvalue is zero.

On the other hand $u \otimes 1$ is an unitary matrix and if we use it in order to perform a transformation of the A_m., which doesn't change Lyapunov exponents, we only change A_m in the corresponding equivalent diagonal matrix.

It turns out that any 4 x 4 matrix A_m for which A_m is diagonal is equivalent to a direct sum of two 2 x 2 matrices each one being of the form (1). This explains, acording to [5], why, in this case, the maximun Lyapunov exponent corresponding to s_m scales with a 1/2 power law and the second Lyapunov exponent corresponding to r_m scales instead with power 2/3.

The case of diamond flushes is similar since then any matrix A_m is diagonalised by $u \otimes u \otimes ... \otimes u$ and, therefore, all but one of the eigenvalues are of zero mean.(eigenvalues are the products of sums and differences of the elements of the diamonds used to build the diamond flush) Note that it is possible to introduce more than one eigenvalue with non zero mean if we change the means of some of the elements of the matrix in order to destroy the initial simmetry of the random process and create differences with non zero means.

Again, for diamond flushes, we use the equivalence with a direct sum of independent processes in order to conclude.

Obviously, in the general case, we can not diagonalise each of the matrices of the product in a n-independent way. But we can see, thanks to the assymptotic behaviour of the product of this type of matrices that, for large m, the behaviour of the product becomes like the one of diamonds.

3. LYAPUNOV, ROTATION AND TRANSIENT.

The rotation angle is an important tool in studing stability problems in dynamical systems since it is well known, after Kolmogoroff, that the robustness of an invariant circle under perturbations is closely related to the diophantine properties of the rotation number of the motion along the circle.

On the other hand, Lyapunov exponents are certainly good indicators of the appearance of stochastic behaviour in the dynamics and therefore a relation of the two notions might bring some enlightment in the transition to chaos.

In [2] we use the rotation angle introduced by Ruelle, [6], for diffeomorphisms on a two-dimensional manifold, which preserve a finite measure of compact support.

The rotation angle is essentially the mean of the differences of the angles appearing in the polar decomposition of the tangent map matrices, for which integer multiples of 2π have been taken into account during the matrix multiplication of the maps.

We have found a dispertion relation between the two quantities in that case where they are reasonable(at least measurable) functions of the parameter.

For the square of the Hénon map:

$$\binom{x}{y} \rightarrow \binom{1 + bx - k(1 + y - kx^2)^2}{b(1 + y - kx^2)} \tag{5}$$

the relation is particulary simple along the bifurcation road:

$$\{\lambda(k) - \log b\} = \pm 2\, P \int \frac{\omega(k') - q/2p}{k' - k}\, dk' \tag{6}$$

Fig. 1 shows that, comparing the actual value of the Lyapunov exponent with the value coming from the formula above, via the rotational angle, one finds a reasonable agreement.

It is an interesting and somewhat surprising consequence of this result that the rotation angle stays constant on the two sides of each period-doubling bifurcation (the point where the Lyapunov exponent is zero) and changes only when this exponent becomes constant (equal logb).

Fig. 2 shows how the transient motion around the fixed points, which are atractors in this case, undergoes a bifurcation in between the correspondent doubling period bifurcation. This is also clearly conected to the values of the parameter for which the eigenvalues are real or immaginary

The situation is much more complicated in the case of hamiltonian systems since in this case the dependence of the invariant measure with respect to the change of the parameter is a very intricate question.Nevertheless it is known that, even after the breaking of an invariant circle, the rotation angle doesn't change for the motion on the remaining invariant Cantor set, the so-called cantorus. And this is clearly not in contradiction with the possible persistence of a similar behaviour in that case.

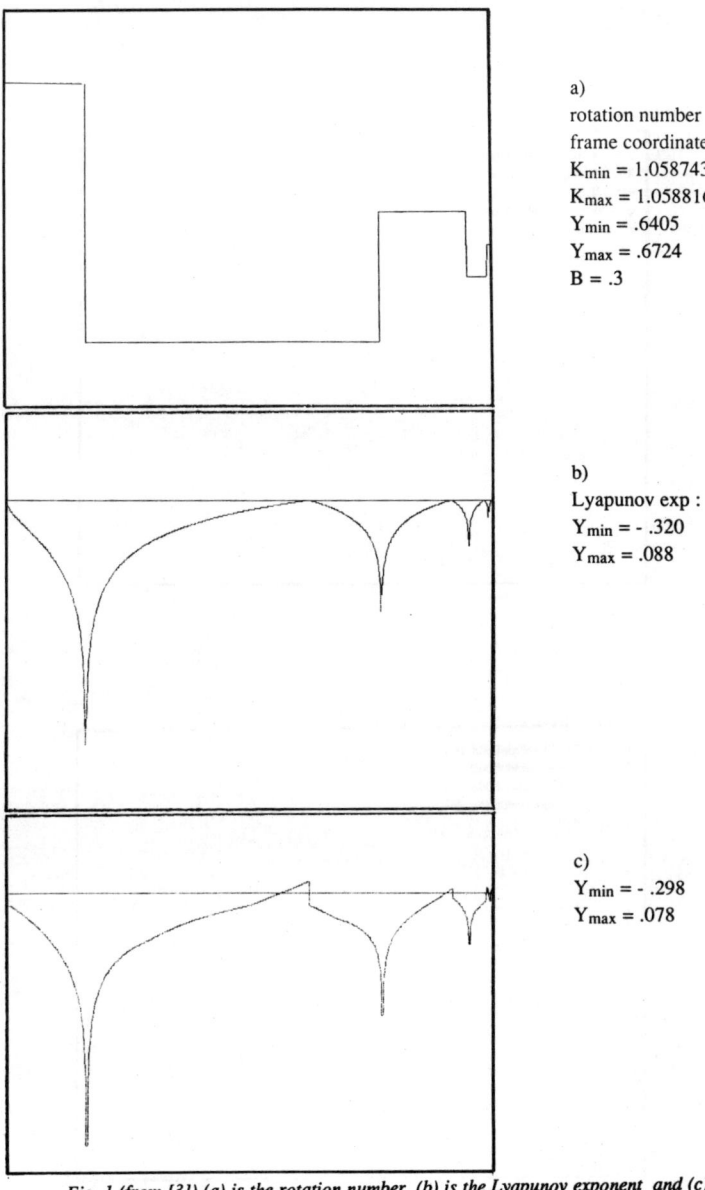

a)
rotation number :
frame coordinates
$K_{min} = 1.058743$
$K_{max} = 1.058816$
$Y_{min} = .6405$
$Y_{max} = .6724$
$B = .3$

b)
Lyapunov exp :
$Y_{min} = -.320$
$Y_{max} = .088$

c)
$Y_{min} = -.298$
$Y_{max} = .078$

Fig. 1 (from [3]) (a) is the rotation number, (b) is the Lyapunov exponent and (c) is the computed lyapunov exponent as it cames from the dispersion relation.

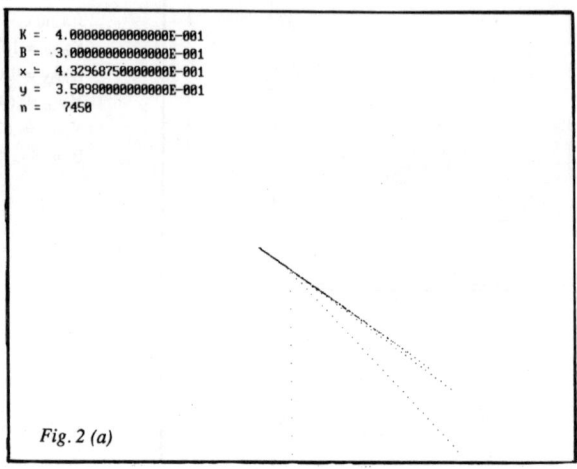

Fig. 2 (a)

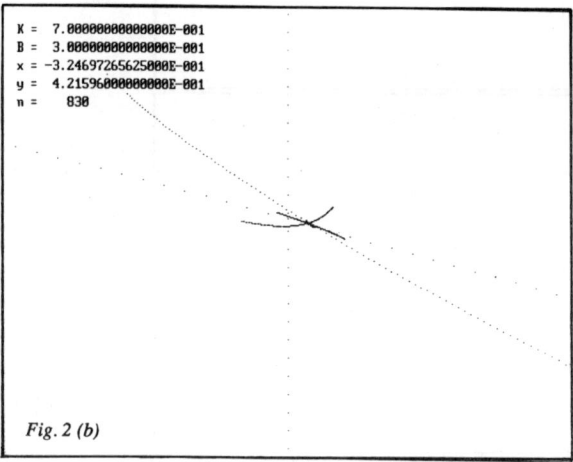

Fig. 2 (b)

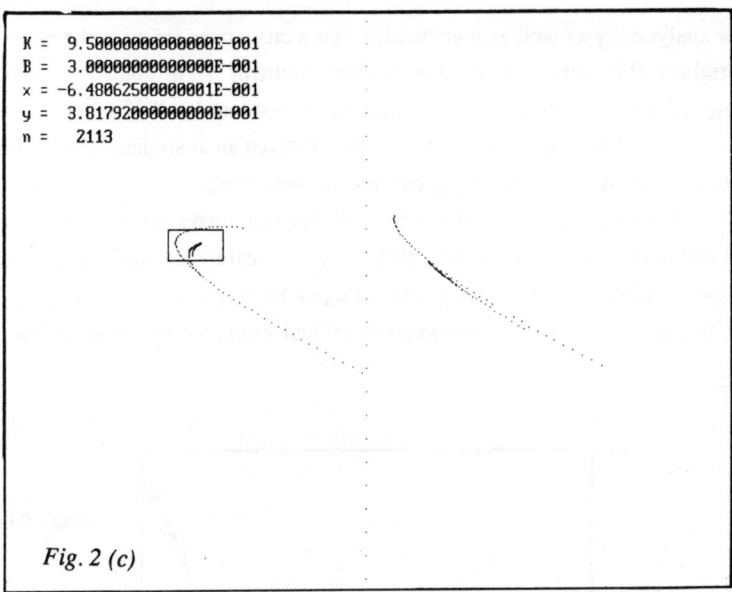

Fig. 2 (c)

Fig.2 (from [3]) Period doubling bifurcation separeted by a modification of the transcient related with the change of the rotation angle.

4. SCALING MORE SUBTLY

In section 2 the scaling laws of Lyapunov exponents are studied in terms of the size of the perturbation.

However such a simple scheme is not completely satisfactory from a physical point of view. The reason being that the statistical properties of the perturbation are far from being described only in terms of their size. This is clear, at least in the case mentioned above of anomalous transport of particles in plasma physics since in this case the second moment of the distribution.of the magnetic fluctuation is related to the widing number of the resonance under consideration.

In order to keep track of the subtle complexity of such stochastic behaviour we take in account the first and the second moment of the process.With M. Rahibe, in [4], we show analytically as well as numerically, that a cross over exists in the plane of the two variables (the mean α and the second moment γ) in such a way that for $\gamma < \gamma_{critic}(\alpha)$ we recover the same scaling as in section 2 but if $\gamma \geq \gamma_{critic}(\alpha)$ we discover a scaling law with respect to γ. We have derived an analytical formula for γ_{critic} as a function of α, as well as the exponents for this new scale.

Fig. 3 gave a numerical ilustration of this result.here the mean α is keeped constant and only γ varies. For small γ only enters the scaling law with respect to α and we can see a constant Lyapunov exponent,its value being fixed by the above described relation. Instead as soon as γ goes beyond its critical value, the new scaling law clearly appears.

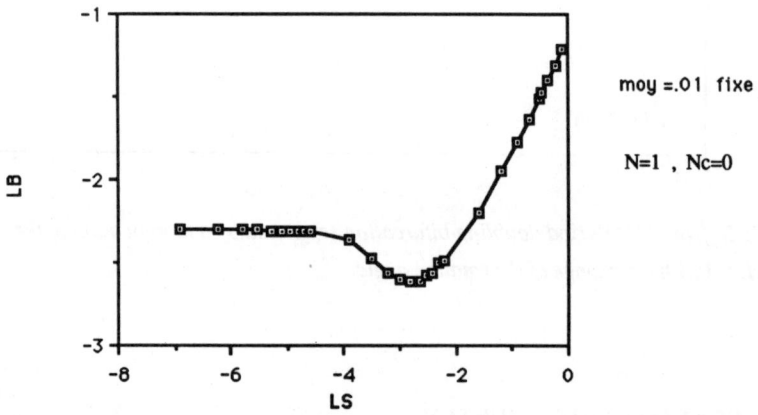

Fig. 3 (from [4]) Lyapunov exponent as a function of γ, in log-log scale, for α constant.

5. CONCLUSIONS

Further conclusions, if any is avaiable, would require a considerable amount of energy.

REFERENCES

[1] Lima, R."Problems Related to Dynamical Systems", in Proceedings of "Renormalization Group-86", Dubna, U.R.S.S..

[2] Lima, R. and Ruffo, S., "Scaling Laws for all Lyapunov Exponents : Models and Measurements"Journ. of Statistical Physics 52 1/2 (1988).

[3] Lambert, A., Lima, R. and Vilela Mendes, R., "Rotation Number and Lyapunov Exponent in Two Dimensional Maps", Preprint Marseille 1988.

[4] Lima, R. and Rahibe, M., to appear soon.

[5] Parisi, G. and Vulpiani, A., J. Phys. A : Math. Gen. 19 (1986).

LYAPUNOV EXPONENTS FOR PRODUCTS OF RANDOM MATRICES IN CONDENSED MATTER AND DYNAMICAL SYSTEMS

Angelo VULPIANI

Dipartimento di Fisica, University of Roma "La Sapienza"
P.le A.Moro, 2 - 00185 Roma
and
GNSM-CISM - Unità di Roma

Abstract

Lyapunov exponents for products of random matrices, related to problems of condensed matter and dynamical systems, are analized both numerically and analitically. The maximum Lyapunov exponent exhibits a power-law behaviour as function of a coupling constant; moreover the spectrum of all the Lyapunov exponents shows a limit distribution in the limite of large size of the matrices. These results are in agreement with those obtained for hamiltonian systems and symplectic maps.

1. - Why Study Products of Random Matrices?

Many problems in physics can be reduced to products of random matrices [1,2,3]. Let us briefly discuss some examples. For a disordered Ising chain with hamiltonian

$$H = - \sum_i J_i \sigma_i \sigma_{i+1} - \sum_i h_i \sigma_i \qquad (1)$$

the random transfer matrix M_i has the following form

$$M_i = \begin{pmatrix} e^{\beta(J_i + h_i)} & e^{\beta(-J_i + h_i)} \\ e^{\beta(-J_i - h_i)} & e^{\beta(J_i - h_i)} \end{pmatrix} \qquad (2)$$

where J_i and h_i are respectively the random coupling and the random field and $\beta = 1/KT$. The free energy is

$$f(T) = -KT \lim_{n \to \infty} \frac{1}{n} \ln(Tr(\prod_{i=1}^{n} M_i)) \qquad (3)$$

It is easy to see that

$$f(T) = -KT \lambda_1 \qquad (4)$$

where λ_1 is the maximum Lyapunov exponent for the product of random matrices M_i [3,4]:

$$\lambda_1 = \lim_{n \to \infty} \frac{1}{n} \ln \frac{|\zeta(n)|}{|\zeta(0)|} = \lim_{n \to \infty} \frac{1}{n} \ln (Tr(\prod_{i=1}^{n} M_i)) \qquad (5)$$

where $\zeta(n) = M_n M_{n-1} .... M_1 \zeta(0)$ and $\zeta(0)$ is a generical vector $\in \mathbf{R}^2$.

Another example is given by the discretized Schroedinger equation on a one-dimensional lattice with a random potential εV_i on each site i:

$$\psi_{i+1} + \psi_{i-1} + \varepsilon V_i \psi_i = E \psi_i \qquad (6)$$

This equation can be written in a recursive form:

$$z(i) = A_\varepsilon(i) z(i-1)$$

where

$$A_\varepsilon(i) = \begin{pmatrix} E - \varepsilon V_i & -1 \\ 1 & 0 \end{pmatrix} \qquad (6')$$

and

$$z(i) = \begin{pmatrix} \psi_{i+1} \\ \psi_i \end{pmatrix}$$

In this case λ_1 for the product of the matrices $A_\varepsilon(n)$ is related to the characteristic length ξ of the localized wave functions [5]:

$$\lambda_1 = \xi^{-1}$$

Let us now consider a sequence of 2Nx2N symplectic random matrices $A_\varepsilon(i)$:

$$A_\varepsilon(i) = \begin{pmatrix} \mathbb{1} & \mathbb{1} \\ \varepsilon\,a(i) & \mathbb{1} + \varepsilon\,a(i) \end{pmatrix} \qquad (7)$$

where $\mathbb{1}$ is the NxN identity matrix and a(i) is a symmetric NxN random matrix.

One can see that products of matrices with form (7) are related to the following symplectic map:

$$\begin{aligned} q(i+1) &= q(i) + p(i) \qquad \mod 2\pi \\ p(i+1) &= p(i) + \varepsilon\,\nabla F(q(i+1)) \end{aligned} \qquad (8)$$

where $q, p \in \mathbb{R}^N$ and $\nabla = (\partial/\partial q_1, \ldots \partial/\partial q_N)$.

Indeed in the computation of λ_1 for the map (8) one has to consider the product of matrices with the following form:

$$B_\varepsilon(i) = \begin{pmatrix} \mathbb{1} & \mathbb{1} \\ \varepsilon\,b(i) & \mathbb{1} + \varepsilon\,b(i) \end{pmatrix} \qquad (9)$$

where $b_{ij}(k) = \partial^2 F(q(k))/\partial q_i \partial q_j$ with $q(k)$ given by the evolution equation (8). The matrices $B_\varepsilon(i)$ can be considered in a certain sense "random" when the dynamical system (8) is chaotic. Therefore in a first crude (but not trivial!) approximation we can consider, for the features of the Lyapunov exponents, the product of the random matrices $A_\varepsilon(i)$ instead of $B_\varepsilon(i)$. The randomness of $A_\varepsilon(i)$ mimics the chaoticity of the trajectory generated by (8).

2. - Scaling laws and thermodynamic limit

Let us consider matrices with the form (7) where a_{ij} will have non-zero value only for $|i-j|\leq 1$ and $(i,j)=(1,N)$ or $(N,1)$ [6]. We have chosen this particular form in order to simulate nearest-neighbour coupled maps. The non-zero elements of $a(i)$ are generated according to different probabilistic laws:

$$a_{ij} = \frac{1}{2} (X_m)^\alpha + \bar{x} \tag{10}$$

where $\bar{x}$ is a fixed value, α is an odd integer and x_m is a random number which is uniformly distributed in the interval $(-1,1)$ when $m=1$, and has a gaussian distribution with zero mean and unit variance when $m=2$. Moreover, one can impose the analogous of the conservation of total momentum $\Sigma_i p_i(k)$=const by requiring the constraint

$$\sum_{j=1}^{N} a_{ij} = 0 \tag{11}$$

The maximal Lyapunov exponent λ_1 displays scaling laws for small ε [6], see Fig.(1):

$$\lambda_1(N) = C(N) \varepsilon^\beta \tag{12}$$

where $\beta=1/2$ if $\bar{x}>0$ and $\beta=2/3$ if $\bar{x}=0$.

This result has been already obtained by Benettin [3] in the case $N=1$. Moreover it is relevant to stress that scaling laws with $\beta=1/2$ or $2/3$ have been found for some two-dimensional billiards [3] and in some conservative maps [7].

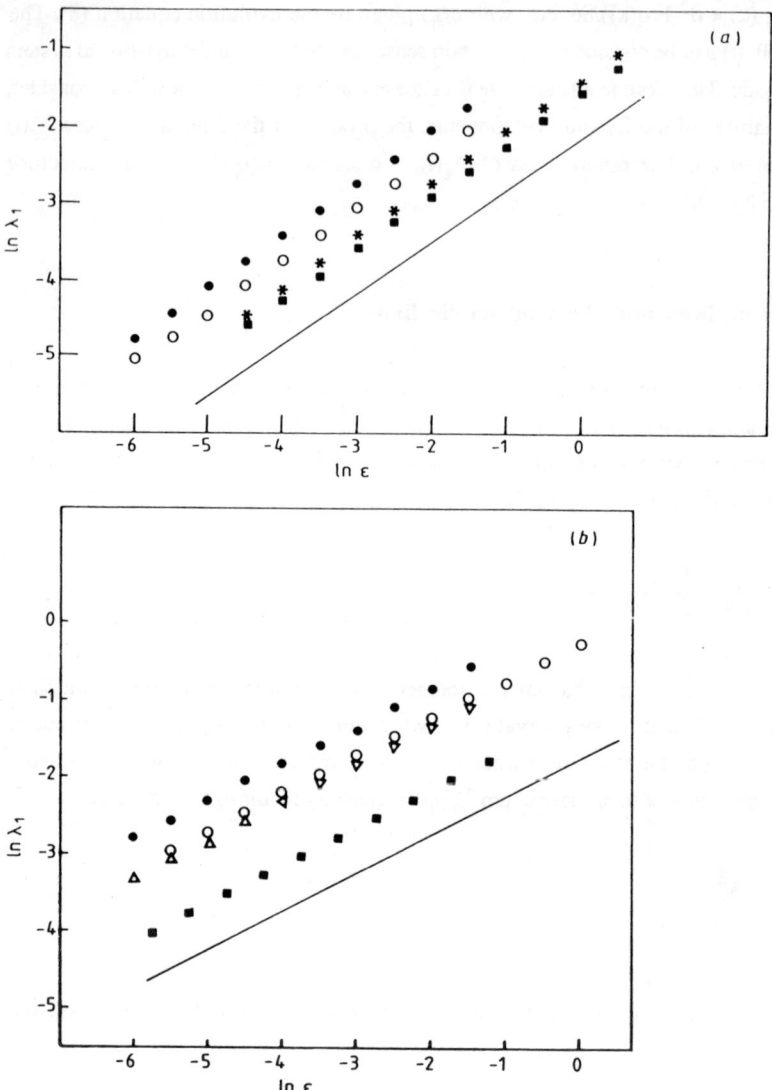

Figure 1. $\ln \lambda_1$ plotted against $\ln \varepsilon$.: (a) $\bar{x} = 0$; ○: $N = 4$, $\alpha = 1$, $m = 1$, constraint (11) imposed; ●: $N = 6$, $\alpha = 1$, $m = 2$, constraint (11) not imposed; ∗: $N = 4$, $\alpha = 3$, $m = 1$, constraint (11) not imposed; ■: $N = 4$, $\alpha = 5$, $m = 1$, constraint (11) not imposed. The line indicates the slope $\frac{2}{3}$; (b) $\bar{x} \neq 0$; ●: $N = 5$, $\alpha = 3$, $m = 2$, $\bar{x} = 0.5$, constraint (11) not imposed; △: $N = 7$, $\alpha = 3$, $m = 2$, $\bar{z} = 0.5$, constraint (11) imposed; ○: $N = 4$, $\alpha = 3$, $m = 1$, $\bar{x} = 0.2$, constraint (11) not imposed; ■: $N = 8$, $\alpha = 1$, $m = 1$, $\bar{x} = 0.1$, constraint (11) imposed. The line indicates the slope $\frac{1}{2}$.

Before going on with the discussion we want to remark that the same scaling law (12) with $\beta=2/3$ holds for the matrix (6') at the band edge (i.e. E=2). Moreover eq.(12) has been obtained analitically for matrices with form (7) [9,10].

One can see that, for a given value of ε, λ_1 quickly approaches to an asymptotic value with increasing N, see Fig.(2). This result may indicate the existence of a sort of "thermodynamic limit" for many properties of the "dynamics" generated by products of random matrices.

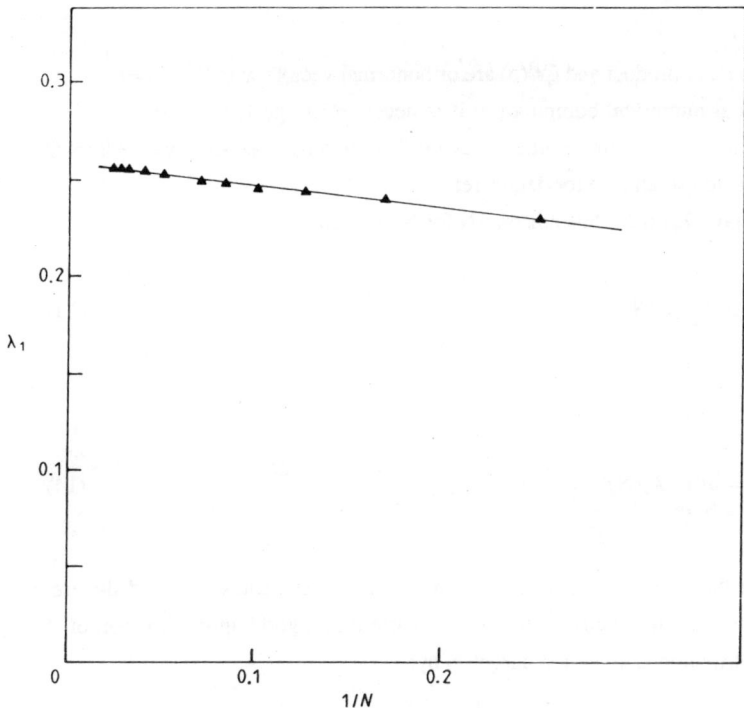

Figure 2. λ_1 plotted against $1/N$ without imposing constraint (11). $\alpha = 3$, $\bar{x} = 0$, $\varepsilon = 0.1$ and $m = 2$.

In order to check this we compute the set of all the Lyapunov exponents $\{\lambda_i\}$ which gives a good (even if not complete) description of a dynamical system. We briefly recall the definition of $\{\lambda_i\}$ for products of random matrices [3,11] where $\lambda_1 \geq \lambda_2 \geq \ldots$:

$$\sum_{i=1}^{m} \lambda_i = \lim_{n \to \infty} \frac{1}{n} \ln |\zeta^{(1)}(n) \wedge \ldots \wedge \zeta^{(m)}(n)|$$

where

$$\zeta^{(i)}(n) = \left[\prod_{k=1}^{n} A_\varepsilon(k) \right] \zeta^{(i)}(o) \quad ,$$

$\wedge$ indicates the inner product and $\zeta^{(i)}(o)$ are orthonormal vectors with $|\zeta^{(i)}(o)|=1$.

For practical numerical computation it is necessary to perform a Gram-Schmidt orthonormalization procedure in order to avoid that the angle between two vectors $\zeta^{(i)}$ and $\zeta^{(j)}$ become too small (see for details ref.[11]).

We find the set of $\{\lambda_i\}$ only depends on i/N for N large enough:

$$\lambda_i \simeq \lambda_1^* \, f(i/N) \tag{13}$$

where

$$\lambda_1^* = \lim_{N \to \infty} \lambda_1(N) \tag{14}$$

In Fig.(3) one can see λ_i plotted against i/N for different values of ε and different probability laws for the matrix elements a_{ij}. Note that a good approximation of the asymptotic distribution is given for "large" N by:

$$\lambda_i \simeq \lambda_1^* \, (1 - i/N) \tag{15}$$

i.e.

$$f(x) = 1 - x.$$

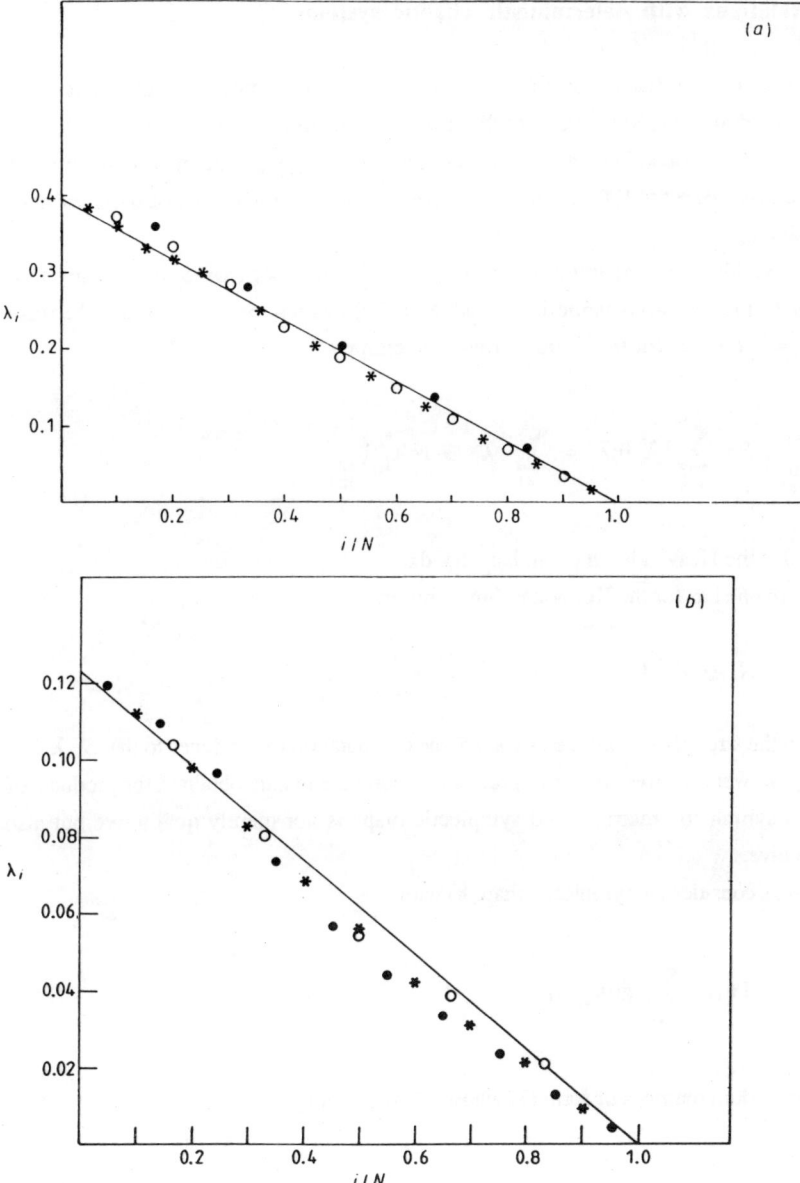

Figure 3. λ_i plotted against i/N at different N imposing constraint (11). (a) $\alpha = 1$, $\bar{x} = 0.5$, $\varepsilon = 1$ and $m = 1$; ●: $N = 6$; ○: $N = 10$; *: $N = 20$. (b) $\alpha = 1$, $\bar{x} = 0$, $\varepsilon = 1$ and $m = 1$; ○: $N = 6$; *: $N = 10$; ●: $N = 20$.

3. - Relations with deterministic chaotic systems

We have seen that the scaling laws (12) are rather common in hamiltonian and symplectic dynamics. Moreover also the asymptotic distribution (13) with a linear shape for f(x) has been found in many hamiltonian systems and symplectic maps in the limit of strong chaos (see ref.[12] for a numerical study and ref.[13] for an analitical treatment of the problem).

The validity of (13) implies the existence of a Kolmogorov-Sinai entropy [14] density in the thermodynamic limit. Indeed using Pesin's formula [15] and because $\lambda_{2N-i+1} = -\lambda_i$ one has for the Kolmogorov-Sinai entropy S:

$$S = \sum_{i=1}^{2N} \lambda_i \, \theta(\lambda_i) = \sum_{i=1}^{N} \lambda_i \simeq N \lambda_1^* \, I$$

where θ is the Heaviside function, $I = \int_0^1 f(x)dx$.
Therefore one has for the Kolmogov-Sinai entropy density

$$s = (S/N) \simeq \lambda_1^* \, I.$$

Since in the strongly chaotic cases $I \simeq 0.5$, the computation of s reduces to that of λ_1^*. Finally we want to stress that the agreement among the results obtained for products of random symplectic matrices and symplectic maps is not merely qualitative but also quantitative.

Let us consider the symplectic map (8) with

$$F(q) = \sum_i g(q_{i+1} - q_i) \tag{16}$$

and the random matrices of form (7) whose a_{ij} are given by

$$a_{ij} = \frac{\partial^2}{\partial x_i \, \partial x_j} F(x) \tag{17}$$

with F(x) as in (16) and x_i uniformly distributed in the $(0, 2\pi)$.

Comparing the $\{\lambda_i\}$ obtained with the true dynamics (i.e. eq.s.(8) and (16)) with those obtained by the product of random matrices, one can see [12] that in the limit of large ε (i.e. strong chaos) the results agree also quantitatively.

This report as been based on ref.s.[6,8,9,12]. I thank the coworkers of these papers for the friendly continuous exchange of ideas. A particular thank goes to Giovanni Paladin.

References

1] C.De Calan, J.M.Luck, T.Nieuwenhuizen and D.Petritis, J.Phys. 18A, 501 (1985)

2] B.Derrida, K.Mecheri and J.L.Pichard, J.Physique 48, 733 (1987).

3] G.Benettin, Physica 13D, 211 (1984).

4] V.I.Oseledoc, Trans. Moscow Math. Soc. 19, 197 (1968).

5] B.Derrida and E.Gardner, J.Physique 45, 1283 (1984).

6] G.Paladin and A.Vulpiani, J.Phys. 19A, 1881 (1986).

7] A.B.Rechester, M.N.Rosembluth and R.B.White, Phys.Rev.Lett. 42, 1247 (1979).

8] G.Paladin and A.Vulpiani, Phys.Rev. 35B, 2015 (1987)

9] G.Parisi and A.Vulpiani, J.Phys. 19A, L425 (1986).

10] R.Lima and S.Ruffo, to be published (1987).

11] G.Benettin, L.Galgani, A.Giorgilli and J.M.Strelcyn, Meccanica 15, 9 and 21 (1980).

12] R.Livi, A.Politi, S.Ruffo and A.Vulpiani, J.Stat.Phys. 46, 197 (1987).

13] J.P.Eckmann and C.E.Wayne, to be published (1987).

14] V.I.Arnold and A.Avez, "Ergodic problems of classical mechanics" (Benjamin 1968).

15] Ya B.Pesin, Dokl. Akad. Nauk 226, 774 (1976).

A COMPLETELY CLASSICAL MECHANISM FOR THE "FREEZING" OF THE HIGH-FREQUENCY DEGREES OF FREEDOM

Giancarlo Benettin

Dipartimento di Fisica "G. Galilei", Università di Padova
Via F. Marzolo, 8 — 35131 Padova, Italy.

1. Introduction

At variance with most talks presented at this conference, my talk is devoted to conservative dynamical systems, precisely to autonomous Hamiltonian systems. Hamiltonian dynamics dominates, in the physical world, whenever friction is absent, in particular in astronomy and in microphysics. I'll restrict myself to microphysics, and will reconsider, in the light of some recent results on Hamiltonian dynamics (in particular, on classical perturbation theory) the ergodic problem, and the related deep conflict between classical statistical mechanics and some crucial experiments, which led, as is well known, to the conclusion that classical mechanics is not adequate to microphysics, and to the birth of quantum mechanics.

From a physical point of view, the most relevant consequence of ergodicity is the so-called principle of equipartition of energy. As is well known, equipartition means essentially that, if one gives (or subtracts) energy to a system, for example by making some work on it, eventually energy goes uniformly distributed among (or subtracted from) all of the degrees of freedom. This property has a central role in the edification of classical statistical mechanics: in particular, it is at the basis of the dynamical interpretation of temperature and of internal energy, and allows one to predict the specific heat of any system, by simply counting the number of degrees of freedom.

However, as is well known, the classical values of the specific heats are apparently in strong conflict with the experimental data. For example:

i. *Diatomic gases:* one expects 7 contributions to the specific heat for each molecule (diatomic molecules have 3 translational degrees of freedom, 2 rotational degrees of freedom, and one vibrational degree of freedom, which however counts twice, as the kinetic and the potential energy contribute separately). Correspondingly, one expects a specific heat $C_V = \frac{7}{2}R$ (in fact, even a much higher value, if one considers also the internal structure of the atoms). As is well known, experiments give instead, at ordinary temperatures, $C_V = \frac{5}{2}R$, as if diatomic molecules were rigid bars, which can only rotate and translate, and even $C_V = \frac{3}{2}R$ at low enough temperature, as if the molecules were points, with no internal structure at all.

ii. *Solids:* let us consider N atoms on a three dimensional lattice; if the interatomic forces are exactly harmonic (linear), then, as is well known, the system is equivalent to a set of $n = 3N$ uncoupled harmonic oscillators (normal modes). However, any anharmonicity, no matter how small, produces a coupling among the oscillators, thus makes possible the energy exchanges among them, and is expected to lead eventually the system to equipartition. The number of contributions to the specific heat is, for such model, two (kinetic plus potential) per oscillator, thus six per particle, so one expects $C_V = \frac{6}{2}R = 3R$ (Doulong-Petit law) at any temperature. On the contrary, low temperature experiments show a non-constant value of C_V, with $C_V \to 0$ for $T \to 0$, as if, at low temperatures, only a few degrees of freedom (precisely, only those of sufficiently low frequency) were effectively taking part in the energy sharing.

iii. *The blackbody:* here we have the same difficulty as for solids; in this case however, because of the infinite number of degrees of freedom per unit volume, the difficulty turns into a real paradox: no equilibrium is now possible, unless the energy per unit volume is infinite, or the temperature is zero. This is perhaps the most serious difficulty, which is usually interpreted by saying that classical statistical mechanics not only is in conflict with experiments, but is also intrinsically unable to deal with systems having infinitely many degrees of freedom per unit volume, i.e. with fields.

Let us stress a relevant fact: all of the above difficulties came from the fact that some degrees of freedom, namely the high frequency ones, apparently do not take part in the energy sharing; they are said to be "frozen". Classical statistical mechanics is usually referred to as being unable to explain the phenomenon of the freezing of the high frequency degrees of freedom.

As is well known, since the celebrated numerical study by Fermi, Pasta and Ulam in 1954,[1] a lot of numerical work has been devoted to investigate the ergodic properties of simple models of solids, with the surprising result that, although the models are completely classical, nevertheless equipartition apparently does not take place, unless the non-linearity (or the energy) is sufficiently high. It would be very interesting to follow with some detail all of these studies, including a few numerical papers devoted to the blackbody problem: but of course, here we have not enough space (for a short review on the subject, let me propose a paper of mine, namely ref. [2]). Here I'll concentrate its attention on the problem of the specific heats of polyatomic gases: precisely, I'll first revisit, in the next section, an old fascinating conjecture advanced by Boltzmann and Jeans at the very beginning of the century, where they propose a completely classical interpretation of the freezing phenomena; then, in the following section, I'll show how this conjecture can be today supported, on the basis of the most recent results in classical perturbation theory.

2. The Boltzmann-Jeans' conjecture

If one considers a model of dilute gas, with short range intermolecular forces, then the energy exchanges among translational, rotational and vibrational degrees of freedom are due either to collisions with the walls, or to two- or few-molecules collisions: many body collisions are, for statistical reasons, highly unlike, and (as is usual in gas dynamics) can be disregarded (thus, in particular, there is no need to consider the dynamics of a system in the thermodynamic limit, as is necessary, in principle, for FPU-like models). The basic question is then whether few-body collisions do produce an efficient energy exchange among the different kinds of degrees of freedom, or instead, in spite of collisions, the vibrational and rotational energy remain essentially constant, or frozen.

As a matter of fact, this is precisely the question proposed a long time ago by Boltzmann and Jeans, within the general discussion on the failure of classical statistical mechanics. Boltzmann[3,4] concentrates the attention on the freezing of rotations, and starts by considering a gas of spherically symmetric rotating molecules (for example, perfectly smooth hard spheres). If the spherical symmetry is exact, then the forces are exactly central, the angular momentum remains constant, and the rotational energy is also constant, and does not contribute at all to thermodynamics; in particular, any measurement of the specific heat in this system would give $C_V = \frac{3}{2}R$, as for molecules with no internal structure. Let us now add a small asymmetry. The rotational energy is no more constant, however it will appreciably change only on a characteristic time scale T, depending on the asymmetry, and larger for smaller asymmetry. Clearly, only if the time scale of the experiments is larger than T, then the rotational degrees of freedom are expected to contribute to the thermodynamics, in particular to the specific heat: otherwise, they would still appear to be frozen, as for symmetric molecules. The intuition of Boltzmann is that, in realistic models, T could be very large ("days or years", in his very words): this would explain the freezing phenomena, saving on the same time the equipartition principle and classical statistical mechanics.

A very important point is the following: even if molecules are non-symmetric, the symmetry can be, so to speak, constructed dynamically, if they rotate sufficiently fast. This fact is illustrated in the figure below, which represents the collision of a molecule with permanent dipole moment P, with a charged wall: clearly, if the center of mass of the molecule does not move appreciably during a complete turn of the molecules, the non-central forces almost exactly compensate; one then expects to have, for fast rotating molecules, something like an "effective dipole moment", $P_{\text{eff}} = P/\omega\tau$, where τ is a natural time-scale associated to the motion of the center of mass (say that, during a time-interval τ, collision forces change slowly due to the movement of the center of mass). For very large ω, the effective coupling constant P_{eff} becomes very small, and molecules are expected to better and better behave as symmetric ones.*

At variance with Boltzmann, Jeans[5,6] considers the freezing of vibrations. The question is in fact quite similar: if ω is very large with respect τ^{-1}, τ being as above, then the molecule is well approximated by its "average configuration", i.e., by a rigid bar, and the internal vibrational motion doesn't play any role in the collision. Now, in Jeans' 1903 paper[5] one can find a fundamental, beautiful intuition (also supported by apparently weak, in fact illuminating heuristic considerations): there are some general mechanisms in dynamics, according to which the effect of a small coupling constant, of order $1/\omega\tau$, is not an energy exchange of

---

* These last considerations are not really explicit in Boltzmann's papers; I toke some freedom to interpret his mind, also in the light of the more explicit discussion by Jeans.

Illustrating the compensation of forces during a collision of a fast rotating molecule with a wall.

the same order $1/\omega$, but a much smaller one, precisely of order $e^{-\omega\tau}$. According to this intuition, an enormous number of collisions, of order $e^{\omega\tau}$, is necessary, in order that a fast vibrating molecule appreciably changes its interval vibrational energy; consequently, the overall vibrational energy of the gas remains constant, or frozen, for a rather large time scale

$$T \sim T_0 \, e^{\omega\tau}, \qquad (1)$$

where T_0 denotes the average time between collisions, and is expected to contribute to the specific heat only for experiments extending over time intervals larger than T. Taking T_0, ω and τ from ordinary gases, Jeans arrives to the impressive conclusion that T can be of the order of "billions of years". It is worthwhile to mention that similar considerations are also invoked by Jeans, to explain classically, still as a non-equilibrium phenomenon, the lack of the ultraviolet catastrophe in the blackbody problem. We shall refer to the exponential law (1) as to the Jeans conjecture.

The Jeans conjecture did not receive much attention, and was no more discussed after the 1911 Solvay Conference[7]. An exponential law appeared again several years later, still on the basis of heuristic considerations, in a paper by Landau and Teller[8] in 1936, which later was widely used in the literature of chemical physics (see, for example, ref. [9]). All of this literature is apparently unaware of Jeans' ideas, but shares, in a sense, Jeans' aim of studying the energy exchanges, during molecular collisions, by means of classical dynamics in place of quantum dynamics. Concerning the approximation scheme used in these papers, let me say that, at least in my opinion, it is essentially of the same level as Jeans' heuristic considerations, although more detailed, and apparently more rigorous.

3. Exponential estimates in perturbation theory

As a matter of fact, the exponential laws appeared as rigorous results in perturbation theory, about ten years ago, within the Arnol'd school, in particular in some papers by Nekhoroshev[10,11] and, more recently, by Neishtadt[12,13] (see also ref.[14-19]). Let me roughly recall some basic ideas: one deals with a nearly-integrable Hamiltonian system, say of the form

$$H(I,\varphi) = h(I) + \varepsilon f(I,\varphi), \qquad (I,\varphi) = (I_1,\ldots,I_n,\varphi_1,\ldots,\varphi_n) \qquad (2)$$

and looks for a near to the identity canonical transformation $(I,\varphi) = \mathcal{C}_\varepsilon(I',\varphi')$, such that the new Hamiltonian $H' = H \circ \mathcal{C}_\varepsilon$ assumes the form

$$H'(I',\varphi') = h(I') + \varepsilon g(I',\varphi',\varepsilon) + \varepsilon e^{-(\varepsilon/\varepsilon_0)^a} f'(I',\varphi',\varepsilon), \qquad (3)$$

with suitable constants $a, \varepsilon_0 > 0$. From this form one sees that I' is constant, up to the small "noise" due to f', which however is negligible up to exponentially large times; correspondingly, the motion of I is the superposition of this "noise" and of the "deformation" $|I - I'|$, due to the canonical transformation; this however is small with ε *uniformly in time*, so that I too is, like I', almost constant for an exponentially long time scale,

$$T \sim T_0\, e^{-(\varepsilon/\varepsilon_0)^a}. \qquad (4)$$

The exponential law (4) differs from Jeans' conjecture (1) essentially because of the exponent a, which, unfortunately, in general is much less than 1 ($a \sim 1/n$ or $\sim 1/n^2$, depending on the problem at hand, for systems with n degrees of freedom).

As shown in a joint work with L. Galgani and A. Giorgilli,[20,21] Nekhoroshev's techniques can be adapted to the Jeans problem. Consider, for simplicity, the collinear collision of two diatomic molecules, and denote by $x = (x_1,x_2)$, $\xi = (\xi_1,\xi_2)$ the coordinates of the centers of mass of the molecules, and respectively of the internal vibrational degrees of freedom. The Hamiltonian has then the form

$$H = h_\omega(\pi,\xi) + \hat{h}(p,x) + f(x,\xi), \qquad (5)$$

with

$$h_\omega = \frac{1}{2}\sum_{j=1}^{2}(\pi_j^2 + \omega\xi_j^2); \qquad \hat{h} = \frac{1}{2}\sum_{j=1}^{2} p_j^2 + V(x,0) \qquad (6)$$
$$f = V(x,\xi) - V(x,0),$$

where V is the interaction potential, which is assumed to decay exponentially with the distance of the interacting molecules. A little reflection shows that the form (7) is in fact obtained quite generally, in the collision of any number of molecules, also in two or three dimensions, and that the (p, x) coordinates may include, beside translations, also rotations.

Roughly speaking, our result can be described as follows: under suitable, reasonable assumptions, one finds a canonical transformation $(p, x, \pi, \xi) = C_\omega(p', x', \pi', \xi')$, such that the new Hamiltonian $H' = H \circ C_\omega$ has the form

$$H' = h_\omega(\pi', \xi') + \frac{1}{\omega \tau} g_\omega(p', x', \pi', \xi') + \mathcal{O}(e^{-\omega \tau}), \qquad (7)$$

and g_ω is such that the Poisson bracket $\{h_\omega, g_\omega\}$ vanishes. It follows that the vibrational energy in the new variables, namely $E'(t) \equiv h_\omega(\pi'(t), \xi'(t))$, is almost constant, and denoting by $t = 0$ the beginning of the collision, where the molecules are far apart, and by T the end of the collision, when the molecules are again far apart, one obtains

$$E'(T) - E'(0) = \mathcal{O}(e^{-\omega \tau}). \qquad (8)$$

Now, at the beginning and at the end of the collision, one could see (in fact, it is quite evident) that the canonical transformation C_ω reduces to the identity, so that the deformation vanishes. This means that one can drop the primes from (8), and get from it Jeans' exponential law.

Let me stress that everything can be stated in a completely rigorous mathematical form. In fact, the simpler case of the collision of a molecule with an atom is studied in ref. [20], where however one proves only the weaker exponential law

$$E(T) - E(0) \sim \mathcal{O}(e^{-(\omega \tau)^a}), \qquad (11)$$

with $a = \frac{1}{2}$. Many molecules collisions are instead studied in ref. [21], by a more powerful (but much more complicated) technique, which also gives (for collisions of identical molecules) $a = 1$.

As is well known, estimates in perturbation theory are always pessimistic, thus in particular the numerical constants appearing in the final results of ref. [20,21] are quite far from being optimal. Nevertheless, one can say that, at least in principle, the exponential law conjectured by Jeans is now definitely proven.

As a remarkable fact, numerical experiments give a rather striking evidence of Jeans' exponential law (1); in particular, for a very rough one-dimensional model of diatomic gas[22], one finds constants t_0 and t_1 which are compatible with the time–scales of "billions of years" invoked by Jeans. Numerical work is also in progress for the problem of "freezing" of fast rotations; preliminary results show a behavior in good agreement with Jeans' exponential law[23]. One can also mention a numerical work on a one-dimensional model of radiant cavity[24], where no reference is made to Nekhoroshev exponential estimates, but some evidence is produced of a "freezing" of the high–frequency modes of the electromagnetic field, for times rapidly increasing with the frequency of the modes.

All of these numerical studies should be considered with care, and possibly repeated on more realistic models; a great effort should also be made to improve the techniques in perturbation theory. Anyhow, on the basis of the already available results, I am quite confident that the exponential laws conjectured by Jeans at the very beginning of the century, today better understood within classical perturbation theory, are an essential deep element of classical mechanics, which could be very relevant for a proper understanding of the relations between classical and quantum physics, and provide in particular a completely classical interpretation of some phenomena, which are usually considered to be purely quantum ones. For a wider discussion on this point, one can also look at the lectures by L. Galgani, A. Giorgilli and myself[25−27] at the 1987 Noto Summer School.

References

[1] E. Fermi, J. Pasta and S. Ulam: *Los Alamos Report* No. LA-1940 (1955), later published in *E. Fermi: Collected Papers* (Chicago, 1965), and *Lect. Appl. Math.* **15**, 143 (1974).

[2] G. Benettin: *Ordered and Chaotic Motions in Dynamical Systems with Many Degrees of Freedom*, in *Molecular-Dynamics simulation of Statistical Mechanical Systems*, Rendiconti della Scuola Italiana di Fisica "E. FERMI", G. Ciccotti and W.G. Hoover editors (North–Holland, Amsterdam 1986).

[3] L. Boltzmann: *Nature* **51**, 413 (1895).

[4] L.Boltzmann: *Lectures on gas theory*, translated by S.G.Brush, University of Cal. Press (1966); see especially section 45:*Comparison with experiments*.

[5] J.H. Jeans: *Phil. Mag.* **6**, 279 (1903).

[6] J.H. Jeans: *Phil. Mag.* **10**, 91 (1905).

[7] *La Theorie du Rayonnment et le Quanta*, proceedings of the 1911 Solvay Conference, M.M. Langevin and M. de Broglie editors (Gauthier – Villars, Paris 1912).

[8] L. Landau and E. Teller: *Physik. Z. Sowjetunion* **11**, 18 (1936).

[9] D. Rapp: *Journ. Chem. Phys.* **32**, 735 (1960).
[10] N. N. Nekhoroshev: *Usp. Mat. Nauk* **32**, (1977) [*Russ. Math. Surv.* **32**, 1 (1977)].
[11] N. N. Nekhoroshev: *Trudy Sem. Petrows.* No. 5, 5 (1979).
[12] A.I. Neishtadt: *Prikl. Matem. Mekan.* **45**, 80 (1981) [*PMM U.S.S.R.* **45**, 58 (1982)].
[13] A.I. Neishtadt: *Prikl. Matem. Mekan.* **48**, 197 (1984) [*PMM U.S.S.R.* **45**, 133 (1984)].
[14] G. Benettin, L. Galgani, A. Giorgilli: *Celestial Mechanics* **37**, 1 (1985).
[15] G. Gallavotti: *Quasi-Integrable Mechanical Systems*, in *Critical phenomena, Random Systems, Gauge Theories*, K. Osterwalder and R. Stora editors, Les Houches, Session XLIII, 1984 (North-Holland, Amsterdam 1876).
[16] A. Giorgilli and L. Galgani: *Celestial Mechanics* **37**, 95 (1985).
[17] G. Benettin and G. Gallavotti: *J. Stat. Phys.* **44**, 293 (1985).
[18] A. Giorgilli, A. Delshams, E. Fontich, L. Galgani and C. Simò: *Effective Stability for a Hamiltonian System near an Elliptic Equilibrium Point, with an Application to the Restricted three Body Problem*, preprint.
[19] A. Giorgilli, *Rigorous Results on the Power Expansions for the Integrals of a Hamiltonian System near an Elliptic Equilibrium Point.*, preprint.
[20] G. Benettin, L. Galgani and A. Giorgilli: *Comm. Math. Phys.* **113**, 87 (1987).
[21] G. Benettin, L. Galgani and A. Giorgilli: *Realization of Holonomic Constraints and Freezing of High-Frequency Degrees of Freedom in the Light of Classical Perturbation Theory, part II*, preprint.
[22] G. Benettin, L. Galgani and A. Giorgilli: *Phys. Lett.A* **120**, 23 (1987).
[23] O. Baldan and G. Benettin: paper in preparation.
[24] G. Benettin and L. Galgani: *J. Stat. Phys.* **27**, 153 (1982).
[25] L. Galgani: *Relaxation Times and the Foundations of Classical Statistical Mechanics in the Light of Modern Perturbation Theory*, lectures given at the Noto School "Non-Linear Evolution and Chaotic Phenomena", G. Gallavotti and A.M. Anile Editors (to appear).
[26] A. Giorgilli: *Relevance of Exponentially Large Time Scales in Practical Applications: Effective Fractal Dimension in Conservative Dynamical Systems*, lectures given at the Noto School "Non-Linear Evolution and Chaotic Phenomena", G. Gallavotti and A.M. Anile Editors (to appear).
[27] G. Benettin: *Nekhoroshev-like Results for Hamiltonian Dynamical Systems*, lectures given at the Noto School "Non-Linear Evolution and Chaotic Phenomena", G. Gallavotti and A.M. Anile Editors (to appear).

Complexity and Relaxation in Nonlinear Chains and in a Model of Radiant Cavity

C. Alabiso§ and M. Casartelli‡

Dipartimento di Fisica dell'Università di Parma
Sezione Teorica - §INFN - ‡GNSM

We present here the results of numerical experiments on the connections between the local complexity of the phase space of hamiltonian systems and dynamical properties such as relaxation, approach to equilibrium, sensitiveness to initial conditions during finite observations, etc. Before going into the more recent studies regarding a model of radiant cavity and the infinite degrees of freedom problem, we recall some previous results on nonlinear dynamical systems with N degrees of freedom, precisely on unidimensional chains of particles interacting via anharmonic potentials (Lennard-Jones, Hénon-Heiles type, Fermi-Pasta-Ulam, Toda...). The problem faced in that case consists in the connection between the features of the trajectory and the sensitiveness to the initial conditions in *the stochastic domain*, in such a way to link the difference between low and high stochasticity to the geometrical properties of the phase space [1-3]. This is done by introducing suitable observables. Precisely, using the variance

$$\mathrm{Var}(x_1, ..., x_M) = (\bar{x^2} - \bar{x}^2)/\bar{x}^2, \quad \bar{x} = \frac{1}{M}\sum_{k=1}^{M} x_k$$

we introduce the following three quantities:

- $S_T = \mathrm{Var}(c_1, ..., c_M)$, where the $c_i = c(t_i)$ are the values computed along the trajectory of the curvature $c(t)$ up to $t_M = T$.
- $G_T = \mathrm{Var}(\rho_1, ..., \rho_M)$, where the $\rho_i = \rho(t_i)$ are the corresponding values of the microcanonical density function ρ.
- $\mathcal{X}_T = \mathrm{Var}(\langle \xi_1 \rangle_T, ..., \langle \xi_N \rangle_T)$, where $\xi_k = |\dot{J}_k|$ are the absolute values of the time derivatives of the *harmonic* action variables J_k, and $\langle \cdot \rangle_T$ denotes the time average up to time T. In the following, we also use $\langle \cdot \rangle$ for $\langle \cdot \rangle_{T \to \infty}$.

All these quantities confirm, with different numerical evidence, a trapping action by residual invariant surfaces in low stochastic domain, during finite time observations. The simplest of the introduced indicators, i.e. $\mathcal{X}_T$, gives the neater results and provides moreover supplementary informations. Hereafter we refer to it esclusively.

What is important to our investigations is that $\mathcal{X}_T$ proves to be an excellent stochastic indicator. Precisely, it happens that when the dynamical system with stochastic transition (the LJ chain in our experiments) overpasses a threshold, constituted by a critical value u^c of the specific energy u, then $\mathcal{X}_T \to 0$. In terms of single modes behaviour, such relation may be expressed as an *equipartition* property of the absolute variations of the harmonic actions:

$$\langle |\dot{J}_k| \rangle = const. \tag{1}$$

or, by taking into account that $E_k = \omega_k J_k$ is the energy of the k-th mode, as a *proportionality* relation for the absolute energy exchanges:

$$\langle |\dot{E}_k| \rangle \propto \omega_k \tag{2}$$

These two and equivalent relations may be read in a geometrical sense the first, in a dynamical sense the second. Indeed, since the action variable J_k reapresents the k-th radius of the unperturbed torus of the purely harmonic system, its absolute velocity is plausibly linked to the increasing instability of the tori. On the other side, $|\dot{E}_k|$ represents the absolute rate of energy exchange of the k-th mode, and it is therefore an index of its activity. Now, a common feature of these nonlinear chains is given by the fact that the frequency spectrum has an upper bound ω_∞ which is rapidly approached for finite N. We may ask then what can happen for a system showing a stochastic transition and an unbounded spectrum of frequencies, for example a linear spectrum.

In lack of theoretical arguments for a general answer, we have tried numerical experiments on a definite model, the Bocchieri-Crotti-Loinger model of radiant cavity introduced in 1972 [4] and studied after them by a number of people [5-7]. It consists of a ZY plate oscillating along the Z axis between two mirrors located at $-l$ and l in the X direction, and driven by an external force $F(z)$. The starting equations of motion are

$$\begin{cases} \dfrac{\partial^2 A_z}{\partial x^2} - \dfrac{1}{c^2}\dfrac{\partial^2 A_z}{\partial t^2} = -\dfrac{4\pi}{c}\sigma \delta(x)\dot{z} \\ m\ddot{z} = -\dfrac{\sigma}{c}\dfrac{\partial A_z(0,t)}{\partial t} + F(z) \end{cases} \tag{3}$$

with $A_z(-l,t) = A_z(l,t) = 0$ and $F(z) = -m\alpha z^3$. The introduction of the normal modes of the field, and a suitable number of substitutions and redefinitions of variables and parameters (which will be presented in detail elsewhere [8]) lead to the final form

$$\begin{cases} \ddot{c}_n + \tilde{\omega}_n^2 c_n = \sqrt{2\gamma}\dot{w} \\ \ddot{w} = -\sqrt{2\gamma}\sum_{n=1}^{\infty}{}'\dot{c}_n^0 - \gamma[\dot{w} + 2\sum_{k=1}^{N(\tau/2)}{}'(-1)^k \dot{w}(\tau - 2k)] - \varepsilon^4 w^3 \end{cases} \quad (4)$$

where $\tilde{\omega}_n = n\pi/2$, $\tau = ct/l$ and the energies of the modes are proportional to the quantities $E_n = \frac{1}{2}(\dot{c}_n^2 + \tilde{\omega}_n^2 c_n^2)$. The parameter ruling the energy of the system is ε, while γ gives the intensity of the coupling. Even if the present model is only a highly idealized model of matter-field interaction, its features are sufficient for a first qualitative answer to our problem.

What is already known is the following: 1) testing the onset of stochasticity by the energy equipartition criterion, it results that *within a fixed observation time* a threshold may be defined for each mode, in the sense that if the energy is not sufficiently large to overpass a certain critical value then that mode cannot reach the equipartition. The value of these thresholds grows approximately with n. It is an open problem if these thresholds depend or not on time. 2) As a consequence, a given amount of energy is shared more easily when the excited modes are those with a lower threshold, i.e. the lower modes. 3) Energy flows more easily from high modes to low modes.

These general features, studied for instance by Benettin and Galgani [6], have been recovered with an improved numerical precision in our experiments, just to fix a clear reference frame for the successive experiments on the rate of energy exchanges. Examples of the results are shown in the Figs.[1-3]. Turning now the analysis to the behaviours of $|\dot{E}_k|$, we observe that 1) a clear trend to equipartition takes place and 2) this trend is almost perfectly coherent with the equipartition of energy. In other words, even if $\langle E_k \rangle_T$ and $\langle |\dot{E}_k| \rangle_T$ are in principle very different quantities, they behave in the same way with respect to the parameters $(T, \varepsilon, \gamma...)$. (Figs.[4-6]) (Of course, a more complete and realistic pattern would appear in the Figs.[1-6] by adding the flatter and flatter curves of the higher modes). As a first consequence, the absolute rate of energy exchanges may be used again as a stochasticity index, to renforce the equipartition criterion. But there are now also new informations: instead of eq.(2) it holds that

$$\langle |\dot{E}_k| \rangle = const \tag{5}$$

and, instead of eq.(1):

$$\langle |\dot{J}_k| \rangle \propto \frac{1}{\bar{\omega}_k} \tag{6}$$

If we consider the subspace spanned by the action variables J_k, the equipartition relation (1) says that in the mean the absolute value of the velocity is there isotropic: stochasticity entails therefore a substantial uniformity of the space. The relation (6), on the contrary, indicates that the thermalization of the modes of the field takes place with growing difficulty for a sort of progressive rigidity of the space. Thus, both on geometrical and dynamical sides, eqs. (5) and (6) give evidence to the deep difference of stochasticity between systems with a finite number of degrees of freedom and the field.

In conclusions, numerical experiments on the rate of energy exchanges prove to be effective in getting informations which would require, otherwise, more sophisticated and difficult approaches.

REFERENCES

[1] M. Casartelli, in *Advances in Nonlinear Dynamics and Stochastic Processes*, R. Livi and A. Politi eds. (Singapore 1985)
[2] M. Casartelli and S. Sello: *Phys. Lett.* **112A**, *249* (1985);
[3] M. Casartelli and S. Sello: *Nuovo Cim.* **97B**, *183* (1987).
[4] P. Bocchieri, A. Crotti and A. Loinger: *Lett. Nuovo Cim* **4**, *341* (1972).
[5] G. Casati, I. Guarneri and F. Valz-Gris: *Phys. Rev.* **16A**, *1237* (1977)
[6] G. Benettin and L. Galgani: *Jour. Stat. Phys.* **27**, *153* (1982)
[7] R. Livi, M. Pettini, S. Ruffo and A. Vulpiani: *Jour. Phys. A*, **20**, 577 (1987).
[8] C. Alabiso and M. Casartelli, in preparation.

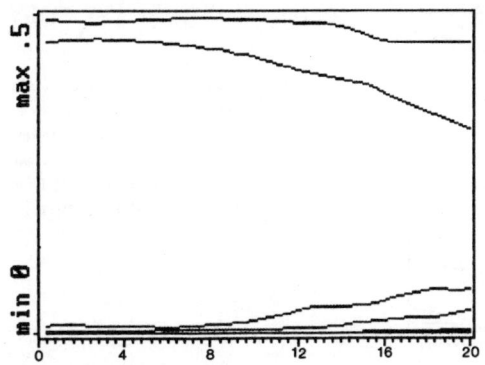

FIG.1 Time behaviour of $\langle E_k \rangle_T$ with $\varepsilon = 6$, time in integration steps/10^6. Modes 9, 7, 3, 5, and 11-17 (from the top) at the first mark, modes 9,7,3,5,11 and 13-17 at the end. The higher modes are still frozen, and no one reaches the equipartition.

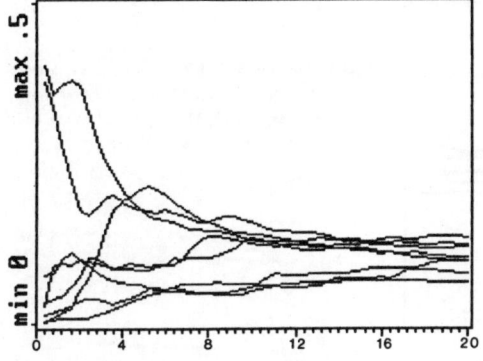

FIG.2 The same as in FIG.1 with $\varepsilon = 14$. Modes 9, 7, 3, 5, 11, 15, 17, 13 at the first mark, modes 5,3,13,7,11,9,15,17 at the end. The trend to equipartition is very clear.

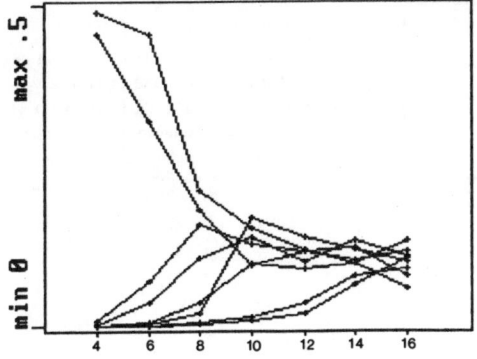

FIG.3 Values for $\langle E_k \rangle_T$ at final T plotted vs. ε. Modes 9, 7, 3, 5 and 11-17 at $\varepsilon = 4$, modes 9, 7, 5, 17, 13, 15, 3, 11 at $\varepsilon = 16$. The role of ε in the thermalization process is clearly shown. The wider spreading at $\varepsilon = 16$ is only apparent, due to the increasing influence of the higher modes not drawn in the figure.

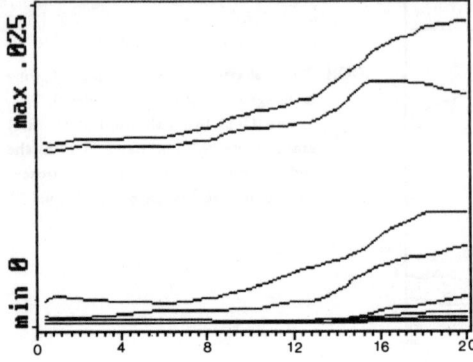

FIG.4 Time behaviour of $\langle |\dot{E}_k|\rangle_T$ with $\varepsilon = 6$, time in integration steps/10^6. Modes 9, 7, 3, 5, and 11-17 (from the top) at the first mark, modes 9,7,3,5,11 and 13-17 at the end. To be compared with FIG.1 for the qualitative features of the pattern.

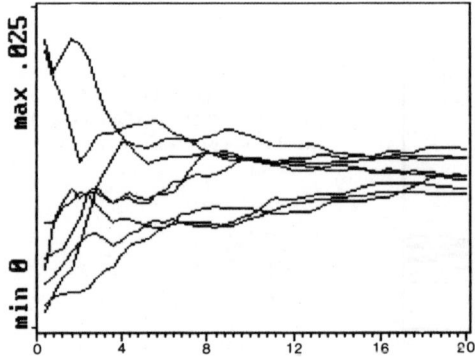

FIG.5 The same as in FIG.4 with $\varepsilon = 14$. Modes 9, 7, 3, 5, 11, 15, 17, 13 at the first mark, modes 5, 13, 3, 11, 7, 9, 15, 17 at the end. To be compared with FIG.2.

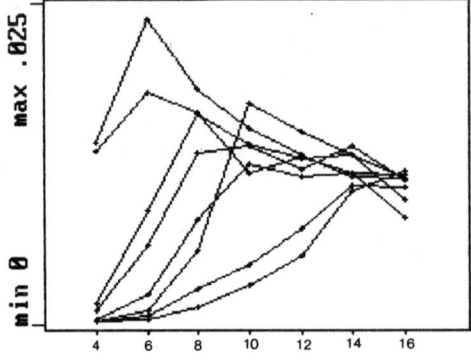

FIG.6 Values for $\langle |\dot{E}_k|\rangle_T$ at final T plotted vs. ε. Modes 9, 7, 3, 5 and 11-17 at $\varepsilon = 4$, modes 17, 7, 9 and 13, 5, 15, 3, 11 at $\varepsilon = 16$. The comments to FIG.3 apply also here.

TRANSITIONS IN SYSTEMS FAR FROM EQUILIBRIUM: A GINZBURG-LANDAU APPROACH

P. Coullet*, L. Gil, and J. Lega
Laboratoire de Physique Théorique.
Parc Valrose. 06034 Nice Cedex. France

Introduction

The study of systems driven far from equilibrium by an external parameter has recently attracted a considerable interest [1]. In the late seventies, it was shown that such physical systems with small aspect ratios behaved like dynamical systems, and were displaying the various ways leading to chaos [2]. The universality of these behaviours is related to the fact that near a bifurcation threshold, the dynamics of the system can be reduced to generic equations, which only depend of the nature of the bifurcated solutions. When spatially extended, such systems can be seen as an infinity of coupled dynamical systems, and they naturally show complex spatio-temporal behaviours [3]. Their dynamics are extremely rich, and their study may give some clue in the understanding of hydrodynamic turbulence.

We shall study the spatio-temporal disorganization of a spatial, temporal or spatio-temporal periodic structure. When the control parameter is varied, the system undergoes a bifurcation, giving rise to a static, temporal or spatio-temporal pattern. Near the threshold, the behaviour of the system is described by an order parameter which obeys a partial differential equation [4,5]. The form of such an equation is only related to the symmetries of the problem, so that various physical systems may exhibit an analogous behaviour. This description can be seen as a generalization of the Landau's theory of equilibrium systems [6], and the order parameter may have singularities which are topological defects [7]. These play an important role in phase transitions [8,9], giving rise to a loss of correlations. By analogy, we will show that the defects are responsible of the desorganization of the structure of spatially extended systems. The striking difference with classical defect-mediated phase transitions is that the transition to a turbulent state in the systems we are studying occurs without external noise. In fact, for a given range of the value of the external parameter, the system presents a phase instability [10,11] which provides a deterministic noise, giving rise to quickly dissociated pairs of topological defects [12].

---

* also Observatoire de Nice. Mont Gros.

In the first part of this paper, we shall give a description through order parameters and amplitude equations of three physical systems undergoing a bifurcation and giving rise respectivelly to a static, a temporal, and a spatio-temporal pattern. In the second part, we shall study the stability of the bifurcated corresponding patterns, and in the third part, we shall consider the existence of more singular solutions, which are topological defects. The mechanisms of creation of defects will be discussed in the last part, and we shall caracterize a form of turbulence due to the presence of defects.

1. Transitions in Systems far from Equilibrium

In a Rayleigh-Benard experiment [13], a dilatable fluid lays among two horizontal plates, and is submitted to a temperature gradient. The control parameter is the Rayleigh number $\mathcal{R}a$, which is proportional to the difference of temperature between the two plates. When increasing $\mathcal{R}a$ beyond a critical value $\mathcal{R}a_c$, the systems goes from a conductive state to a convective one, and a static roll pattern appears in the box. The temperature of the fluid in a median plane between the two plates can be written as:

$$T = T_0 + A(x,y,t)\exp ik_0 x + c.c. + ...,$$

where x is the direction perpendicular to the rolls, $\frac{2\pi}{k_0}$ is the period of the pattern and corresponds to two counter rotating rolls, A is the small and slowly varying complex order parameter, and the dots stand for higher order terms. Near the bifurcation threshold, where such a description is valid, the order parameter A follows the amplitude equation:

$$\frac{\partial}{\partial t}A = \mu A + (\frac{\partial}{\partial x} - \frac{i}{2k_0}\frac{\partial^2}{\partial y^2})^2 A - |A|^2 A, \tag{1}$$

where μ measures the distance to the instability threshold. The symmetries of this equation reflect those of the physical system. This one is assumed to be isotropic and invariant under space translations ($\vec{r} \to \vec{r} + \vec{a}$), and under time translations ($t \to t + \tau$). After the bifurcation, the roll structure induces a spatial order, and thus breaks some of these symmetries. Nevertheless, the bifurcated pattern remains invariant under a class of symmetries of the initial system which are $t \to t + \tau$, $y \to -y$ and $x \to x \pm \frac{2\pi}{k_0}$. Besides, since the initial system is invariant under space translations, any translated roll structure is also a possible solution for the bifurcated pattern. This explains why Eq.(1) is invariant under the transformation $A \to A\exp i\phi$, which corresponds in the expression of T to a translation of $\frac{\phi}{k_0}$ in the x direction. In the same way, since the system is invariant under parity ($x \to -x$), the amplitude equation has real coefficients, i.e. it is left unchanged by the transformation $A \to \bar{A}$.

Instead of a spatial order, the bifurcated pattern may induce a temporal order [14]. This is the case for instance in chemical reactors where beyond a critical value f_c of the flux f of reactives into the reactor, a reference concentration begins oscillating periodically in time. This concentration can be written as:

$$C = C_0 + A(x,y,t) \exp i\omega_0 t + c.c. + ...,$$

where $\frac{2\pi}{\omega_0}$ is the period of the temporal oscillation. Here again, A is slowly varying in space and time. The symmetries of the physical system are assumed to be the same as in the previous case, but since the bifurcated solution does not break any spatial symmetry, the amplitude equation for the order parameter A has to be isotropic and invariant under space translations. Besides, since any time translated solution must also be a solution, this equation has to be invariant under the transformation $A \to A \exp i\phi$. Thus, the amplitude equation reads:

$$\frac{\partial}{\partial t} A = \mu A + (1 + i\alpha)\Delta A - (1 + i\beta)|A|^2 A. \tag{2}$$

The existence of complex coefficients is related to the oscillatory nature of the bifurcated solution: since this solution breaks the invariance under time translations, and since the inversion of time is not a symmetry, the amplitude equation has no reason to be left unchanged by the transformation $A \to \bar{A}$.

Our last example will describe the bifurcation towards a wavy structure [15]. This occurs for instance in a Rayleigh-Benard experiment with a binary mixture [16,17]. Beyond a critical value of the difference of temperature between the two plates, a system of propagating rolls appears. The temperature in a median plane between the two plates can be written as:

$$Q = Q_0 + A \exp i(\omega_0 t + k_0 x) + B \exp i(\omega_0 t - k_0 x) + c.c. + ...,$$

where the order parameter A corresponds to a left-propagation in the x direction, and B is associated to a right-propagation. These two order parameters obey the following amplitude equations:

$$\frac{\partial}{\partial t} A + c\frac{\partial}{\partial x} A = \mu A + (1+i\alpha)(\frac{\partial}{\partial x} - \frac{i}{2k_0}\frac{\partial^2}{\partial y^2})^2 A + i\epsilon(\frac{\partial^2}{\partial x^2} + \zeta\frac{\partial^2}{\partial y^2})A \\ - (1+i\beta)|A|^2 A - (\gamma+i\delta)|B|^2 A \tag{3a}$$

$$\frac{\partial}{\partial t} B - c\frac{\partial}{\partial x} B = \mu B + (1+i\alpha)(\frac{\partial}{\partial x} + \frac{i}{2k_0}\frac{\partial^2}{\partial y^2})^2 B + i\epsilon(\frac{\partial^2}{\partial x^2} + \zeta\frac{\partial^2}{\partial y^2})B \\ - (1+i\beta)|B|^2 B - (\gamma+i\delta)|A|^2 B \tag{3b}$$

where μ measures the deviation from the critical situation, c, α, ϵ and ζ describe dispersive effects, β and δ are associated with nonlinear renormalization of the temporal frequency, and γ is the competition parameter between travelling and standing waves, corresponding to non trivial homogeneous solutions of Eqs.(3). These equations are invariant under the following transformations: $A \to A \exp -i\Phi$, $B \to B \exp i\Phi$ which reflects the initial invariance under space translations, $A \to A \exp i\Psi$, $B \to B \exp i\Psi$ which reflects the invariance under time translations, $x \to -x$, $A \to B$, $B \to A$ and $y \to -y$ which reflect the parity symmetry.

These three examples have shown that, near the threshold of a bifurcation, one can describe various systems in an analogous way, and that the form of the associated amplitude equations is only related to the nature of the bifurcation.

2. The Bifurcated Solutions and their Stability

Each of the above amplitude equation possesses a class of homogeneous solutions, which correspond to the perfect bifurcated pattern. When varying the external parameters, such a solution may become unstable. We shall now study these solutions, and their stability.

The homogeneous stationary solutions of Eq.(1) are $A_0 = \sqrt{\mu} \exp i\phi$, where the phase ϕ is arbitrary. In order to study their stability, we look for a solution of Eq.(1) under the form $A = (\sqrt{\mu} + a(x, y, t)) \exp i\phi$, where a is a small perturbation. The associated linear problem in the Fourier space has two eigenvalues, which read:

$$\sigma_\pm = -\mu - (k_x^2 + \frac{k_y^4}{4k_0^2}) \pm \mu$$

The first one σ_- is negative and the associated eigenvector corresponds to an amplitude mode, which is stable. The second one is negative for $|k| \neq 0$, and is marginal for $k = 0$. The existence of the marginal mode is related to the invariance of the physical system under space translations, and corresponds to a phase translation for the order parameter. Therefore, the perfect bifurcated solution is always stable. We shall see that the oscillatory nature of the two others physical systems described above will cause the loss of this stability.

Equation (2) admits a class of stationary solutions which read: $A_0 = \sqrt{\mu} \exp i(-\beta \mu t + \phi)$, where ϕ is arbitrary. As before, we study its stability with respect to small perturbations. The associated eigenvalue problem has two solutions, which read in the limit of small $|k|$:

$$\sigma_\pm = -\mu - (k_x^2 + k_y^2) \pm \mu[1 - \frac{\alpha\beta}{\mu}(k_x^2 + k_y^2) + ...]$$

The eigenvalue σ_- is negative and corresponds to a stable amplitude mode. But if $1 + \alpha\beta$ is negative, σ_+ is positive, and the associated eigenvector will be exponentially increased under the dynamics of Eq.(2). Thus, the solution A_0 may become unstable if $1 + \alpha\beta < 0$. Since the eigenvalue σ_+ is associated to the phase mode, one says that a phase instability may occur.

Now we will consider the homogeneous solutions of Eqs.(3). They are of two kinds: the travelling waves:

$$A = \sqrt{\mu} \exp i(-\beta\mu t + \phi_a), \qquad B = 0$$

and

$$A = 0, \qquad B = \sqrt{\mu} \exp i(-\beta\mu t + \phi_b),$$

and the standing waves:

$$A = \frac{\mu}{1+\gamma} \exp i(-\frac{\beta+\delta}{1+\gamma}\mu t + \phi_a), \qquad B = \frac{\mu}{1+\gamma} \exp i(-\frac{\beta+\delta}{1+\gamma}\mu t + \phi_b).$$

The former are stable with respect to homogeneous perturbations when $\gamma > 1$, the latter when $-1 < \gamma < 1$. Assuming that the range of external parameters has been chosen so that $\gamma > 1$, we can now study the stability of travelling waves with respect to any small perturbation. As described above, we look for the eigenvalues of the linearized problem near the travelling wave solution. They read:

$$\sigma_\pm = -\mu - (k_x^2 + \frac{k_y^4}{4k_0}) + i(-ck_x - \alpha\frac{k_y^2}{k_0^2}k_x) \pm \mu[1 - \frac{\beta\epsilon\zeta}{k_0\mu}k_y^2 - \frac{\beta\epsilon}{k_0\mu}k_x^2$$
$$- \frac{(\epsilon\zeta)^2}{2\mu^2}(1+\beta^2)k_y^4 - \frac{\alpha\beta}{\mu}(k_x^2 + \frac{k_y^4}{4k_0^2}) + i\frac{\beta}{k_0\mu}k_x k_y^2 + ...]$$

Since the real part of σ_- is always negative, the amplitude mode is stable. But the real part of σ_+ will become positive if $1 + \alpha\beta + \frac{\epsilon\beta}{k_0}$, or $\beta\epsilon\zeta$, or $\frac{1}{k_0^2}[1 + \alpha\beta + (1+\beta^2)\frac{2(\epsilon\zeta k_0)^2}{\mu}]$ are negative, and thus a phase instability may occur.

The previous examples show that the existence of a phase instability for the perfect bifurcated pattern is related to the imaginary parts of the coefficients of the amplitude equations. This is a fundamental difference between Eq.(1) and Eqs.(2) and (3). An other way to see this is that since Eq.(1) has only real coefficients, it can be written as:

$$\frac{\partial A}{\partial t} = -\frac{\delta \mathcal{F}}{\delta \bar{A}},$$

where the Lyapunov functional

$$\mathcal{F} = \int [-\mu|A|^2 + |(\frac{\partial}{\partial x} - \frac{i}{2k_0}\frac{\partial^2}{\partial y^2})A|^2 + \frac{1}{2}|A|^4] dx dy$$

is a real quantity. Hence, it is the analog of a free energy. It is not the case for Eqs.(2) and (3) where $\mathcal{F}$ would be a complex quantity. Thus, non variational systems, i.e. systems which cannot be described through a real free energy, may present a phase instability, which will turn out to be a noise generator. Namely, we shall see in the last part of this paper that this phase instability leads to the creation of topological defects in the system, and thus is responsible for a loss of correlations. We shall now describe the defects associated to the ordered structures we have considered in this second part.

3. Topological Defects

Besides the homogeneous solutions described above, the amplitude equations possess some singular solutions, which are defects [3]. Their existence and their stability may be understood through topological arguments [18]. Here, we shall give only a brief numerical description of these solutions, and a geometrical argument for their stability. The core of a defect is a region where the order parameter varies quickly in space. Far from the core, the defect reaches the periodic solution, so that one recovers the perfect bifurcated pattern. It may be surprising to study such solutions through the amplitude equations which have been derived in the approximation of small spatial variations. But the existence and the nature of a defect are only related to the symmetries of the problem, and to the state of the order parameter far from the core. Only the precise shape of the core of a defect cannot be obtained through an amplitude equation, but it is irrelevant here.

The defect of a static roll pattern -described by Eq.(1)- is a dislocation. It corresponds to the insertion of a pair of two counter rotating rolls. Such a defect also occurs in travelling or in standing wave patterns -described by Eqs.(3). The Fig.(1) shows two dislocations of travelling waves. Travelling wave patterns possess an other class of defects, which are domain walls between counter propagating waves.

In the case of temporal periodic patterns -described by Eq.(2)- the defects take the form of spiral waves [14], which propagate out of the core. They are the analogs of the dislocations of roll patterns. The phase of the order parameter turns by 2π around the core of a dislocation or of a spiral wave, that is:

$$\int_\Gamma A\overrightarrow{dl} = 2\pi,$$

where Γ is an oriented path around the core of the dislocation. Therefore, the order parameter has to vanish at the core of the defect, and the lines of the plane (x,y)

where $\Re(A) = 0$ and $\Im(A) = 0$ cross themsleves at this point. The Fig.(2) shows a spiral wave defect (Fig.(2a)) and the associated lines $\Re(A) = 0$ and $\Im(A) = 0$ (Fig.(2b)).

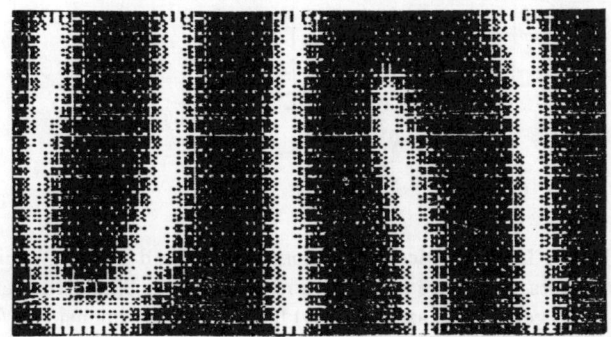

Figure 1: Numerical simulation of Eq.(3) showing two dislocations of travelling wave patterns.

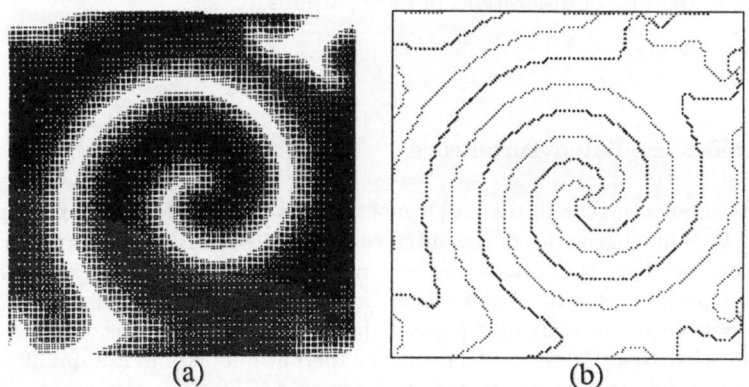

Figure 2: Numerical simulation of Eq.(2) showing a spiral wave defect.

The above geometrical constatation ensures the stability of such defects, since a little perturbation of A will shift and deform the lines $\Re(A) = 0$ and $\Im(A) = 0$, and thus will move the core of the defect, but will not make it desappear. This is not the case in one spatial dimension, since $\Re(A)$ and $\Im(A)$ do not vanish generically at the same point (see Fig.(3)).

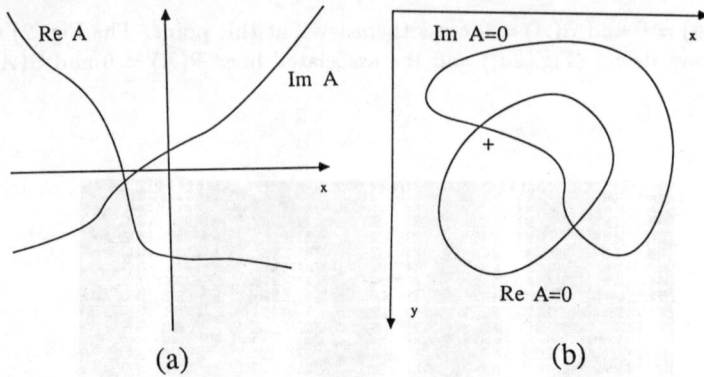

<u>Figure 3</u>: (a) In one space dimension, $\mathcal{R}e(A)$ and $\mathcal{I}m(A)$ vanish at different points which are not the same in general. Thus, the relation $|A|=0$ is not generically satisfied. (b) In two spatial dimensions, $\mathcal{R}e(A)$ and $\mathcal{I}m(A)$ vanish along lines which generally cross one-another at two points, where $|A|=0$ is satisfied.

In the last part of this paper, we shall numerically describe a turbulent state caracterized by a great number of topological defects in the system. We shall restrict our description to the case of spiral wave defects, but analog results could be given in the case of dislocations in wave patterns.

4. A Defect-mediated Turbulence

The simpler answer to the problem of the creation of defects is that they are induced by inhomogeneities of the initial conditions. Namely, in real experiments, defects can rise from the edges of the box, and even in the bulk if the experimental set is not perfect. But the system itself can also create its own defects. In fact, as it is well known in statistical physics, defects play an important role in phase transitions [9]. Near the transition point, the thermic fluctuations are amplificated, and pairs of defects may be created, and dissociated [8]. An analogous scenario can occur in far from equilibrium systems. But there is no need to add an external noise which would be the analog of thermic fluctuations. Near the threshold, the phase instability can be considered as a deterministic noise generator, and after some transient, pairs of topological defects are created. These pairs are quickly dissociated, and the defects move rapidly throughout the box; new pairs appear

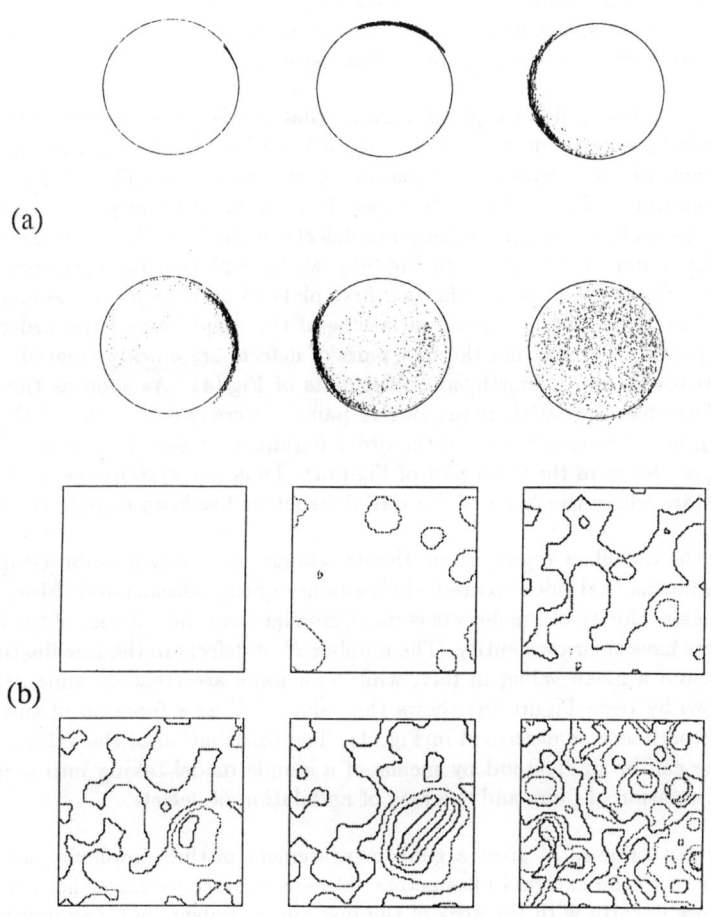

Figure 4: Numerical simulation showing the transition between the phase turbulence and the topological turbulence. (a) Distribution in the complex plane of the complex order parameter throughout the box, at t=50, 75, 100, 105, 110, and 500. The circle $|A| = \sqrt{\mu}$ has been plotted. (b) Lines $\Re(A) = 0$ and $\Im(A) = 0$ in the plane (x,y), at the same times.

while other anihilate. We give here a numerical description of this turbulent state termed "topological turbulence", in the case of a physical system which is described by Eq.(2). The initial conditions are chosen so that there are only little fluctuations around the homogeneous solution. The external parameters satisfy the inequality $1 + \alpha\beta < 0$, and we study the dynamics of the system.

The system first follows a phase regime, that is the phase of the order parameter spread out through the box. In Fig.(4a), we have plotted in the complex plane the value of the complex order parameter at each of the collocation points, for successive times. The perfect pattern would correspond to only one point following the circle of radius $\sqrt{\mu}$. If there are defects in the box, there will be some points at the center of this circle. In Fig.(4b), we have plotted the corresponding lines $\Re(A) = 0$ and $\Im(A) = 0$. The two first plots of each figure are associated with a phase regime. Then, strong variations of the amplitude of the order parameter appear in the box, and the first pairs of defects are quickly created. This corresponds to the third, fourth, and fifth plots of Fig.(4). As soon as the first pair of defects has appeared, many others pairs are created (see the sixth plot of Fig.(4b)), and the distribution of the order parameter covers the whole disk of radius $\sqrt{\mu}$, as shown in the sixth plot of Fig.(4a). Thus, the system has reached a turbulent state where the order of the initial structure has been completely lost.

When the turbulent regime is on, the defects are uniformly distributed in the box, which means that once created, the pairs are quickly dissociated. Moreover, since the mean velocity of the defects is much smaller than the velocity of the field, these objects have their own entity. The number N of defects in the box fluctuates in time around a mean value; in fact, while new pairs are created, some defects anihilate two by two. Figure (5) shows the value of N as a function of time for the same experiment as described in Fig.(4). The distribution of the values of N during time can be understood by means of a simple model taking into account the rate of creation of pairs and the rate of anihilation of defects.

The mean value of N gives a good measurement of the complexity, since it is only related to the external parameters. Moreover, since we have checked that $< N >$ grows linearly with the area of the box, one can define a mean density of defects d. In Fig.(6), we have plotted the value of d as a function of α, for a fixed value of $\beta = 2$. As a general feature, d grows as $|\alpha|$ is increased, but its growth is bounded by the lessening of the density of linearly unstable modes. This explains the decrease of d for large values of $|\alpha|$.

The existence of such singular objects rising, moving, and disappearing in the box, leads necessarily to a loss of correlations in the system. The correlation function of the field is defined as $\mathcal{C} = < A(x_0, y_0, t)\bar{A}(x, y, t) >$, where $< . >$ means the average over time. Figure (7) shows the behaviour of $\mathcal{C}$ as a function of x. The loss of correlation is clearly exponential, and thus the initial ordered structure has

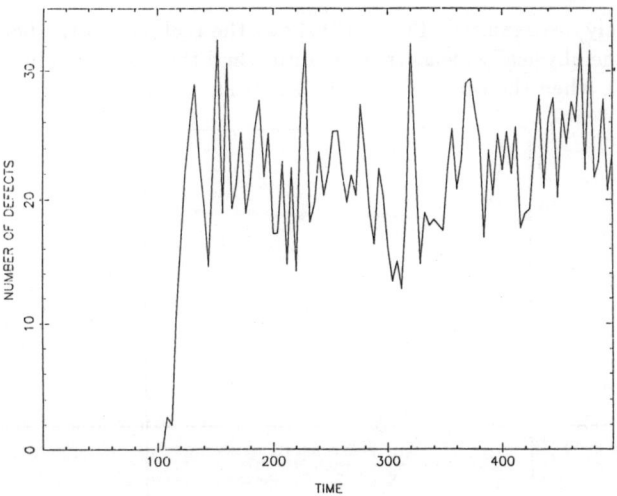

Figure 5: Number of defects in the box, corresponding to the experiment described in Fig.(4), and averaged over a time period of 40 units of time. The topological turbulent regime is characterized by a mean number of defects.

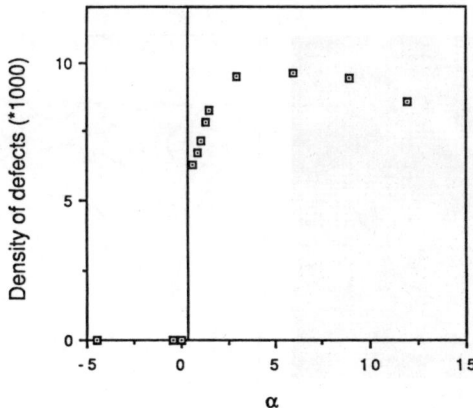

Figure 6: Diagram showing the density of defects as a function of α.

been completely disorganized. Figure (8) shows the real part of the field A, which is related to the physically measurable quantity, and the associated lines $\Re(A) = 0$ and $\Im(A) = 0$, when the turbulent regime has been reached.

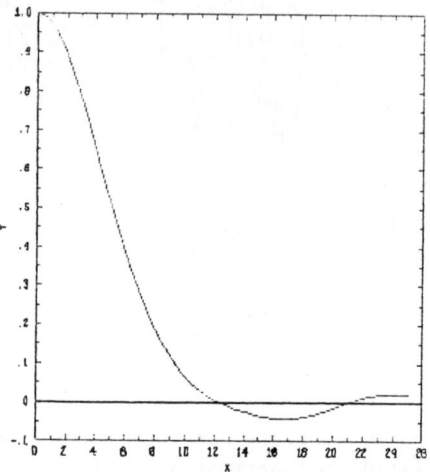

Figure 7: Correlation function $\mathcal{C}$ as a function of x.

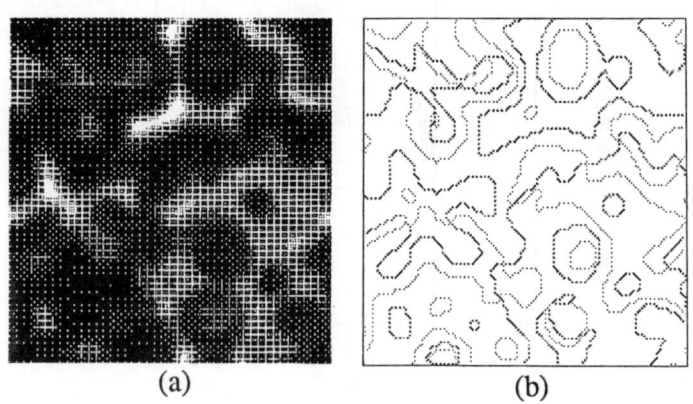

Figure 8: Numerical simulation of a topological turbulent field, showing in the plane (x,y), (a) the real part of the complex order parameter, and (b) the associated lines $\Re(A) = 0$ and $\Im(A) = 0$.

Conclusion

The numerical simulations described above have shown that the phase instability can lead to a disorganization of the system characterized by the presence of numerous topological defects in the box. These singular objects, which have a complex motion throughout the box and a finite lifetime, are responsible for the turbulent state of the system, which is termed "topological turbulence".

Our results may be related to the experiments made about ten years ago by E.Guyon and al. [19], who observed a defect-mediated disorganization of a convective structure, in the case of a nematic liquid crystal submitted to an elliptical shear.

Besides, such phenomena are observed nowadays in wave patterns in nematic liquid crystals, where the spatio-temporal ordered structure may be destroyed by the presence of rising, moving, and disappearing dislocations [20].

All the numerical simulations described in this paper have been made on CRAY1 and CRAY2 machines at the CCVR. We have used a pseudo-spectral code with periodic boundary conditions and a "slaved-frog" temporal scheme.

Acknowledgements

We thank U.Frisch, J.L.Meunier, J.P.Provost, D.Repaux, S.Ruffo, and D.Walgraef for fruitful discussions. We acknowledge the CCVR (Centre de Calcul Vectoriel pour la Recherche) where the numerical simulations have been performed, the NCAR, the CPAI (Centre Pilote d'Analyse d'Images) of the Observatoire de Nice, and the DRET (Direction des Recherches Etudes et Techniques) for a financial support under contract n^0 86/1511.

References

1. *Cellular Structures in Instabilities*, Eds. J.E.Wesfreid and S.Zaleski, Springer Verlag (1984)

2. *Chaotic Behaviour of Deterministic Systems*, Eds. G.Iooss, H.G.Hellerman, and R.Stora, North-Holland (1981)

3. *Propagation in Systems far from Equilibrium*, Eds. J.E.Wesfreid, H.R.Brand, P.Manneville, J.Albinet, and N.Boccara, Springer-Verlag (1988)

4. L.A.Segel, J. Fluid Mech. **38**, 203 (1969)

5. A.Newell, et J.Whitehead, J. Fluid Mech. **38**, 279 (1969)

6. L.Landau and E.Lifshitz, *Statistical Physics*, Pergamon Press (1959)

7. *Physics of Defects*, Eds. R.Balian, M.Kleman, and J.P.Poirier, North-Holland, Amsterdam (1980)

8. J.M.Kosterlitz, and D.J.Thouless, Prog. Low Temp. Phys. **78**, 371 (1978)

9. For a review, see B.I.Halperin in Ref.7, and D.R.Nelson in *Phase Transitions and Critical Phenomena*, Eds. C.Domb and M.S.Green, Vol.7, Academic Press, London (1983).

10. Y.Pomeau and P.Manneville, J. Phys. Lettres **40**, L-609 (1979)

11. Y. Kuramoto, Prog. Theor. Phys. **71**, 1182 (1984).

12. P.Coullet, L.Gil, and J.Lega, *Defect-mediated Turbulence*, preprint, submitted to Phys. Rev. Lett.

13. F.H.Busse, Rep. Prog. Phys. **41**, 1929 (1978)

14. Y.Kuramoto, *Chemical Oscillations, Waves, and Turbulence*, Ed. H.Haken, Springer, New York (1984)

15. P.Coullet, S.Fauve, and E.Tirapegui, J. Phys. Lett. **46**, 787 (1985)

16. I.Rehberg and G.Ahlers, Phys. Rev. Lett. **55**, 500 (1985)

17. E.Moses and V.Steinberg, Phys. Rev. A **34**, 693 (1986)

18. P.Coullet, C.Elphick, L.Gil, and J.Lega, Phys. Rev. Lett. **59**, 884 (1987)

19. J.M.Dreyfus, and E.Guyon, J. Physique **42**, 283 (1981), and E.Guazelli, and E.Guyon, C. R. Hebd. Séan. Acad. Sci. **292 II**, 141 (1981)

20. I.Rehberg and V.Steinberg, private communication

A Complex Behavior of the Gas of Particles Moving According to the Equation $x_{n+1} = (Ax_n + B) \bmod C$

M. Wolf

INSTITUTE OF THEORETICAL PHYSICS
PL 50-20 WROCLAW, CYBULSKIEGO 36, POLAND

Let us imagine the circle with nodes regularly distributed along it and denoted by $0, 1, \ldots, C-1$, see fig.1. Around this circle the material point is jumping from one node to the another one acording to the following rule : The position of the ball at the instant of time n we will denote by x_n, so $x_n \in \{0, 1, \ldots, C-1\}$. During an elementary interval of time a ball shifts by $Ax_n + B$ nodes in e.g. clockwise direction. Here A and B are natural numbers smaller then C : $0 < A < C$, $0 < B < C$. This rule can be written as the following equation of motion :

$$x_{n+1} = (Ax_n + B) \bmod C. \qquad (1)$$

It turns out that there exists a big variety of possible behaviors of the particle according to the particular values of driving parameters A, B and C. On the one hand the particle can perform a random walk around the circle. On the other hand it is possible that the particle will fall after at most *two* jumps into such a node in which it will remain for ever, *regardless of the initial value* x_0.

The following theorem asserts when it is possible that the sequence generated by iterations of (1) will possess a maximal period equal to C (for proof see e.g. [1]):

Theorem 1 : The period T of the sequence generated by the iterations of equation (1) is equal C if and only if :

i) B and C are mutually prime ;

ii) $A-1$ is a multiple of each prime p being a divisor of C;

iii) $A-1$ is a multiple of 4 if C is a multiple of 4.

Among the integers A, B and C fulfilling the requirements of the above theorem there are such ones that the succesive x_n will be weakly correlated in statistical sense [1]. In such a case the equation (1) is used for generation of random numbers in computers. So, with appropriately chosen A, B and C the particle will perform a random walk around the circle. It is a case of disorder in our system : each site of the phase space will be occupied after each C units of time - the motion is ergodic and the Poincare's recurrence theorem holds (with time od recurrence equal C).

Now let us focus on the exactly opposite case, namely let us look for the fixed points of (1). Let us introduce the function

$$f(x) = (Ax + B) \bmod C. \qquad (2)$$

Now we can write

$$x_{n+1} = f(x_n). \qquad (3)$$

The n-th element x_n can be expressed by the first one x_0 as the following superposition :

$$x_n = \underbrace{f \circ f \circ \ldots \circ f}_{n \text{ times}}(x_0) \equiv f^n(x_0) . \qquad (4)$$

Let x^* denote the fixed point of (2) :

$$x^* = f(x^*) . \qquad (5)$$

The above equation can be written in the following form :

$$x^* = Ax^* + B - Cq^*, \qquad (6)$$

where

$$q^* = [(Ax^* + B)/C]. \qquad (7)$$

The existence of fixed points of the mapping (2) is equivalent to the existence of integer solutions x^* and q^* of equation (6). The equations of the form

$$ax + by = c, \qquad (8)$$

where a, b and c are the given integers for which one seeks solutions x, y in integers are called *diophantine equations* (see e.g. [2]). Such an equations can have finite number of solutions, infinitely many or no one. As is well known the equation (8) has the solutions in integers if and only if c is the multiple of the Greatest Common Divisor (GCD) of numbers a and b. It is common to denote the GCD of a and b as (a,b). Writting (6) in the form

$$Cq^* - (A-1)x^* = B$$

we see that the following lemma holds :

Lemma 1: The necessary condition for the existence of fixed points of the mapping (3) is that B should be a multiple of $(C, A-1)$.

It can be proved [3] that it is possible to have d different fixed points :

Theorem 2 : The sequence

$$x_{n+1} = ((ad+1)x_n + bd) \bmod cd$$

possesses d fixed points. Denoting the smallest one by $x^{*(0)}$ all remaining fixed points are given by the formula

$$x^{*(r)} = x^{*(0)} + rc, \qquad r = 0,1,\ldots,d-1 .$$

In particular it is possible to have for rather pathological choice $A=1$ and $B=C$ C fixed points as from

$$x_{n+1} = (x_n + C) \bmod C \quad (\equiv x_n \bmod C)$$

it follows that $x_{n+1} = x_n$.

Now let us look more closely to the case when only *one*

single fixed point exist. It is natural to consider the fixed point of (2) as an attractor. Each attractor has its own domain of attraction called the *basin*. For the case of natural numbers we take the following

Definition : The basin $\mathcal{B}(x^*)$ of the attractor x^* is the set of such initial points $x_0^{(i)}$ ($i=1,2,\ldots,b$) of the sequences (1) for which there exist such numbers $N^{(i)}$ that $f^{N^{(i)}}(x_0^{(i)}) = x^*$.

Let us remark that for real numbers (continous case) the fixed point is usually reached after infinite number of iterations. The following theorem can be proved [3]

Theorem 3 : $\mathcal{B}(x^*) = \{0,1,\ldots,C-1\}$ if and only if there exist such a natural number N that A^N is divisible by C :

$$(A^N \equiv 0) \bmod C. \qquad (9)$$

If (9) is satisfied then the unique fixed point of (2) is given by the formula

$$x^* = \left(\frac{A^n - 1}{A - 1} B\right) \bmod C \qquad (10)$$

It turns out that if C is the prime then the equation (9) does not have integer solutions for A and N, for the proof see also [3].

Finally let us consider a gas of particles moving according to eq.(1). Let us remark that such a gas consists of *noninteracting* particles. The figures 2,3,4 and 5 show some examples of the possible evolutions of the gas. On the top of each figure the initial positions of balls are plotted (the time is directed downwards). Succesive rows correspond to succesive time steps in evolution. The starting positions as well as the number of particles were chosen randomly. In the example on Fig.2 all particles fall into one node, so for A and B fulfilling requirements of the Theorem 3 the particles resemble bosons and condensate to one state. On the contrary for A, B and C fulfilling requirements of the theorem 1 all the time each site will be occupied by at most one particle and the gas behaves like a set of fermions. Let us remark that Figures 2-5 resemble the ones obtained for the behavior of cellular automata [4], so it is possible that there is a connection between them and the model described by (1).

REFERENCES :

[1] D.Knuth, *The art of computer programming*, vol.2, *Seminumerical Methods*, Addison-Wesley,New York, 1981

[2] R.Courant, H.Robbins, *What is Mathematics*, Oxford University Press, many editions, e.g.1947, 1961

[3] M.Wolf, preprint ITP UWr 87/680, will be published in Int.Journ.Theor.Phys. 27 (1988), issue 1

[4] see e.g.D.Farmer, T.Toffoli and S.Wolfram edts. *Cellular Automata*, Physica **10D**, 1984, pp.1-248

Figure captions :

Fig.1 The illustration of the model.

Fig.2 Example of the attractor with maximal basin. Sites occupied by the particles are black. Here $C = 65536 = 2^{16}$, $A = 1154 = 2 \cdot 557$, so all particles fall into the attractor after at most 16 iterations.

Fig.3 Here $A = 78$, $B = 33$, $C = 195 = 3 \cdot 5 \cdot 13$. All particles fall after one unit of time either into the 4-cycle (33,72,189,150) or into the fixed point $x^* = 111$.

Fig.4 It is an example of chaos for A, B and C fulfilling the conditions of Theorem 1. Here $C = 243 = 3^5$, $A = 16$, $B = 17$. All the time each site is occupied by at most one particle.

Fig.5 It is an example for Theorem 2. Here $a=2$, $b=3$, $d=10$, $c=17$, so there are ten fixed points. Besides fixed points $x^*=7, 24,...,160$ there are also cycles seen.

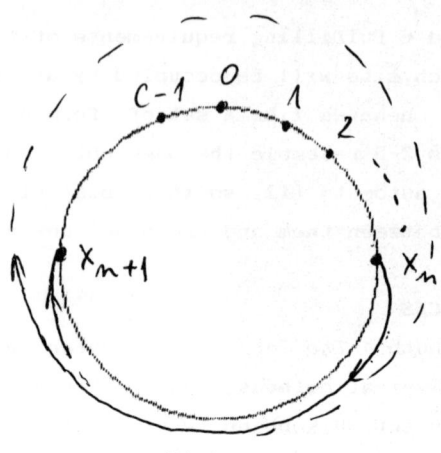

Fig. 1

Fig. 2

Fig. 3

Fig. 4

Fig. 5

Experiments on Spatio-Temporal Chaos

SHIL'NIKOV CHAOS IN LASERS

F.T. Arecchi

Istituto Nazionale di Ottica

and

Dept. of Physics - University of Firenze - Italy

ABSTRACT

The onset of deterministic chaos in lasers is studied by referring to low dimensional systems, in order to isolate the characteristics of chaos from the random fluctuations due to the coupling with a thermal reservoir. For this purpose, attention is focused on single mode homogeneous line lasers, whose dynamics is ruled by a low number of coupled variables. In the examined cases, experiments and theoretical model are in close agreement. In particular I describe Shil'nikov chaos, how it can be characterized, and the strong resulting coupling between nonlinear dynamics and statistical mechanics.

1. COHERENCE AND CHAOS IN LASERS

Quantum optics from its beginning was considered as the physics of coherent and intrinsically stable radiation sources. Lamb's semiclassical theory /1/ showed the role of the e.m. field in the cavity in ordering the phases of the induced atomic dipoles, thus giving rise to a macroscopic polarization and making possible a description in terms of very few collective variables. In the case of a single mode laser and a homogeneous gain line this means just five coupled degrees of freedom, namely, a complex field amplitude E, a complex polarization P, and a population inversion N. A corresponding quantum theory, even for the simplest laser model, does not lead to a closed set of equations, however the interaction with other degrees of freedom acting as a thermal bath (atomic collisions, thermal radiation) provides truncation of high order terms in the atom-field interaction /2,3,4/. The problem may be reduced to five coupled equations (the so-called Maxwell-Bloch equations) but now they are affected by noise sources to account for the coupling with the thermal bath /5/. As they are stochastic, or Langevin, equations, the corresponding solution in closed form refers to a suitable weight function or phase space density. In any case the average motion matches the semiclassical one, and fluctuations play a negligible role if one excludes the bifurcation points where there are changes of stability in the stationary branches. Leaving out the peculiar statistical phenomena which characterize the threshold points and which suggested a formal analogy with thermodynamic phase transitions /6/ the main point of interest is that a single mode laser provides a highly stable or coherent radiation field.

From the point of view of the associated information, the standard interferometric or spectroscopic measurements of classical optics, relying on average field values or on their first order correlation functions, are insufficient. In order to characterize the statistical features of quantum optics it was necessary to make extensive use of photon statistics /7,8/.

From a dynamical point of view, coherence is equivalent of having a stable fixed point attractor and this does not depend on details of the nonlinear coupling, but on the number of relevant degrees of freedom. Since such a number depends on the time scales on which the output field is observed, coherence becomes a question of time scales. This is the reason why for some lasers coherence is a robust quality, persistent even in presence of strong perturbations, whereas in other cases coherence is easily destroyed by the manipulations common in the laboratory use of lasers, such as modulation, feedback or injection from another laser.

Here I give a general presentation of low dimensional chaos in lasers. For a more complete approach to the problem, I refer to a recent monograph on the subject /9/.

We focus on those situations in quantum optics which permit close comparison between experiments and theory. By purpose I do not tackle the vast class of inhomogeneously broadened lasers, where it is extremely difficult to derive close correspondences between experiments and theory because of the large number of coupled degrees of freedom.

If we couple Maxwell equations with Schrödinger equations for N atoms confined in a cavity, and expand the field in cavity modes, keeping only the first mode which goes unstable, this is coupled with the collective variables P and Δ describing the atomic polarization and population inversion as follows (E being the slowly varying mode amplitude):

$$\dot{E} = -kE + gP$$

$$\dot{P} = -\gamma_\perp P + g E \Delta \qquad (1)$$

$$\dot{\Delta} = -\gamma_\parallel (\Delta - \Delta_o) - 4g\, PE \; .$$

For simplicity we consider the cavity frequency at resonance with the atomic resonance, so that we can take E and P as real variables and we have three coupled equations. Here k, $\gamma_\perp$, $\gamma_\parallel$ are the loss rates for field, polarization and population, respectively; g is a coupling constant and Δ_o is the population inversion which would be established by the pump mechanism in the atomic medium in the absence of coupling. While the first equation comes from Maxwell's equations, the other two imply the reduction of each atom to a two-level atom resonantly coupled with the field.

The presence of loss rates means that the three relevant degrees of freedom are in contact with a "sea" of other degrees of freedom. In principle, Eqs. (1) could be deduced from microscopic equations by statistical reduction techniques /5/.

The similarity of the Maxwell-Bloch equations (1) to the Lorenz equations /10/ would suggest the easy appearance of chaotic instabilities in single-mode, homogeneous-line lasers. Indeed the Lorenz model is a suggestive example of the general fact that a nonlinear coupling of at least three dynamical degrees of freedom may induce instabilities in the motion, which in such cases becomes irregular. However time scale considerations rule out the full dynamics for most of the available lasers. The Lorenz equations have damping rates within one order of magnitude. In contrast, in most lasers the three damping rates are wildly different from one another.

The following classification has been introduced /11/

<u>Class A</u> (e.g., He-Ne, Ar$^+$, Kr$^+$, dye): $\gamma_\perp \simeq \gamma_\parallel \gg k$.

The two last equations can be solved at equilibrium (adiabatic elimination procedure) and one single nonlinear field equation describes the laser. N=1 means a fixed point attractor, hence coherent emission.

<u>Class B</u> (e.g., ruby, Nd, CO_2): $\gamma_\perp \gg k \simeq \gamma_\parallel$.

Only the polarization is adiabatically eliminated and the dynamics is ruled by two rate equations for the field and population. N=2 allows also for periodic oscillations.

<u>Class C</u> The complete set of eqs. (1) has to be used, hence Lorenz-like chaos is feasible, whenever $\gamma_\perp \simeq \gamma_\parallel \simeq k$.

We have carried out a series of experiments on the birth of deterministic chaos in CO_2 lasers (Class B). In order to increase by at least 1 the number of degrees of freedom, we have tested the following configurations.

(i) Introduction of a time dependent parameter to make the system non autonomous /12/. Precisely, an electro-optical modulator modulates the cavity losses at a frequency near the proper oscillation frequency Ω provided by a linear stability analysis, which for a CO_2 laser happens to lie in the 50-100 KHz range, making it easy to take an accurate set of measurements.

ii) Injection of signal from an external laser detuned with respect to main one, choosing the frequency difference near the above mentioned Ω . With respect to the external reference the laser field has two quadrature components which represent two dynamical variables. Hence we reach N = 3 and observe chaos /11/.

(iii) Use a bidirectional ring, rather than a Fabry-Perot cavity /13/. In the latter case the boundary conditions constrain the forward and backward waves, by phase relations at the mirrors, to act as a single standing wave. In the former case the forward and backward waves have just to fill the total ring length with an integer number of wavelengths but there are no mutual phase constraints, hence they act as two separate variables. Furthermore, when the field frequency is detuned with respect to the center of the gain line, a complex population grating arises from interference of the two counter-going waves, and as a result the dynamics becomes rather complex, requiring $N > 3$ dimensions.

(iv) Add an overall feedback, besides that provided by the cavity mirrors, by modulating the losses with a signal provided by the output intensity /14/. If the feedback has a time constant comparable with the population decay time, it provides a third equation sufficient to yield chaos.

Notice that while methods (i), (ii) and (iv) require an external device, (iii) provides intrinsic chaos. In any case, since feedback, injection and modulation are currently used in laser applications, the evidence of chaotic regions cautions against optimistic trust in the laser coherence.

Of course, the requirement of three coupled nonlinear equations does not necessarily restrict the attention to just Lorenz equations. In fact none of the explored cases i) to iv) corresponds to Lorenz chaos.

2. SHIL'NIKOV CHAOS, THE METHOD OF RETURN TIME, AND FLUCTUATION ENHANCEMENT

Of the whole phenomenology explored in the past years, I select a single topic of particular relevance. I report experimental evidence of quasi homoclinic behavior characterized by pulses with regular shapes but chaotic in their time sequence /15/. The regularity in the shape means that the points at any Poincaré section are so closely packed that extremely precise measurements of their position would be required to yield relevant features of the motion. On the contrary, return times to a Poincaré section close to the unstable point display a large spread, due to the sensitive dependence of the motion upon the intersection coordinate. Based on such a consideration, we introduce the spread in return times as the specific indicator of homoclinic chaos. Our experimental data yield iteration maps of return times in close agreement with those arising from the theory of Shil'nikov chaos /16,17/. Thus, the test introduced in Ref. 15 appears as the most direct one to characterize chaotic dynamical situations associated with pulses almost equal in shape but having fluctuating occurrence times. Furthermore, at variance with the theory, the experimental return maps show a variable degree of thickening independent from the measurement accuracy. This is the phenomenon of fluctuation enhancement already described in the decay of the unstable state of a macroscopic system /18/. This phenomenon introduces unavoidable statistical features in the nonlinear dynamics of a macroscopic system. In such case the collective description in terms of a few dynamical variables breaks down, because of large fluctuations. This was first observed in the switch-on of a laser /18a/ and then in many quenching phenomena such as spinodal decomposition /18b/.

Our experimental evidence of Shil'nikov type instability is based on a quantum optical system, namely a laser with an overall feedback. Precisely, we work on a single mode CO_2 laser with an intracavity electro-optic modulator yielding cavity losses proportional to the laser output intensity (Fig. 1) /14/. If the time scale of the feedback loop is of the same order of the other two relevant variables, the system becomes three dimensional. Such a system is described by three first order differential equations for the laser intensity $x(t)$, population inversion $y(t)$ and modulation voltage $z(t)$ as follows /14/:

$$\dot{x} = -K_o \, x \, (1 + \alpha \sin^2(z) - y),$$

$$\dot{y} = -\gamma_{\parallel} (y + x y - A), \qquad (2)$$

$$\dot{z} = -\beta (z - B + r \, x),$$

where $K_o = (c/L) \, T$ is the non-modulated cavity loss parameter, L is the cavity ength, T is the effective transmission of the cavity, $\gamma_{\parallel}$ is the population decay rate. The intensity $x(t)$ is normalized to the saturation intensity $I_s = \gamma_{\parallel}/2G$, with G the field-matter coupling constant. The population inversion $y(t)$ is normalized to the threshold inversion K_o/G, $z(t)$ is the modulation voltage normalized to the $\pi/V_{\lambda/2}$, with $V_{\lambda/2}$ the $\lambda/2$ modulator voltage, A is the normalized pump parameter, β is the damping rate of the feedback loop, r is a coupling coefficient between the detected intensity $x(t)$ and the normalized $z(t)$ voltage, B is the bias voltage appied to the EOM, $\alpha = (1 - T)/T$.

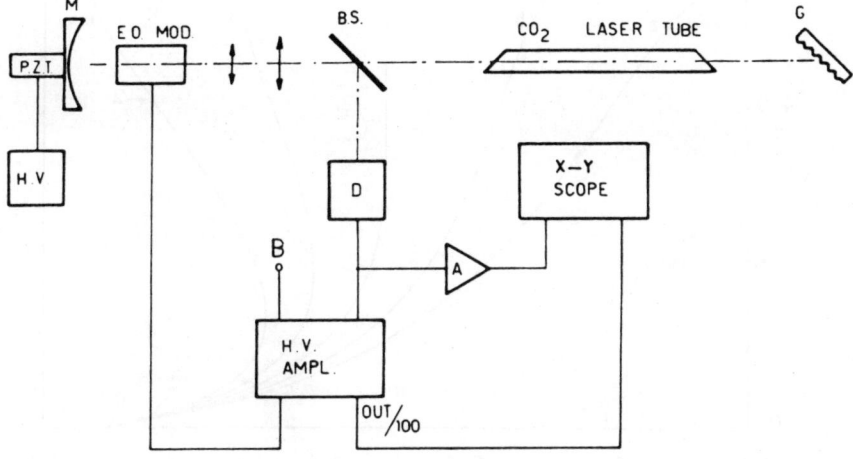

Fig. 1 Experimental set up. M - total reflecting mirror mounted on a PZT drive. E.O.MOD - electro-optic modulator. BS - ZnSe beam splitter. G - grating. D - HgCdTe detector. B - bias voltage.

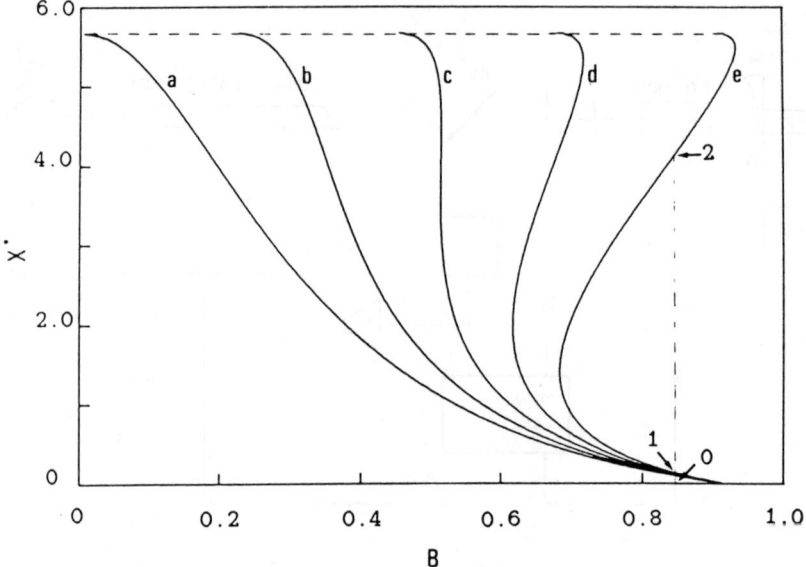

Fig. 2 Plot of the normalized stationary intensity x* versus B for a given pump rate. Call r the gain of the feedback loop. Curves a), b), c), d) and e) refer to r = 0.0, 0.04, 0.08, 0.12 and 0.16 respectively. Points 0,1 and 2 indicated by arrows, are the stationary points for B = 0.838 and r = 0.16.

The stationary solution (x*, y*, z*) for the system (2) implies the condition

$$B = r\ x^* + \arcsin\left[(A/1 + x^*) - 1)/\alpha\right]^{1/2}. \quad (3)$$

In Fig. 2 we report the stationary laser intensity versus one of the control parameters (B) for different values of the second one (r). This shows the coexistence of three fixed points for a wide range of r values.

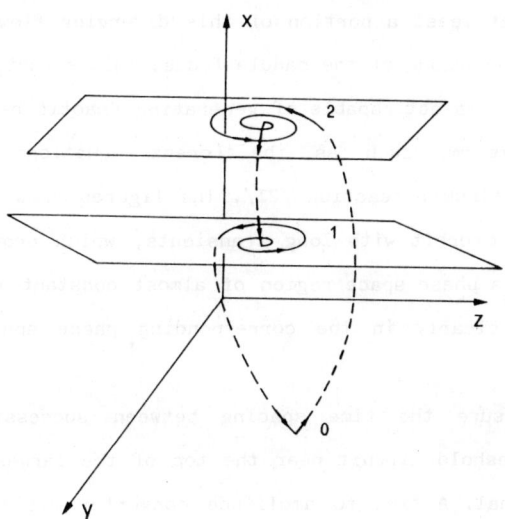

Fig. 3 Schematic view of a trajectory in the phase space (x = laser intensity, y = population inversion, z = feedback voltage) when the dynamics is affected by all the three unstable stationary points.

In Fig. 3 we present a schematic view of the trajectory in the three dimensional space, obtained by a linear stability analysis of the motion around the stationary points, and qualitative connections between the linear manifolds (dashed lines) /19/. Ref. 19 describes the competition of the three instabilities in controlling the global features of the motion. We adjust the control parameter in order to have a dominance of the saddle focus and reduce the influence of the other two instabilities. This way, the motion consists of a homoclinic orbit asymptotic to the saddle focus. This instability provides an exponential divergence within the flow while the homoclinic orbit ensures that at least a portion of this diverging flow is reinjected into the neighbourhood of the saddle focus. This structure of the flow is one of the simplest capable of generating chaotic behavior in many automonous systems such as the Lorenz equations /20/ or the Belousov-Zhabotinskii reaction /21/. The figures show clear evidence of a homoclinic orbit with long transients, which provide a lengthy permanence in a phase space region of almost constant intensity. This appears more clearly in the corresponding phase space projections (Figs. 4, 5).

We measure the time spacing between successive orbits by setting a threshold circuit near the top of the largest peak of the intensity signal. A time to amplitude converter yields the sequence τ_i of successive time spacings, which are then classified as a statistical distribution by a multichannel pulse height analyser, or stored in a digitizer, so that correlation functions or return maps can be sorted out.

The statistical distribution of return times is a broad featureless curve which does not offer clues on the ordering if τ_i. In contrast, the return map displays an extremely regular structure (Fig. 6). To check if we are seeng a one-dimensional (1D) return map where the remaining thickness is due to the observation technique or whether the map is more than 1D and its thickness hides new

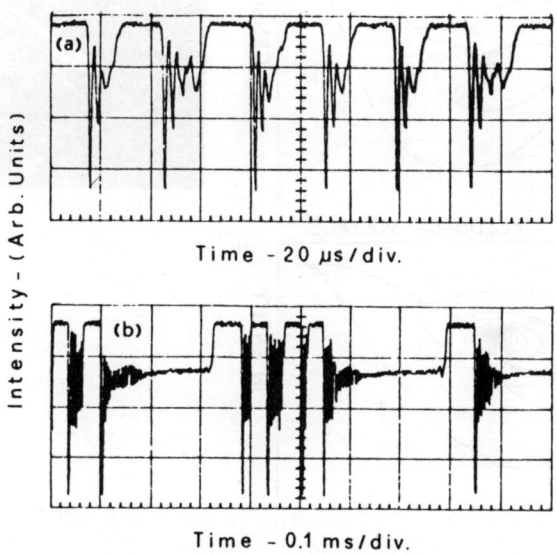

Fig. 4 Time plots of the laser intensity x(t) in the regime of Shil'nikov chaos. Intensity increases downward. a) and b) refer to the same B value, but two different gains r of the feedback loop. Fig.4b) shows two long transients corresponding to a large number of small spirals around the saddle focus (see also Fig. 5b)).

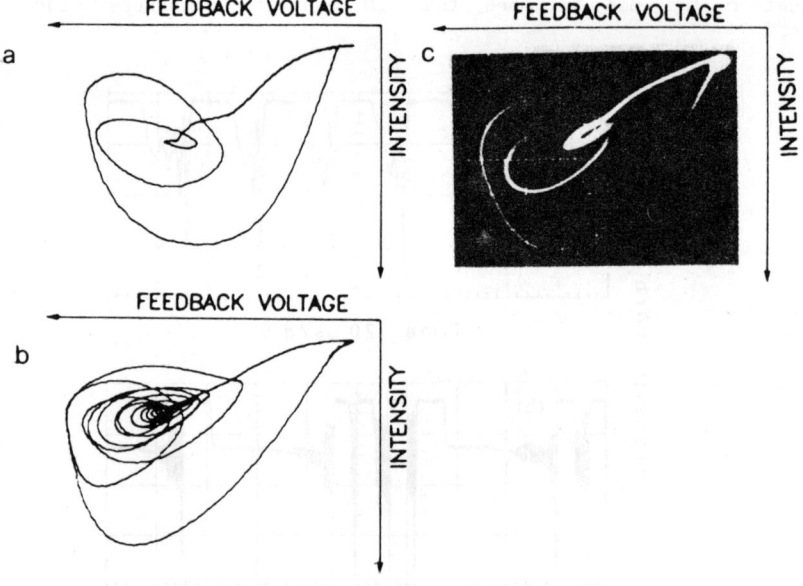

Fig. 5 Phase space projections x-z (laser intensity-feedback voltage). a) and b) are single orbits obtained by a digitizer, while c) is a photographic exposure over 1sec. a) and b) refer respectively to the same parameter situation of Fig. 4a), and b).

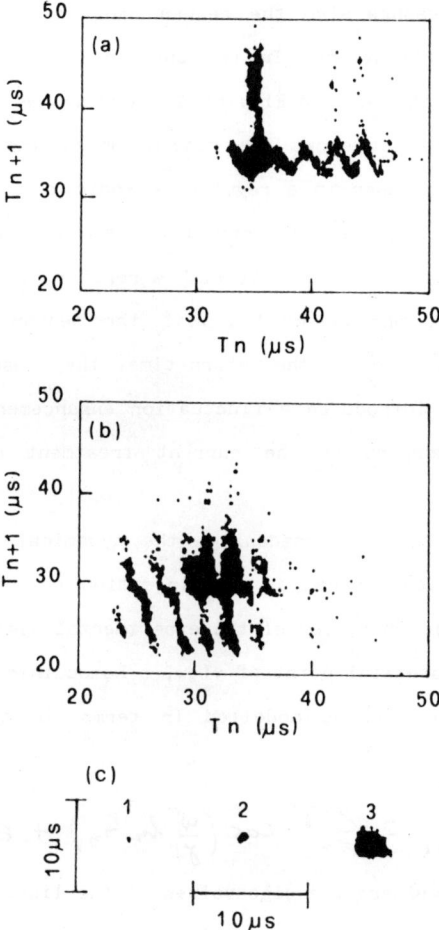

Fig. 6 Experimental return time maps. a) and b) refer to the same gain and to B values of 0.459 and 0.427 respectively. c) shows the maps corresponding to regular periodic situations, namely, 1) an electronic oscillator, 2) the laser in a regular periodic regime and 3) the laser just at the onset of the instability but still with a regular period.

information, we measure also the return maps corresponding to three regular periodic situations. In the absence of fluctuations in τ_i they should be point like. In Fig. 6c 1) corresponds to an electronic oscillator and it just shows the resolution of the measurement, 2) corresponds to the laser in a regular periodic regime away from the Shil'nikov instability, and 3) corresponds to the laser just on the verge of the instability but still with a regular period. In this last case, the fluctuations associated with the nearby transition show that, even without chaos in the return time, the close approach to an instability point introduces a fluctuation enhancement, which has no theoretical counterpart in the current treatment of deterministic chaos.

To deal with this broadening, the dynamical equations should include a statistical spread in the injection coordinate to account for the macroscopic character of the experimental system.

From a theoretical point of view, a homoclinic orbit asymptotic to a saddle focus can be modelled in terms of the following 1D iteration map:

$$\zeta_{n+1} = \zeta_n^{\frac{\lambda}{\gamma}} \cos\left(\frac{\omega}{\gamma} \ln \zeta_n\right) + \epsilon \qquad (4)$$

where γ and $-\lambda \pm i\omega$ are the eigenvalues of the linearized flow at the saddle focus, ζ is the coordinate along the unstable manifold and ϵ is the deviation along ζ from the homoclinic orbit at the Poincaré section in the neighbourhood of the saddle point ($\epsilon = 0$ corresponds to the homoclinic condition).

If we build a small cubic box of unit side centered at the saddle focus and oriented along the eigenvectors, any tiny difference in the entrance coordinate along the expanding axis ζ will strongly influence the residence time inside the box and hence the spacing from the next re-injection.

Observing that most of the time is spent in the box around the saddle point, we relate the return time τ to the coordinate ζ of the unstable manifold by $\zeta = \zeta_o \exp(\gamma \tau)$, thus obtaining an iteration map for the return times:

$$\tau_{n+1} = -\ln\left[e^{-\frac{\lambda}{\gamma}\tau_n} \cos\frac{\omega}{\gamma}\tau_n + \epsilon \right] \quad ,(5)$$

where λ, γ, ω and ϵ are the same as above.

The main difference between the experimental maps reported in Fig. 6 and the theoretical ones (Fig. 7), is related to the finite thickening independent of the accuracy of the measurements. This spread is due to a transient fluctuation enhancement, peculiar in the decay of an unstable state of a macroscopic system as already stressed in Ref. 18. This phenomenon unavoidably introduces further statistical fluctuations into the chaotic dynamics of a macroscopic system.

As it was shown in Ref. 18, even though this spread has no relevance on the average dynamics, it contributes a large fluctuation enhancement whenever the system slides downhill. In Ref. 18 this was observed in the transient decay of an unstable state, here we report the same feature repeated at each Poincaré cycle.

As a conclusion, low dimensional chaos is described by a small number of coupled deterministic equations (as e.g. Eq (1)), that we write in general as

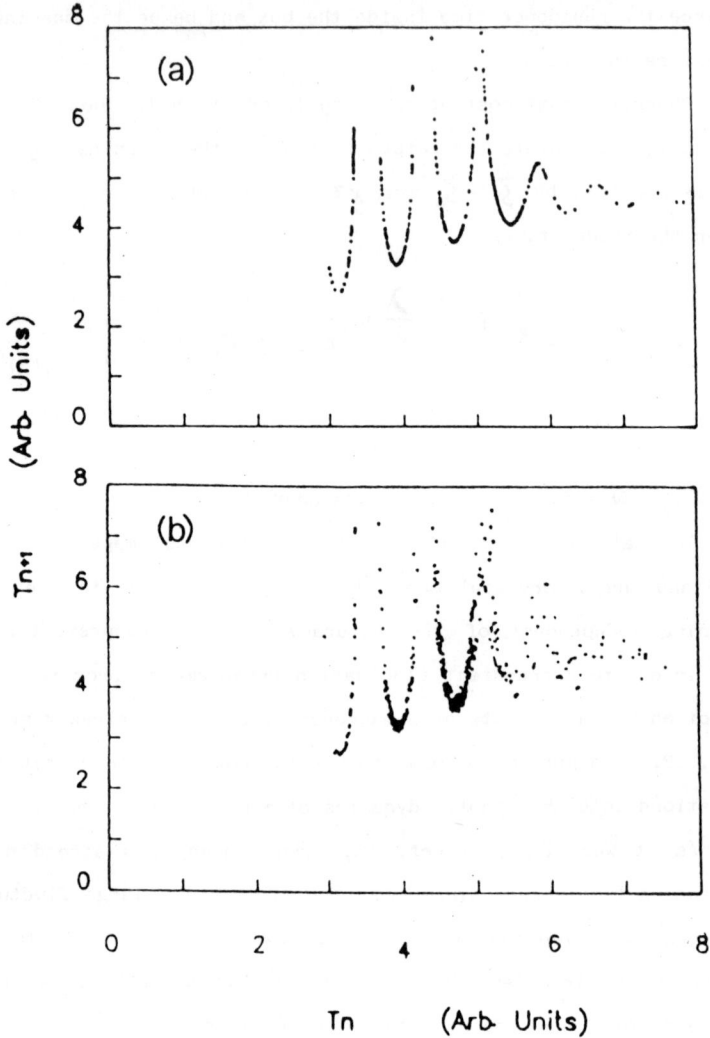

Fig. 7 Numerical return time maps for Shil'nokov chaos. a) no addition of noise or anharmonic contributions. b) with the addition of noise and 1% of second and third harmonic contributions.

$$\dot{x}_i = F_i(\{x_j\}) \quad (i,j = 1, 2, 3) \tag{6}$$

where F_i are nonlinear functions whose power expansion implies terms as $x_i \, x_j$ and higher order. But whenever this low dimensional chaos is a contracted description of a large system, then x_i are collective variables corresponding to macroscopic dynamics, and the nonlinearities of Eq (2) depend critically on whether

$$\langle x_i \, x_j \rangle \simeq \langle x_i \rangle \langle x_j \rangle . \tag{7}$$

Relation (7) fails to hold in the decay situations typical of Shilnikov chaos. In such a case, we have a strong coupling between nonlinear chaotic dynamics and statistical mechanics.

REFERENCES

/1/ W.E. Lamb, Jr., Phys. Rev. 134, A1429 (1964).

/2/ H. Haken, Laser Theory, in Encyclopedia of Physics vol. XXV/2c, ed. S. Flügge, Springer 1970.

/3/ M. Scully and W.E. Lamb, Jr., Phys. Rev. Letters 16, 853 (1966), Phys. Rev. 159, 208 (1967) and 166, 246 (1968).

/4/ J.P. Gordon, Phys. Rev. 161, 367 (1967).

/5/ H. Haken, Synergetics, 3rd Ed., Springer 1983.

/6/ F.T. Arecchi, in Order and Fluctuations in Equilibrium and Nonequilibrium Statistical Mechanics (Proc. XVII Solvay Conf. on Physics, ed. G. Nicolis et al.), J. Wiley 1981.

/7/ R.J. Glauber, in Quantum Optics and Electronics, (ed. C. De Witt et al.) Gordon and Breach, 1965.

/8/ F.T. Arecchi, in Quantum Optics (ed. R.J. Glauber), Academic Press 1969.

/9/ F.T. Arecchi and R.G. Harrison (eds.), Instabilities and Chaos in Quantum Optics, Springer Verlag 1987.

/10/ E.M. Lorenz, J. Atmos. Sci. 20, 130 (1963).

/11/ F.T. Arecchi, G.L. Lippi, G.P. Puccioni and J.R. Tredicce, Optics Comm. 51, 308 (1984).

/12/ F.T. Arecchi, R. Meucci, G. P. Puccioni and J.R. Tredicce, Phys. Rev. Lett. 49, 1217 (1982).

/13/ G.L. Lippi, J.R. Tredicce, N.B. Abraham and F.T. Arecchi, Optics Comm. 53, 129 (1985).

/14/ F.T. Arecchi, W. Gadomski and R. Meucci Phys. Rev. A 43, 1617 (1986).

/15/ F.T. Arecchi, A. Lapucci, R. Meucci, J.A. Roversi and P. Coullet (to be published).

/16/ L.P. Shil'nikov, Dokl. Akad. Nauk. SSSR 160, 558 (1965).
L.P. Shil'nikov, Mat. Sbornik 77, (119) 461 (1968) and 81, (123) 92 (1970).

/17/ A. Arneodo, P.H. Coullet, E.A. Spiegel and C. Tresser, Physica 14D, 327 (1985).

/18/ a) F.T. Arecchi, V. Degiorgio and B. Querzola, Phys. Rev. Lett. 19, 1168 (1967); b) F. Haake, Phys. Rev. Lett. 41, 1685 (1978); c) F.T. Arecchi and A. Politi, Phys. Rev. lett. 45, 1215 (1980); F.T. Arecchi, A. Politi and L. Ulivi, Nuovo Cimento 71B, 119 (1982).

/19/ F.T. Arecchi, R. Meucci and W. Gadomski, Phys. Rev. Lett. 58, 2205 (1987).

/20/ P. Glendinning and C. Sparrow, Jour. Stat. Phys. 35, 645 (1984).

/21/ F. Argoul, A. Arneodo and P. Richetti, Phys. Lett. A120, 269 (1987).

ADVANCES IN EXPERIMENTAL NONLINEAR DYNAMICS: SPACE/TIME PATTERNS

J.P. Gollub, F. Simonelli, and J. D. Goldberg

Haverford College and The University of Pennsylvania
Haverford, PA 19041
USA

ABSTRACT

We discuss various improved methods for experimental studies of nonlinear dynamical systems. These methods offer the promise of improved contact with theory, and include real–time modal decomposition, systematic transient studies, automated measurements of stability boundaries, and techniques for characterizing the spatial complexity of systems with many interacting modes.

1. INTRODUCTION

How can complex, time–varying patterns be characterized and described? How can experimental data on the dynamics of complex systems be effectively compared with mathematical models? It is apparent that there is a large gap between the kind of information provided in most experiments on nonlinear systems, and what is required for a proper comparison with theory. The purpose of this note is to outline some of these difficulties and to summarize (qualitatively) preliminary efforts to overcome them, in the context of experiments on space/time patterns at fluid interfaces. A detailed description of this work appears elsewhere[1,2].

2. CONSTRUCTING PHASE SPACE TRAJECTORIES FOR EXPERIMENTAL SYSTEMS

The basic paradigm of nonlinear dynamics is the assertion that the behavior of deterministic nonlinear systems can be understood by projecting the dynamics into a finite dimensional phase space. This fundamental step is surprisingly difficult to achieve experimentally. A considerable amount of work has been done using the "embedding method", in which time delayed versions of a single time series (obtained from a local measurement or a global average property) are used to construct trajectories in a multidimensional phase space.[3] This method is based on the idea that a single time series and its derivatives or delayed replicas will typically contain all of the important dynamical information. Is this assertion actually plausible for a typical nonlinear continuum of experimental interest, such as a fluid system?

The answer to this question depends on the number of dynamical variables that are important. If this number is sufficiently small (probably under six or so), as often occurs in systems where the basic dynamical scale is not much smaller than the size of the system, then the embedding method can be utilized to obtain trajectories in phase space and determine the properties of the resulting attractors[3]. If the motion is chaotic, the dimension and entropy of the chaotic states can be determined.

However, this approach is still generally unsuitable to achieving any understanding of the nonlinear processes occurring in the system. The reason is that there is no clear physical interpretation of the dynamical variables being used to describe the motion. A more desirable approach is to make a modal decomposition in terms of the basic spatial modes that would be used in an economical theoretical description. This concept has been discussed and emphasized by Newell[4], but experimenters have not yet taken heed, for the most part. One reason may be that it is often not obvious in advance what a suitable basis set would be in a particular case.

When chaotic dynamics occur close to onset in fluids, however, it may be useful to use as a basis the functions that become unstable at the initial instability threshold. This choice has proven useful, for example, in studies of surface wave dynamics, where the linearly unstable modes are simply two-dimensional Fourier modes.

However, even knowing this fact, it is not immediately clear how to accomplish a proper modal decomposition. Is it necessary to do a full Fourier decomposition at each time step? This can in fact be done with the aid of image processing methods in some cases[5] but it is inefficient if only a few modes are actually involved in the dynamics. In that case, it is possible to obtain the amplitudes of the relevant Fourier modes from a number of local probes equal to the number N of interacting modes. It is only necessary to solve an N x N system of complex linear equations at each time step. We have demonstrated that this can be done quite accurately when N=2, and the method can easily be extended to a larger number of probes.[1,2]

The method of time-resolved modal decomposition has several advantages over embedding techniques. The most important consideration is that the resulting variables have a clear physical interpretation. If a useful theoretical model is expressed in terms of Fourier modes, for example, then it is important to know whether a particular experimental state is a pure mode or a mixture of several spatial modes. Even this fundamental point is inaccessible without a modal decomposition. Furthermore, one can easily avoid the kind of uncertainty that plagues the embedding method, where the time delay is arbitrary and must be optimized.

In the method of modal decomposition, the primary difficulty is not the measurement itself, but rather the development of appropriate ways of visualizing and processing the results. When the phase space has more than two dimensions, then various graphical methods may be helpful, including: multiple projections, oscillating displays; the use of color; return mappings, etc.

3. STUDYING PHASE SPACE STRUCTURE

The behavior of a nonlinear dynamical system is not completely revealed by locating and studying an attractor. First, there may be many attractors; locating all of them gives a much more complete view of the dynamics of the system. Second, it is desirable to determine the structure of the basins of attraction: how large are they? Are they interlaced in complex ways?

Third, the transients leading to the attractors are also important. They can reveal the existence of multiple time–scales. For example, our studies of surface wave dynamics revealed that trajectories typically decayed on a fast time–scale to a well defined manifold, followed by a much slower decay to the fixed points. The transients can also reveal *unstable* fixed points and unstable periodic orbits. These are actually just as important from a dynamical point of view, yet they are rarely seen experimentally. The origin and fate of an attractor is often linked to that of an unstable attractor.

The key to exploring phase space dynamics in this kind of detail is a facility for controlling the initial conditions. For an externally driven system, the frequency ω, amplitude A, and phase ϕ of the external signal can be varied under computer control to allow a variety of initial conditions to be explored. The basic method is to decide on a set of parameters (ω, A, ϕ) at which the behavior of the system is to be explored, but to begin slightly away from that point at $(\omega+\delta\omega, A+\delta A, \phi+\delta\phi)$, followed by a sudden change to (ω, A, ϕ). The increments $(\delta\omega, \delta A, \delta\phi)$ are varied systematically and the resulting transient behavior observed repeatedly. Though it is not possible to obtain all relevant initial conditions in this way, the method does in practice yield considerably more information than the usual techniques.

For systems that are not externally driven, it would be necessary to introduce controlled perturbations in order to carry out this program.

4. EXPLORING COMPLEX PARAMETER SPACES

In order to compare nonlinear behavior with theoretical models, it is usually necessary to know the stability boundaries of the system, that is, the lines or surfaces in parameter space where bifurcations occur that define distinct dynamical regimes. Generally, the stability boundaries are functions of several external parameters. Furthermore, the instabilities are often sensitive functions of one or more of the parameters, so that the stability boundaries are complex sets or even fractals in some cases. For these reasons, it is in practice quite difficult to measure stability boundaries quantitatively.

Therefore, we have explored the possibility of automated determination of stability boundaries. In this method, the parameters are varied under computer control, and a set of conditions is developed to allow automatic recognition of the various dynamical regimes. If a stability boundary is crossed, this fact is detected and smaller steps are subsequently taken to locate the crossing point accurately. There are two basic difficulties in this process. The first is that some stability boundaries involve "critical slowing down," so that the time required to observe the steady state behavior diverges as the instability is approached. The search process must take this fact into account. Second, the presence of hysteresis or multiple stable states can complicate the search for stability boundaries, so that a unidirectional search is required, in which the boundary is always approached from the same direction.

We find that it is indeed possible to explore a complex parameter space in a completely automated fashion.[1,2] The resulting stability boundaries are less subject to experimental bias and are more precise than those that could be obtained manually. Since the locations of the stability boundaries place strong constraints on theoretical models, this information is a critical part of achieving a meaningful comparison between theory and experiment.

5. MANY MODES: SPATIO–TEMPORAL CHAOS

In the previous sections, we have described methods for experimental study of nonlinear systems in which a relatively small number of modes interact. Recently, there has been increasing interest in the opposite limit where many modes participate in the dynamics[6,7] so that the patterns are spatially complex and they vary chaotically in time. In this limit, a phase space description becomes impractical or even meaningless. However, a modal decomposition can still be useful, as we found in our studies of interfacial wave dynamics.

When many modes are involved, the decomposition cannot be done by a small number of local probes. On the other hand, modern digital image processing methods do allow a time–resolved modal decomposition to be made. This process requires a rather time–consuming analysis (such as a two dimensional Fourier transform) at each time step. The availability of array processors on powerful workstations makes this a realistic experimental approach (though it is not inexpensive).

It is of course not sufficient simply to make the modal decomposition. How is that information to be used? Clearly, the answer will depend on the experimental system. In our studies of interfacial wave dynamics, we noted that most of the spectral power resides on a ring in Fourier space. This means that there is a dominant wavenumber, and the various interacting modes differ essentially in their orientation within the plane. (This wavenumber is related to the frequency of the forcing via the dispersion relation of capillary–gravity waves.) The distribution of spectral power on the ring depends on the driving amplitude and frequency, and it can be time–dependent. When the pattern is orderly, all of the spectral power resides in a few discrete spots. On the other hand, in disorganized patterns, the power is dispersed more uniformly on the ring. To illustrate this fact we show in figure 1(a) a digitized shadowgraph image of the surface. The orientational disorder is almost complete, but a fundamental length–scale is clearly visible. In the corresponding power spectrum of figure 1(b), it is clear that most of the power is concentrated in a ring whose radius is the fundamental wavenumber. As expected the power is distributed rather uniformly on the ring in this case.

(a)

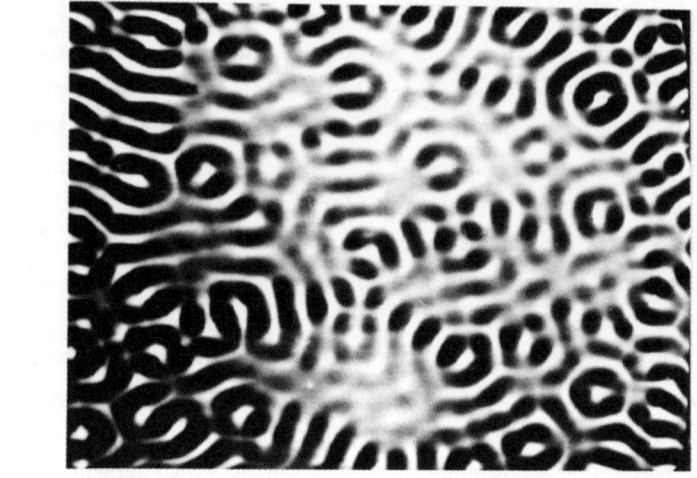

(b)

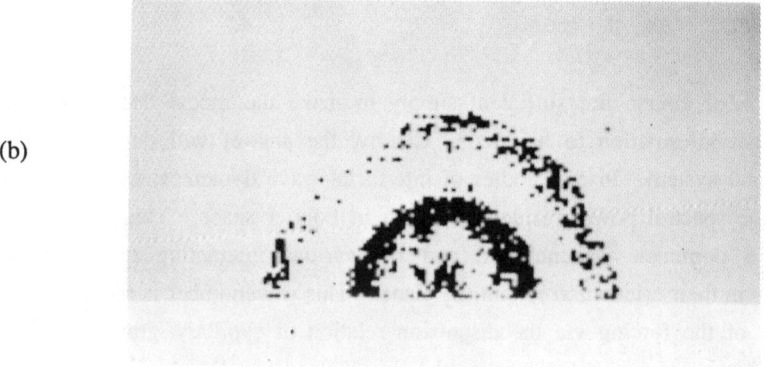

Fig. 1 a) Shadowgraph image of an orientationally disordered pattern, and b) the corresponding two–dimensional Fourier spectrum. Most of the power is concentrated in a ring whose radius is the inverse of the characteristic length scale. The distribution of power on the ring is almost uniform here, but is concentrated in discrete spots for orientationally ordered patterns.

The entire pattern can be described quantitatively by a "spectral complexity" C defined[8] as follows:

$$C = (\Sigma\, P_{ij})^2 / \Sigma\, (P_{ij})^2 \tag{1}$$

where P_{ij} is the spectral power in a particular element of the ring in Fourier space. This quantity goes to 1 when only a single Fourier mode is excited, and it goes to N (the number of discrete elements in Fourier space that fall on the ring) when the pattern is highly disordered. Only the fundamental ring is taken into account, the other (outer ring) being due to harmonics. The spectral complexity C is usually normalized to be independent of N, so that data taken in very different situations can be compared. Other measures of complexity can also be used, such as the spectral entropy,[9] a quantity that would be 0 for an ordered pattern and 1 for a uniform distribution of power in the ring. We find that both methods give similar results although the spectral entropy (due to the logarithm in its definition) appears to be more sensitive to modes with small amplitude.

Spatially complex patterns can be characterized by determining the spectral complexity C as a function of parameters (driving amplitude and frequency). We have found that the average complexity is generally much larger for time–dependent than for time–independent states. By measuring the spectral complexity at each time step, we obtain a time series $C(t)$ that is more closely related to the pattern variations than any local or global signal would be.

Though this work is still in progress, it is worth noting one surprising result. In our studies of surface waves with many interacting modes, we found that spatially simple states (with low values of C) recur aperiodically during the chaotic motion. This intriguing fact, which was not noted in qualitative observations, is an example of the utility of a quantitative measure of pattern complexity. Clearly, the study of spatio–temporal chaos is just beginning, and substantial improvements in experimental approach will be required.

ACKNOWLEDGEMENT

This work was supported by DARPA–URI Contract N00014–85–K–0759 to Princeton University.

REFERENCES

1. F Simonelli and J.P. Gollub, "Stability Boundaries and Phase Space Measurements for Spatially Extended Dynamical Systems", to appear in Revs. Sci. Instr. (1988).

2. F. Simonelli and J.P. Gollub, "Surface Wave Mode Interactions: Effects of Symmetry and Degeneracy", in preparation.

3. H. Swinney and J.P. Gollub, "Characterization of Hydrodynamic Strange Attractors", Physica 18D, 448 (1986).

4. A.C. Newell, "Chaos and Turbulence; is there a connection?", in *Proceedings of the Conference on Mathematics Applied to Fluid Mechanics and Stability* (SIAM, 1985).

5. S. Ciliberto and J.P. Gollub, "Chaotic Mode Competition in Parametrically Forced Surface Waves", J. Fluid Mech., 158, 381, (1985).

6. A.B. Ezerskii, M.I. Rabinovich, V.P. Reutov and I.M. Starobinets, "Spatiotemporal Chaos in the Parametric Excitation of a Capillary Ripple", Sov. Phys. JETP 64, 1228 (1986).

7. J.P. Gollub, "The Onset of Turbulence: Convection, Surface Waves, and Oscillators", in *System Far From equilibrium*, ed. by L. Garrido., Lecture Notes in Pysics Vol. 132 (Springer, Berlin 1980).

8.   J. Crutchfield, D. Farmer, N. Packard, R. Shaw, G. Jones and R.J. Donnelly, "Power Spectral Analysis of a Dynamical System", Phys. Lett. 76A, 1 (1980).

9.   R. Livi, M. Pettini, S. Ruffo, M. Sparpaglione and A. Vulpiani, "Equipartition Threshold in Nonlinear large Hamiltonian Systems: The Fermi–Pasta–Ulam Model", Phys. Rev. A 31, 1039, (1985).

DYNAMIC OF RAYLEIGH-BENARD SPATIAL CONFIGURATIONS IN CONFINED GEOMETRY

M. Dubois
SPSRM - CEN Saclay
91191 Gif-sur-Yvette cedex France

Rayleigh-Bénard convection has been revealed to be a good system to study the approach of turbulence. In large geometries, certain spatial features (defects, frustration effects due to boundary conditions, etc)[1] may lead to spatially turbulent states; while in confined geometries, the evolution of thermal oscillators with the Rayleigh number Ra can give deterministic chaos, i.e. only temporal chaos[2]. In the later case, the spatial configuration intervenes only to determine the specific oscillators related to it, and then has to remain constant in the Ra domain under study. With this condition, experiments have shown the existence of the various scenarios leading to deterministic chaos with the presence of strange attractors[3] and in certain cases the physical mechanisms underlying this dynamics have been evidenced.

Nevertheless, even in confined geometries (horizontal aspect ratios $A_x \simeq 2d$ and $A_y \simeq d$, with d the depth of the fluid layer) many different spatial arrangements of the rolls are available. More particularly, when the convection is achieved with high Prandtl fluids, the Ra values at which the thermal oscillators are present, are very high and at a given Ra number, different structures (then different dynamical states) can be observed[4]. These structures are all solutions of the convective system, each one with its own Ra domain of stability. In the following, we will show briefly some experimental results obtained with high Prandtl fluids and related to dynamical behaviours due to the competition between spatial modes.

1. STABLE LOW SPATIAL TURBULENCE

In the geometry $A_x = 2$, $A_y = 1.2$, widely used to study the routes to deterministic chaos, the sequence of the states generally observed when increasing the Ra number (with Prandtl number > 20) is the following : stationnary, monoperiodic, then chaotic, or biperiodic then chaotic (depending on the scenario). But for slightly different geometries, a direct transition from a stationnary state to a chaotic one was observed to take place. In our experiments (Pr = 38), with a constant value for A_y (A_y = 2) and an adjustable value for A_x (here $2 < A_x < 4$), this transition occured for Ra values around 100 Rac (where Rac is the critical Ra value, Rac = 1708) and for the entire range of A_x values, except windows near A_x = 2, A_x = 3 and A_x = 4 for which a stable configuration is generally observed at these Ra values, even after long transient periods.

What are the characteristics of this chaotic state ? In a mean stable structure, abnormal currents (one or two) appear, irregularly in time; they are not stable and move towards the main descending or ascending streams in the center of the cell (see Fig. 1). It looks like a continuous tendancy for the structure to adjust its local wavenumber to get a more stable configuration, but without success. This behavior is chaotic, as shown by the Fourier spectrum of a signal sensible to the temperature gradient in the center of the cell : this spectrum exhibits a broad band, but there are no characteristic frequencies.

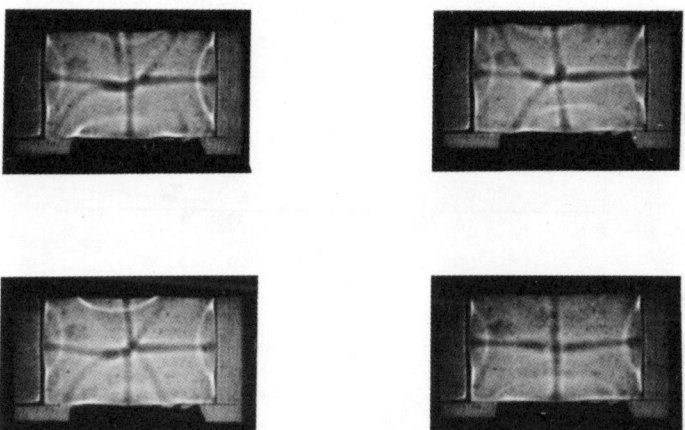

Fig. 1: Shadowgraphic pictures of a convective structure seen from above, (A_x=3.3, A_y=2, Ra/Rac=120).The dark lines represent ascending motions. From one picture to another, we can see extracurrents, which appear erratically in time. They mover towards the main streams which are seen alone, as on the picture n° 4, when no extracurrents are present.

To know more about the number of degrees of freedom involved in this dynamical situation, the correlation dimension has been calculated with the algorithm proposed by P. Grassberger and I. Procaccia[5] in the case of measurements performed with A_x = 2.47 and Ra/Rac ≃ 120. The problem was how to choose the best sampling time T_s, for no characteristic time are really present except the fact that the spread of the Fourier spectrum is 0.1Hz. We constructed two time series successively, one with T_s = 10s and another with T_s = 30s, of 15 000 points each. The results of the calculation are shown on the figure 2, where the calculated correlation dimension ν is reported as a function of the embedding time T_e

$$T_e = (D - 1) * p\, T_s$$

where D is the embedding dimension and $p\, T_s$ is the time delay taken for the reconstruction of the attractor. We can remark that there is a plateau of the ν values in the same T_e range for both time series. This gives a ν value around 6.5, which seems very likely to be related to the number of effective degrees of freedom of the described mechanism.

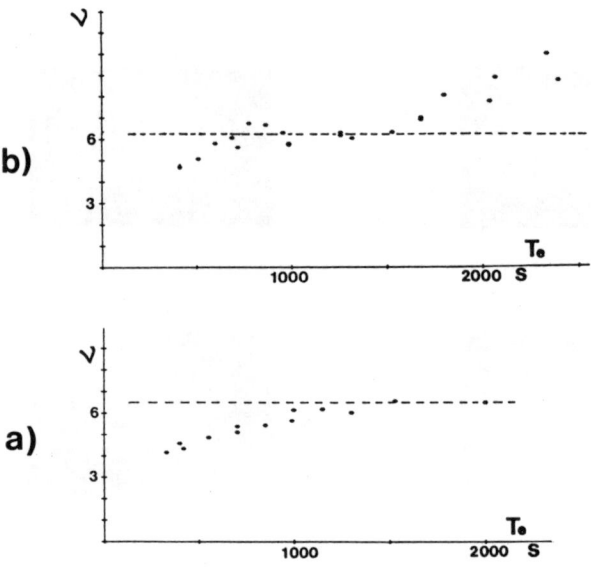

Fig. 2 : Calculated values of the correlation dimension ν versus the embedding time Te :   a) Time series constructed with T_s = 10s.
b) Time series constructed with T_s = 30s.

This low calculated correlation dimension is an indication that in the studied geometry (here $A_x = 2.47$ and Ra/Rac = 120) the observed chaotic state, though due to a spatial instability, is nevertheless restricted to a small number of interactive spatial modes[6]. This fact reminds us of some features observed in the study of the Kuramoto-Shivashinsky equation, with a small lenght for the system[7]. For certain values of this length, complex situtations may be observed, due to the competition between equally probable spatial modes[8]. Although a direct comparison between our experimental situation and the Kuramoto-Shivashinsky model may not be exactly appropriate, our structure being three-dimensional, with the observed dynamics taking place at high value of the Rayleigh number, the two mechanisms may nevertheless be similar.

2. COMPLETE SPATIAL TURBULENCE

The chaotic state, described just previously, is generally stable, i.e., once established, it lasts as long as the conditions (Ra, A_x) remain in its domain of stability. Another type of turbulent state is also observed in the small boxes, but generally as a transient between a structure which becomes unstable and a new one[9]. So the Ra domain, for which it may appear, depend on the Prandtl number of the fluid and on the aspect ratio.

At first sight, this turbulence resembles that observed in large cells, and indeed the corresponding Fourier spectra of the temperature (or velocity) fluctuations exhibit a broad band noise without any sharp peaks. The main streams are continuously moving and no spatial order is present. But from time to time, there is a freezing of the spatial arrangement which remains stable for a while (fig. 3). The spatial chaos then reappears and the alternating sequence of spatial turbulence, followed by a new spatial configuration, continues. These quasi-stable configurations may correspond to stable or metastable structures which can be observed at different Ra values. The lifetime of these temporary lockings varies from one locking to the other. On the other hand, the total duration of this turbulent behaviour which leads to a final stable structure depends on the experiment and may vary from a few hours to a few ways. This turbulent behaviour may be understood as a wandering in the complete phase space of the system, with temporary locking near the points of stable configuration. In this sense, the phenomenon gives an indication on the different stable solutions, whose competition leads to turbulence.

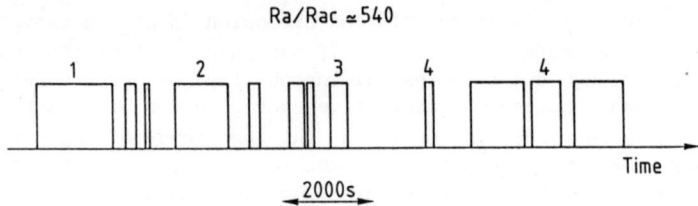

Fig. 3 : Diagram showing a typical succession of chaotic and spatial ordered states in a long transient turbulent behaviour ($A_x = 2$, $A_y = 1.2$, $P_r = 130$, Ra/Rac $\simeq$ 540). The ordered states correspond to within the rectangles. Those rectangles labelled 1, 2 ... refer to well known structures which are stable at the same Ra value.

BIBLIOGRAPHY

1) Pocheau, A., Croquette, V., Le Gal, P., Phys. Rev Lett. $\underline{55}$, 1094 (1985)
   Heutmaker, M.S., Gollub, J.P., Phys. Rev. A $\underline{35}$, 242 (1987)
2) Dubois, M., Bergé, P., J. de Physique $\underline{42}$, 167 (1981)
   Gollub, J.P., Benson, S.V., J. Fluid Mech. $\underline{100}$, 4119 (1980)
3) Dubois, M., Bergé, P., Physica Scripta $\underline{33}$, 159 (1986)
4) Bergé, P., Dubois, M., Croquette, V., "Convective Transport and Instability Phenomena" Euromech 138 Kärlsrühe (1981) ed. by J. Zierep and H. Oertel, p. 123
5) Grassberger, P., Procaccia, I., Phys. Rev. Letters $\underline{50}$, 346 (1983)
6) Bergé, P., in "Physics of Chaos and Systems far from equilibrium" (Chaos 87) ed. by Duong-Van
7) Manneville, P., "The Kuramoto-Sivashinsky equation : a progress report" in "Propagation in systems far from equilibirum" Les Houches - Avril 87 - Springer Proceedings in Physics (in Press)
8) Ciliberto, S., Gollub, J.P., Phys. Rev. Letters $\underline{52}$, 922 (1984)
9) Bergé, P., Dubois, M., Phys. Letters A, 93A, 365 (1983)

TRANSITION BETWEEN LOCALIZED OSCILLATIONS AND TRAVELING WAVES IN THERMAL CONVECTION

M. A. Rubio[+], S. Ciliberto and P. Bigazzi.

Istituto Nazionale di Ottica.
Largo Enrico Fermi 6. 50125 - Arcetri.
Firenze. Italy.

ABSTRACT.

The spatiotemporal behavior in Rayleigh-Benard convection in a small rectangular cell has been experimentally studied. Two main regimes have been found, one characterized by localization in space and the other by a traveling perturbation. The transition between both regimes has been studied and it has been shown that this transition presents many of the features that are usually associated with Shilnikov type homoclinic chaos. Besides, we propose several magnitudes that are useful to quantitatively study the complexity of spatio temporal chaotic regimes.

1.- INTRODUCTION.

During last years Rayleigh-Benard (RB) convection [1], has been widely used to study a large number of nonlinear problems, that range from pattern selection [2] to transition from regular to chaotic behavior [3].

For small cells it has been fully demonstrated by experiments that very often the time dependent states can be described by low dimensional attractors, either periodic or strange, but the problem of the coupling between temporal behavior and spatial structure still remains to be solved. The relationship between spatial and temporal degrees of freedom in a fluid system is extremely important to understand the transition from regular to turbulent flow and it certainly should clarify the actual relevance of some low-dimensional chaotic states as precursors of turbulence.

Due to the intrinsic interest of this problem the relationship between spatial order and temporal chaos has been recently investigated with different approaches. Numerical studies are available in several systems [4].

Here we describe experiments on RB convection in which many of the spatiotemporal features are qualitatively similar to those observed in reference [4].

2. The Experimental Set-up.

The experiments were conducted in a rectangular cell of sizes $lx = 4$ cm, $ly = 1$ cm and $d = 1$ cm. The x and y horizontal axes of the coordinate reference frame are respectively perpendicular and parallel to the roll axis, while the z axis is the vertical one.

The working fluid is silicon oil with Prandtl number $Pr = 30$ at $T = 25\,^{\circ}C$.

The experimental system has been fully described in reference [5]. We only remind that for a quantitative analysis of the spatiotemporal behavior of the fluid we detect the horizontal temperature gradient $u(x,t)=dT(x,z_o,t)/dx$, measured at 64 points on a horizontal line at height $z_o = 0.5$ cm. The length of the line covers the whole length of the cell.

To better identify the different spatio-temporal regimes we have computed the functions $w(x,t) = u(x,t) - \underline{u}(x)$ that accounts for the fluctuations of $u(x,t)$ around the time averaged pattern $\underline{u}(x)$ and the spatio-temporal correlation function $C(\Delta, \tau)$.

To characterize the different structures we have computed time resolved spatial Fourier transforms $\phi(n,t)$, where $2\pi n/lx$ is the wavevector. Then, calling $\Upsilon_n(t) = |\phi(n,t)|^2$, we define an energy as

$$E(t) = \sum_n^{32} \Upsilon_n(t)$$

and a "spectral entropy" function, $\sigma(t)$, as is usually done in the studies of energy spatial distribution in oscillator chains [6]:

$$\sigma(t) = \frac{1}{\sigma_{eq}} \sum_{n=1}^{32} a_n \log a_n$$

where

$$a_n = \frac{\Upsilon_n(t)}{E(t)}$$

and σ_{eq} is the equipartition value.

3. Experimental procedure and results.

The experiment has been performed in the following

way. We start from zero temperature difference between the two plates. Then we increase the temperature of the bottom plate till a maximun difference around 300 Rc. The quasisteady steps in which the temperature difference Δ T has been increased are separated by a sufficient amount of time for the system to relax to a stable state. This type of run has been repeated several times to check for the dependence of the regimes attained on the way in which the control parameter was varied.

Finally we have labeled the different regimes founded using the Rayleigh number[1] normalized to the critical Rayleigh number Rc = 1708, that in our system corresponds to a temperature difference Δ Tc = 0.059 C.

Analysing the fluid behavior as a function of r=R/Rc we first find a steady four roll structure that remain stable up to r = 80. Above this threshold several time dependent regimes and stationary structures alternate up to the transition which is our main concern now.

This transition occurs at r = 182. The spatiotemporal regime prior to this bifurcation can be identifyed as a localized one in figures 1 a) to d), that correspond to a value of r = 174, at which the regime is periodic. Its main features are clearly seen in the spatiotemporal correlation function in figure 1c) and the evolution of

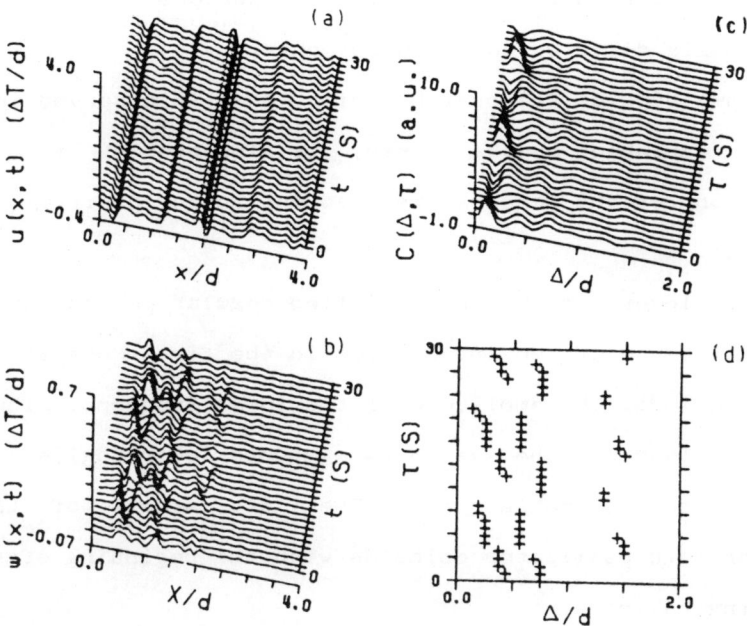

Fig. 1. Space-localized oscillations at r=174. The time evolution is periodic with frequency 77 mHz. Space-time evolution of (a) u(x,t) and (b) w(x,t). (c) corresponding C(Δ, $\hat{\tau}$). (d) Time evolution of the positions of the maxima of C (crosses).

its maxima (figure 1d)).

Upon increasing r over the critical value, $r_i = 182$, there is a great change in the spatiotemporal evolution of the system. At r = 184, the time evolution of the energy E(t), is shown in figure 3a). A detailed analysis of this signal reveals that there are regular periods interrupted by randomly distributed turbulent bursts.

This type of temporal behavior, that is observed in a global variable as the energy, is also present in the evolution of the temperature gradient in any position of the cell.

A closer view to the so-called regular periods shows the existence of two frequencies in the first stages of these periods. The amplitude of the lower frequency slowly decays, leading the system to a monoperiodic regime that appears to be damped too. The last stages of this monoperiodic oscillator coincide with the beginning of the bursting region.

This evolution can be interpreted in the sense that the behavior of the system in this region of parameter space is ruled by five main degrees of freedom. Four degrees of freedom account for the two couples of complex conjugate eigenvalues with negative real part corresponding to the two damped oscillating modes. The last degree of freedom corresponds to the positive real

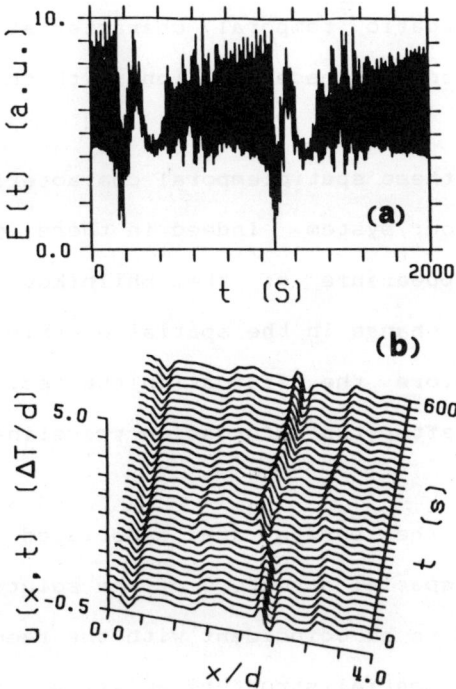

Fig. 2. Transition between localized oscillations and traveling waves at r=184. (a) Time evolution of the energy E(t). (b) Space-time evolution u(x,t) averaged over 4 periods of the fast oscillations.

eigenvalue that allows for the bursting behavior.

This type of evolution is characteristic of Shilnikov type homoclinic chaos [7] that has been observed in numerical integration of simple models of nonlinear partial differential equations [4]. We have carefully compared the spatio temporal characteristics of the numerical solutions of these equations with those found in our experiment.

Several of these spatiotemporal characteristics could be checked in our system. Indeed in these mathematical models, the appearance of the Shilnikov type chaos coincides with a change in the spatial distribution of the oscillators; before the bifurcation the oscillators are localized while after the bifurcation traveling waves are present.

Moreover, the bursts are associated with major changes in the spatial structure of the solutions. These bursts are found to be coincident with the reemergence of the stationary spatial structure existing just prior to this transition. Finally, even wave number spatial modes evolve in counter phase respect to the odd ones.

Turning back to our experimental results, some of these spatio-temporal features can be unambiguously seen in figures 2b) and 3 a,b), corresponding to a value of

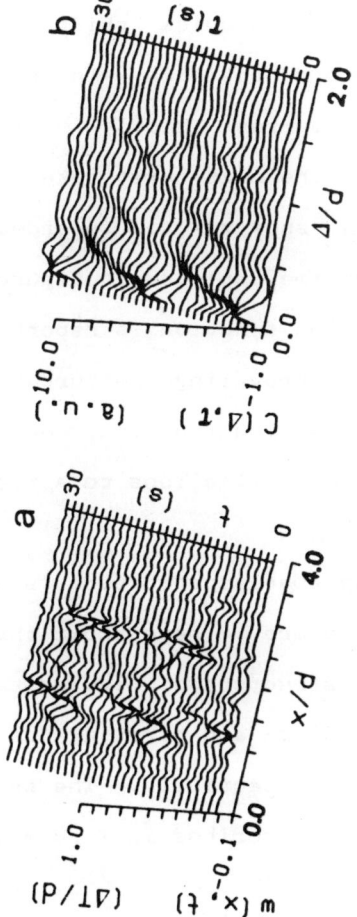

Fig. 3. (a) Space-time evolution of $w(x,t)$ recorded during the laminar period of the regime shown in Fig. 2. (b) Corresponding correlation function $C(\Delta, \tau)$.

r=184.

In figure 2b) we have plotted the evolution of $u(x,t)$ averaged over 4 seconds to eliminate the fast oscillation. There is a great structure change coincident with the burst that corresponds to a quick shift of the main current to a more centered position. In the following regular period, this current slowly returns to its original position, until the next burst takes place.

In figure 3a) we show $w(x,t)$ recorded during a laminar period of the regime depicted in figure 2a). Its spatiotemporal correlation function is reported in figure 3b). The presence of a traveling perturbation can be clearly seen in the correlation function, and therefore a transition from localized oscillations to a traveling wave behavior has taken place.

An analysis of the spatial structure is carried out in figures 4a) to c), where we have shown three relevant time averaged spatial structures. Figures 4a) and b) correspond to the regular and bursting parts of the evolution in figure 2a) respectively. The shift of the central current already described in figure 2b) is seen once again.

However the most striking feature in figure 4 is the similarity between figures 4b) and c). Figure 4c) reports the stationary spatial structure existing just prior to

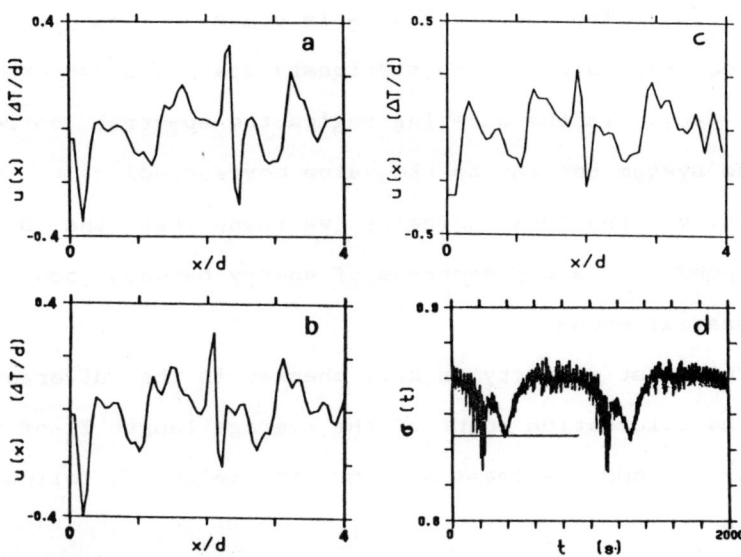

Fig. 4. Time averaged spatial structures at r=184, (a) during a laminar period, (b) during a burst. (c) Spatial structure at r=141. (d) Time evolution of the entropy at r=184; the horizontal line corresponds to the value of the entropy at r=141.

the time dependent regime leading to this transition (r=141).

This check of the similarity between both structures can be performed in a quantitative way using the previously defined spectral entropy. In fact in figure 4d) we show the evolution of this quantity corresponding to figure 2a). The straight line is drawn at the entropy value corresponding to the stationary spatial structure in figure 4c). In the bursting region the spectral entropy of the system returns to the value corresponding to the stationary structure. Finally we found that the burst correspond to a big exchange of energy between odd and even spatial modes.

The last property we have checked is the divergency near the bifurcation point of the average length T_o of the laminar periods. A least squares fit yields the following relation

$$T_o = 850 * (r - r_b)^{-0.11} \text{ Sec.}$$

6. Conclusions.

We have shown that in Rayleigh-Benard convection in a small rectangular cell chaotic regimes may be associated with either localized oscillations or travelling perturbations. Furthermore the transition between these two spatiotemporal regimes takes place via a bifurcation that has several properties of a Shilnikov type homoclinic bifurcation. These properties have been checked by comparison with some characteristic bifurcation schemes that are usually found in numerical simulation of Kuramoto-Sivashinski type equations.

Finally the functions $E(t)$ and $\sigma(t)$ appear to be usefull parameters that allow to make quantitative experimental studies of different spatiotemporal behaviors.

+ Permanent Address: Dpto. Fisica Aplicada. E.T.S. Arquitectura, U.P.M., Av. Juan de Herrera s/n. 28040-Madrid. Spain.

REFERENCES.

1. For the definition of this instability see P. Berge' in this volume.

2. Gollub J.P. and McCarriar A.R., Phys. Rev. A $\underline{26}$, 3470 (1982).
   Pocheau A., Croquette V. and Le Gal P., Phys. Rev. Lett. $\underline{55}$, 1094 (1985).
   Heutmaker M.S., Fraenkel P.N. nd Gollub J.P., Phys. Rev. Lett. $\underline{54}$, 1369 (1985).
   Ahlers G., Cannell D.S. and Steinberg V., Phys. Rev. Lett., $\underline{54}$, 1373 (1985).

3. Gollub J.P. and Benson S.V., J. Fluid Mech. $\underline{100}$, 449 (1980).
   Libchaber A., Laroche C. and Fauve S., J. Physique $\underline{43}$, 221 (1982).
   Dubois M., Rubio M.A. and Berge P., Phys. Rev. Lett., $\underline{51}$, 1446 (1983).

4. Hyman J.M. and Nicolaenko B., Physica 18D, 113 (1986).

5. Ciliberto S. and Rubio M.A., Physica Scripta, $\underline{36}$, 920 (1987).
   Ciliberto S. and Rubio M.A., Phys. Rev. Lett., $\underline{58}$,

2652 (1987).

6. Livi R., Pettini M., Ruffo S., Sparpaglione M. and Vulpiani A., Phys. Rev. A **31**, 1039 (1985).

7. Guckenheimer J. and Holmes T., Nonlinear Oscillations, Dynamical Systems and Bifurcations of Vector Fields, Springer-Verlag, Berlin (1983).

ANALOGUE SIMULATION OF STOCHASTIC PROCESSES BY MEANS OF MINIMUM COMPONENTS ELECTRONIC DEVICES

Leone Fronzoni

Dipartimento di fisica- Universita di Pisa
Piazza Torricelli 2, 56100   Pisa
ITALY

ABSTRACT

Although the precision of the analogue technique by means of electronic devices is modest compared to the computer calculation, it must be remarked that in many cases it is convenient to have available immediate results rather than a more accurate numerical precision. Few basic considerations on the analogue techniques are given and two important applications are discussed.

1. INTRODUCTION

The interest on stochastic processes is increasing in the last years because of the importance that this field involves in many disciplines , moreover some difficulties arise for testing the theory and, often, long times computation need in spite the large use of advanced computers.

In this section it is shown a simple analogue technique based on the "minimum electronic components devices" that allows us to obtain a high speed and low cost simulation of stochastic equations. Of course the analogue experiments does not replace completely computer simulation rather, a comparison of the results obtained by using these two techniques, where it is possible, has proven to be a powerful method.

2. THE MINIMUM COMPONENTS TECHNIQUE

In order to improve a traditional electronic analogue simulation it was developed a new technique which allowed us to reduce the thermic drift and the influence of both internal and external electric noise.

The main feature of analogue simulation via electronic devices have been discussed in many papers ([1,2,3,4,5,6,7,8] . Here, the principles of the "minimum components technique " ([9] are briefly exposed. They can be resumed in three points:

i) We have to write the equations in such a way as to reduce to a minimum the number of electronic components.

ii) We have to renormalize the equations with respect to the time so as to produce characteristic times well within the operating ranges of the electronic devices.

iii) We have to assemble the circuits in only one block. This reduces the parasitic elements and the influence of unwanted external noise.

This method is particularly useful when very long times need to perform statistical distributions and to point out slowing down effects. In the next section two significant examples are presented and the experimental results are given.

3. THE EFFECT OF THE EXTERNAL NOISE ON THE HOPF BIFURCATION-THE BRUSSELATOR

Consider the following system of equations

$$\frac{dx}{dt} = A - (1 - B)x + x^2 y$$
$$\frac{dy}{dt} = Bx - x^2 y \qquad (3.1)$$

This system of equations (Brusselator) [10] describes the dynamics of a particular chemical oscillating reaction where x and y denote the concentrations of two chemical substances. If the parameter B is regarded as a control parameter Hopf bifurcation arises at $B = 1 + A^2$

Then consider

$$B = B_0 + f(t) \qquad (3.2)$$

where f(t) is a Gaussian Colored noise defined by

$$<f(t)f(s)> = (D/\tau) e^{-|t-s|/\tau} \qquad (3.3)$$

D characterize the noise intensity and τ is the correlation time of the noise. Using the " **minimum components design philosophy"** , only two integrators and two multipliers are enough to simulate the Brusselator. Infact, there are only two first order differential equations and two multipicative terms: "z=x y"and "x z=x^2 y". It is important to note that the terms Bx and (1-B)x do not involve multipliers devices but only resistance elements . Moreover the single term " x^2 y" can be added to the two equations without repeating the multiplication operations.

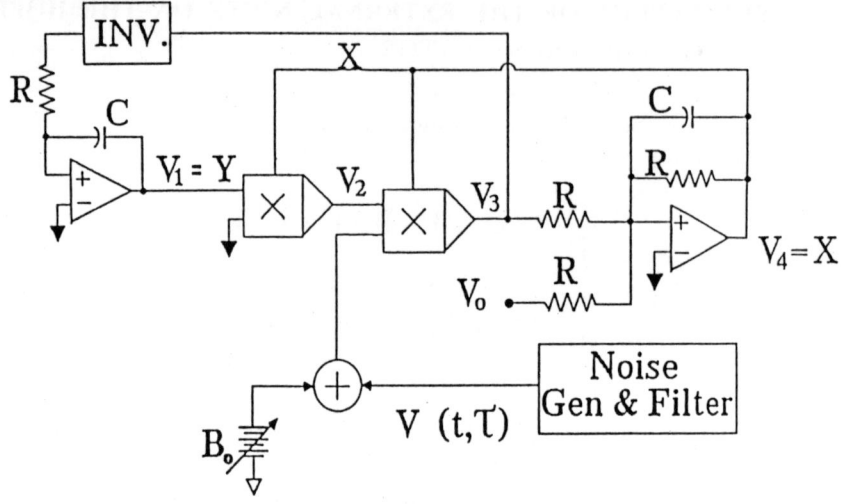

Fig.1 Scheme of the analogue circuit . From Fronzoni L. et al.(11

Fig.1 shows the scheme of the analog circuit built in agreement with the previous consideration. The equations of the electronic circuits are

$$RC \frac{dx}{dt} = V_0 - [1 + (B_0 + V_n)] x + x^2 y$$

$$RC \frac{dy}{dt} = (B_0 + V_n) x - x^2 y \qquad (3.4)$$

where V_n is the voltage value of the noise at the input of the electronic circuit and the correlation function reads

$$< V_n(t) V_n(s) > = (D_n / \tau_c) e^{-|t-s|/\tau_c} \qquad (3.5)$$

D_n and τ_c denote the intensity and the correlation time of the noise respectively. Both the voltage V_n and τ_c must be scaled in such a way to have the complete agreement with the eqs. (3.1) :

$$f(t) = V_n(t)\sqrt{RC} \qquad \tau = \frac{\tau_c}{RC} \qquad (3.6)$$

In the eqs.(3.1). The threshold of the Hopf bifurcation arises when $B_0 = 2$ volts.

When the noise is superimposed to the voltage B_0 the limit cycle becomes a statistical quantity and the bifurcation nondistinct. In fig.2 a three dimensional view of the statistical distribution $P(x,y)$ is shown when the control parameter is above the threshold. Of course it is not clear what it would be the meaning of threshold when a noise is present on the system. However, a statistical definition of the new bifurcation can be made in terms of the topology of the two dimensional stationary statistical density $P(x,y)$[11].

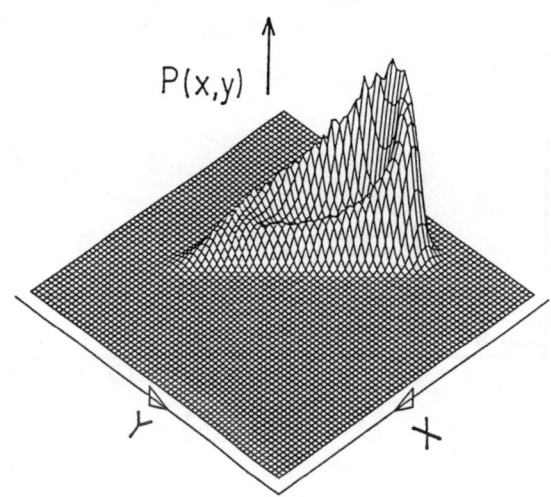

Fig.2 Three-dimensional view of statistical distribution $P(x,y)$. From L.Fronzoni et al. [11

3.1 THE POSTPONEMENT OF THE HOPF BIFURCATION

From fig.2 it appears that the statistical distribution is shaped like a mountain with an irregularly shaped crater at the top. The new threshold can be defined as the value of B^* for which the crater **"fails to hold water"**.

The Lefever-Tuner theory (12 predicts both advancements and postponements of the threshold depending on the value of the correlation time of the noise τ. Within the accuracy of our simulation we observe only postponement of the threshold previously defined, for $0.1 < \tau < 10$.

Fig. 3 shows the results where B^* and B_c denote the threshold with and without noise, respectively. The threshold is posponed by the noise and is a linear function of the noise intensity D. The threshold shift is smaller for large τ values (high colored noise).

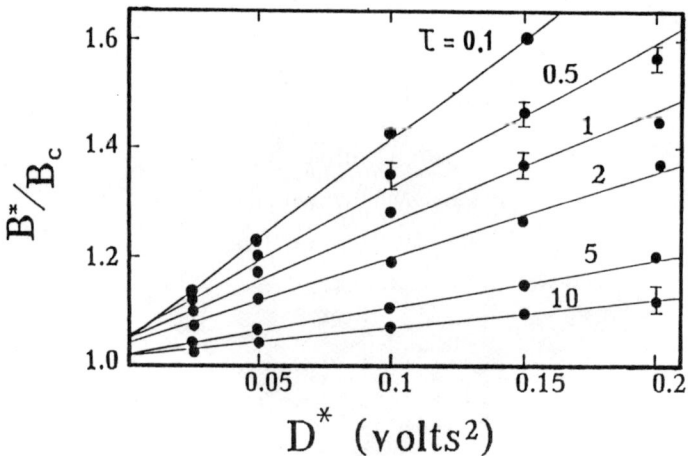

Fig.3 The postponed threshold as funtion the intensity D^* for various noise correlation times. From Fronzoni et al. (11

4. FLUCTUATIONS IN TWO COUPLED OSCILLATORS

This example is interesting because it evidences two important aspects:
a) The statistical fluctuations in a system very far from the equilibrium state.
b) The peculiarity of these fluctuations implies the waste of large amount of time in order to integrate the equations. The analogue simulation allows to bypass this difficulty.

Let consider this system of equations :

$$\frac{d^2x}{dt^2} = -\gamma \frac{dx}{dt} - \omega^2 x - \beta x^3 - \alpha xy$$

$$\frac{d^2y}{dt^2} = -\lambda \frac{dy}{dt} - \omega_0^2 y + \frac{1}{2}\alpha x^2 + f(t) \quad (4.1)$$

This system corresponds to a non-linear oscillator (x) coupled with a linear oscillator (y) in the presence of a stochastic force f(t). The coupling potential is

$$U(x,y) = \frac{1}{2} \alpha x^2 y \quad (4.2)$$

γ and λ are the damping coefficients and f(t) plays the role of a thermal source applied to the linear oscillator (y) with temperature T defined by the correlation function

$$<f(t)f(s)> = 2 \lambda K_B T \delta(t-s) \quad (4.3)$$

The theory developed by the Pisa group [13] shows that there is a critical temperature T_C defined by the following conditions:

a) $<x^2> = 0$ for $0 < T < T_c$

b) $<x^2> = $ const $*(T - T_c)$ for $T > T_c$ (4.4)

This means that the non linear oscillator (x) remains at rest for temperature below a critical value T_c as shown in fig.4.

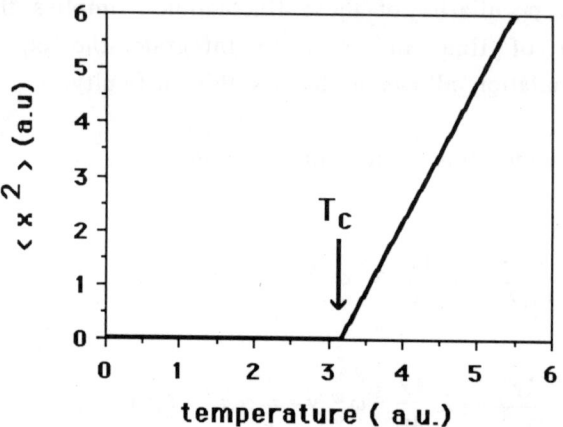

Fig.4 $<x^2>$ v.s. the temperature T of the linear oscillator (y)

This behavior is quite surprising but it is important to note that in the real experiment a little additive noise is present on the non-linear oscillator (x) so that there are some amplitude fluctuations also below the threshold T_c. However the system is far from equilibrium and the energy equipartition is not necessarily ensured. At the Tc value corresponds a threshold value for $<y^2>$. The last theoretical developments leads us to :

$$<y^2>_{thres} = \frac{\omega^2 \gamma (4\lambda^2 \omega^2 + (4\omega^2 - \omega_0^2))}{\alpha^2 \lambda \omega_0^2} \quad (4.5)$$

4.1 EXPERIMENTAL SET UP AND RESULTS

The electronic circuit to simulate the eqs.(4.1) is more complicate than that of the Brusselator and to avoid to make heavy this section a block scheme of the electronic circuit is given in fig. 5. The two oscillators are coupled by means of a multiplier and a noise generator is applied to the input of a linear oscillator. For more details and a complete description of the non-linear oscillator see the papers in refs. (8,9). The experimental results are shown in fig.6.

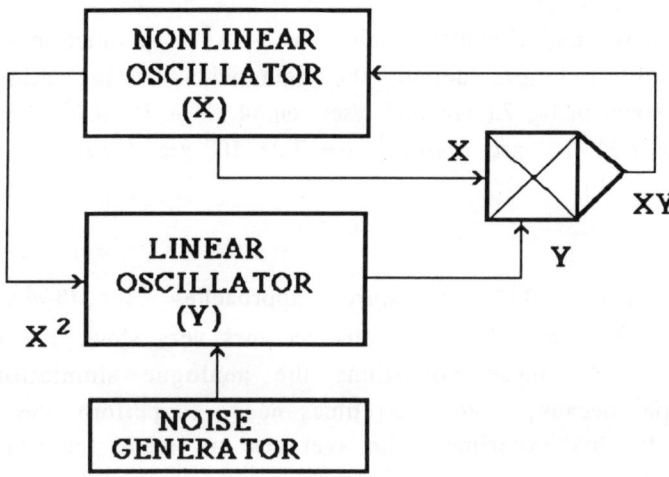

Fig.5 Scheme of electronic circuit to simulate the equations system (4.1).

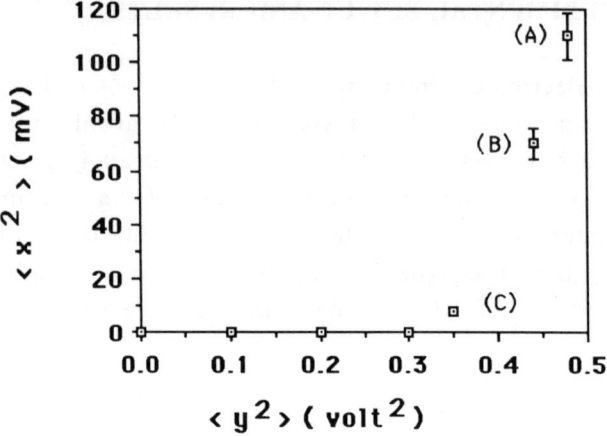

Fig.6 Experimental results for $<\dot{x}^2>$ as function $<y^2>$. Capital letters denote the data where fluctuations are shown in fig 7.(ω= 867 c/sec, ω_0=4.17 ω, λ=10.7 ω, γ=0.1 ω, β=7.21 10^5 sec $^{-1}$/volt ^{2}, α= 7.21 10^5 sec $^{-1}$/volt)

When the temperature approaches the threshold T_c the caracteristic time of fluctuations becomes very slow (slowing down effect). In these conditions the analogue simulation becomes essential because very long times need to perform the averages of x^2. In this experiment the averages are made for times T_{av} = 500/ω.

Fig.7 compares the fluctuations of $<x^2>$ for differents values of $<y^2>$ corresponding to the previous experiments. These results show that the fluctuations become less frequent as $<y^2>$ approaches to the threshold (case C).

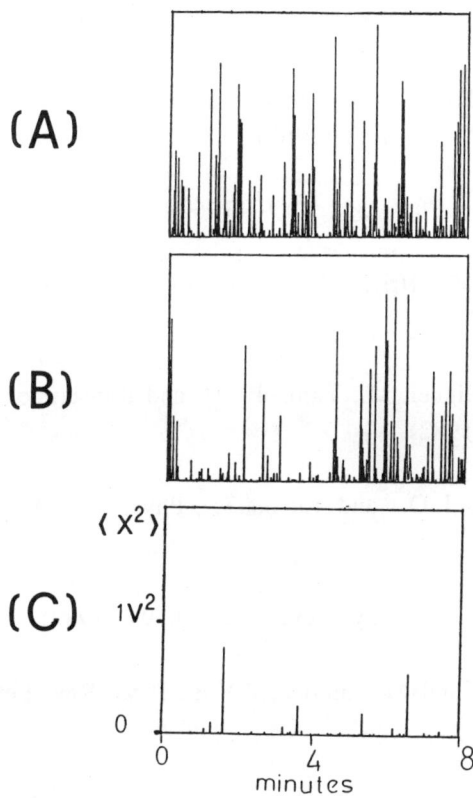

$< x^2 >$ v.s. the time for three different values of $< y^2 >$ over the threshold corresponding at the points (A), (B) and (C) in fig.6.

ACKNOWLEDGMENT

The author would like to thank Prof. Frank Moss and Prof.P.V.E.McClintock for the data acquisition achievement.

REFERENCES

1. Stratonovich, R.L, Topics in the theory of Random Noise. Gordon and Breach, vol 1 (1963), vol 2 (1967).

2. Landauer, R. J.Appl. Phys. 33, 2209 (1962)

3. Testa,J., Perez, J. and Jeffries, C. , Phys .Rev. Lett. , 48 , 714-717 (1982)

4. Sancho,J.M. , San Miguel, M., Yamzoki, H. and Kawakubo, T. , Physica 116A, 560 (1982).

5. Matsumoto, T., Chuu,L.O. and Tanaka,S. , Phys. Rev. A , 30 , 1155-1157 (1982)

6. Arecchi, T. and Lisi, F. , Phys.Rev. Lett. 49, 94 (1982)

7. Smithe, I., Moss, F. and McClintock, P.V.E., Phys. Rev. Lett. 51 , 1062 (1983)

8. Faetti ,S., Festa, C. ,Fronzoni, L. and Grigolini, P. in "Memory Function Approaches to Stochastic Problems " eds Evans, M.W. , Grigolini, P, Pastori Parravicini,G., vol 62,Wiley & Sons(1985).

9. Fronzoni, L. , " Advances in Nonlinear Dynamics and Stochastic Processes" eds Livi, R. and Politi, A. , 201 , World Scientific (1985)

10. Nicolis, G. and Prigogine, I. , "Self-Organization in Nonequilibrium Systems ", Wiley , New York (1977)

11. Fronzoni, L., Mannella, R. , McClintock, P.V.E. and Moss, F. Phys. Rev. A 36 ,834 (1987)

12. Lefever, R. and Turner, J. Wm. ,Phys. Rev.Lett. 56,1631 (1986)
13. Bonci, L., Fronzoni, L. and Grigolini, P., in preparation

Networks

PERIODS, DAMAGE SPREADING, AND MULTIFRACTALITY IN COOPERATIVE SYSTEMS

D. Stauffer, Institute of Theoretical Physics, Cologne University, 5000 Köln 41, West Germany (perm. addr.) and Department of Physics, St. Francis Xavier University, Antigonish N.S., Canada B2G 1C0.

ABSTRACT

This review attempts to bring some order into chaos: Can we understand the transition to chaos in the way we understand second order phase transitions for the last 20 years? In particular, we look for criteria like: an exponential increase of the periods with systems size in the chaotic phase, as opposed to a power law in the non-chaotic phase; spreading of damage (unlimited increase of Hamming distance) in the chaotic phase and its fractal dimension at the transition to chaos; and multifractal versus standard scaling behavior. As examples we use computer simulations in Q2R cellular automata, Kauffman models, Ising models and neural networks.

INTRODUCTION

Thirty years ago, the physics of phase transitions was very confusing: Lots of different materials and models exhibited lots of different properties, and the simplicity of the classical mean field theory (van der Waals equation) or of the Landau ansatz for free energy got lost in numerous deviations found from these theories. About 20 years ago the situation changed dramatically: By defining a suitable order parameter and working with simple scaling assumptions, which suitably generalized mean field and Landau theories, most continuous phase transitions could be described by critical exponents, and from the knowledge of two or three exponents one could predict, via scaling relations, the remaining critical exponents. This phenomenological scaling theory was later justified by the renormalization theory. In this way systems as diverse as superfluid helium, magnets and fluids could be desecribed by a unified formalism. Moreover, within this scaling theory one could find universality classes of different models and materials which had exactly the same critical exponents.

Can we find a similar description for the transition to chaos? I will not give one here, nor will I claim that I am asking the right questions. I will present, however, some numerical evidence for striking similarities in the behavior of different models: Q2R cellular automata, Kauffman models, "neural" networks, and standard Ising magnets. An earlier version of this review was given at the workshop "Geometrical Aspects of Critical Phenomena", April 10-11, 1987 at Antigonish, Nova Scotia, edited by Jan et al.

MODELS AND QUESTIONS

In all models discussed here, each site of a large lattice carries a variable, called a spin by physicists, which can be in one of two states called *up* and *down*. The value of the variable at time $t + 1$ is determined, either completely or with a certain probability, by the values of its nearest neighbor variables on the lattice at time t. Such systems are called deterministic *cellular automata*, if no probability is involved, and probabilistic cellular automata otherwise. Minor modifications of this definition will be made to implement our models. On the square lattice with $K = 4$ neighbors there are $2^K = 16$ different neighbor configurations and thus $2^{16} = 65536$ different rules. According to these rules neighbor configurations can lead to the result up or down. For example, one rule says that the spin is always up, ignoring completely its four neighbors.

The Q2R cellular automat [1] is important since it is a very fast simulation of the Ising magnet in some sort of microcanonical ensemble: We flip a spin if and only if it has as many up as down neighbors. 4200 spin flip attempts per microsecond were reached on a four-processor Cray-2 [2]. Such simulations start with a random distribution of spin orientation and later lead to "correct" (i.e. Ising like) critical temperatures, spontaneous magnetizations and specific heats on the square and simple cubic lattices [1,3]; some Boltzmann factors are known to be wrong in the ferromagnetic phase [3]. To conserve the energy (in the Ising sense) exactly, one separates the lattice into two sublattice, like on a chessboard, and updates one sublattice at time $t + 1/2$, the other at time $t + 1$.

In the Kauffman model [4], each "spin" represents a biological gene which can be on or off. Each lattice site selects randomly which of the 65536 possible rules (on the square lattice) it wants to obey, and stays with this selection for the rest of the simulation. We may also regard Kauffman's cellular automata as models for democracy: Can some degree of order appear out of numerous stubborn individuals with their own values and own friends. To have a continuously varying parameter p oneselects the rules for each site not completely at random, but for each site and each neighbor configuration, one takes with probability p a rule which says *up* and with probability 1-p a rule which says *down*.

Similarly, "neural" networks show order out of disorder: Each spin S_i is up if the sum $\Sigma_k c_{ik} S_k$ over the nearest neighbors S_k of spin i is larger than a certain threshold value. Here the "synaptic strengths" c_{ik} are randomly distributed and can be both positive and negative. Kürten's paper in these proceedings gives more details on this McCulloch-Pitts model, where an up spin corresponds to a firing neuron. (In agreement with biological facts one allows here interactions c_{ik} even between far-away neurons i and k. A related network is known as the Little-Hopfield model).

Finally, in the Ising model we flip a spin with probability

$$\exp(-\beta \Delta E)/(1 + \exp(-\beta \Delta E)),$$

where ΔE is the change in the interaction energy $-J\Sigma S_i S_k$ connected with flipping spin i. This Metropolis algorithm [5] is therefore a probabilistic cellular automat; we follow the tradition of using for each spin the already updated values of those neighbor spins which have been treated before at the same time step, and we go through the lattice like a typewriter.

In these four models we now ask three questions: How large are the periods of the limit cycles into which the deterministic cellular automata must finally end up? What is the "damage" caused by a single error, i.e. how do two different lattices compare with each other if initially they are identical except for the center spin? Is this damage fractal or multifractal, and what are its fractal dimensions?

PERIODS

A lattice with N spins which can either be up or down has exactly 2^N different spin configurations. Thus after at most 2^N time steps it must always go back to a configuration reached earlier, if we work with deterministic and not with probabilistic cellular automata. If the deterministic cellular automata are reversible, we must go back to the initial configuration (which is easy to check numerically); otherwise some initial transient time elapses before the system reaches a limit cycle in configuration space which then is followed again and again. Our period is the length of this limit cycle, that means the number of time steps after which one is reaching the same configuration again. Spins which never move are assigned a period of unity. The whole system has as its global period the lowest common multiple of the local periods of all spins. Probabilistic cellular automata have infinite periods if their random numbers are really random.

The numerically most extensive study was made for Q2R cellular automata, with global periods up to 928 million time steps [6]; few studies involved longer observation times. It turned out, for N up to 10^3, that the median period "always" grows exponentially with the number N of spins in the system, for the paramagnetic region, the ferromagnetic region as well as for the Curie point itself. More precisely, if initially a fraction p of spins is down, and the rest is up, then for $p = 0.5$ (paramagnet), $p = 0.08$ (Curie point), and $p = 1/16$ (ferromagnet) a simple exponential is observed by Schulte et al [6]. It is possible that near $p = 0.03$ some other "cluster period" transition exists [8]; perhaps for smaller p the median periods grow weaker than exponentially with N. Even at $p = 0.5$, the most chaotic choice corresponding to infinitely high temperatures, the periods increase much weaker than 2^N, roughly as only $2^{N/4}$. Thus a simulation covers only an exponentially small fraction of the total configuration space: Q2R is not ergodic.

From the technical point of view this work is interesting since it is based on Zabolitzky's special microprogrammable computer [7] and precedes any conventional mainframe simulations of comparable quality. Also, the median and the geometric mean (logarithmic average) of the global periods are about the same, whereas the arithmetic mean is dominated by the largest observed period. These periods can fluctuate by five orders of magnitude; thus any theory for the periods should define the method of averaging.

In the Kauffman model on the square lattice, the global periods seem to increase exponentially with the number N of sites in the chaotic phase [9]; again these periods are much smaller than 2^N. On the non-chaotic side of the transition the increase with N is much weaker, possibly $N^{1/4}$. The most accurate data were obtained [10] right at the transition to chaos, where on an L*L square lattice the global periods varied as exp (L/2.66). The average *local* periods (for the sites separately) seem to remain finite; the increase of the global period comes from the fact that the lowest common multiple of all local periods is the larger the more sites are involved. For example, if all numbers between 1 and 25 appear as local periods, the global period is already more than 10^{10}, far longer than any simulation known to me.

For "neural" networks, Kürten (these proceedings) found an exponential variation of the period with N for the chaotic phase, a power law for the non-chaotic phase, and a stronger power law increase at the transition to chaos.

For Ising models we do not have periods if the algorithm involves random numbers. During the conference I learned that anharmonic coupled oscillators (Fermi-Ulam-Pasta model) have an ergodicity transition such that periods increase exponentially with system size on one side and with a power law on the other side of the transition [11].

In summary, therefore, several quite different deterministic systems share the property, that at the transition to chaos, the periods also exhibit a phase transition from power law to exponential dependence on system size. The Q2R simulations do not fit into this scheme, and the next section on damage spreading will show that there again Q2R does not do want I want [12].

DAMAGE SPREADING

Damage spreading asks how a single "error" affects the future development of the system. For example, a genetic mutation may change very little in later life, or it may change a lot. Thus Kauffman studied in his genetic model the influence of a single spin flip on the limit cycles. Much earlier, of course, human beings studied the consequences of a single decision like marriage on their later lifes, and in this sense most conference participants know a damage spreading experiment already from Adam and Eve.

What do we (following Adam, Eve and Kauffman) do in practice? We simulate two systems, identical in nearly all their properties, and even use the same set of random numbers for probabilistic rules like in the Ising model. The only difference between the two systems is that in their center one spin is different. It is an up spin in one replica if it points down in the other. We then compare during the time development of both systems which spins in one system differ from their corresponding spins in the other systems. We call the set or the number of different spins in that comparison the *damage*; other words in the literature are Hamming distance, difference picture, or the complement of the overlap. Needless to say, for damage assessment in real applications we might need criteria more refined than just counting the number of different spins in the two replicas.

Damage spreading is different from nucleation: If a droplet is nucleated in a supersaturated vapor, then average quantities like density change, and we have *one* system moving from a metastable to a stable equilibrium. For damage spreading, we look at two systems which both may be in complete equilibrium and which both then have the same average quantities. Damage consists of the difference between two particular configurations, a question usually not asked in Statistical Physics (see Swendsen's talk here for an exception) but very natural for biological mutations [4].

For systems with continuous variables, these problems are very old and are inherent in any stability analysis: Does a small error grow (usually exponentially), or does it remain limited? Chaos deals with those cases where the initial damage spreads over the whole system. In discrete systems, where the amount of damage can only be 0, 1, 2, ..., much less work has been done on damage spreading in this sense. Most of the quantitative studies deal with the Kauffman model whereas Ising models [12-14] were only recently studied in this aspect.

For the *Kauffman model*, it was even possible to determine fractal dimensions (critical exponents) for damage spreading. If the parameter p in the Kauffman model is above a threshold p_c of about 0.3 on the square lattice, then damage spreads over the whole lattice; for $p < p_c$ damage remains limited. More precisely, in the chaotic phase for $1/2 > p > p_c$ there is a nonzero probability that an initial single spin flip eventually affects a finite fraction of the whole lattice, even if the lattice is arbitrarily large.

The damage front propagates on average with a constant velocity through the system; however, within the damaged region not every site is damaged. Obviously, even between two completely uncorrelated configurations the damage affects only half the spins since the others are "accidentally" the same; in the Kauffman model the damaged fraction is slightly smaller even at $p = 0.5$ since the two configurations are still correlated by their identical rules.

For p below p_c, on the other hand, the number of damaged sites is always limited. Close to p_c, a single initial spin flip may lead later to, say, about 20 damaged neighbors, but that number does not increase with lattice size. For smaller p, the damage may even vanish after a few time steps.

Thus close to p_c one may find critical behavior: For p approaching p_c from below the number of sites damaged by one initial spin flip diverges (similar to the mean cluster size in percolation theory); for p approaching p_c from above, the fraction of damaged lattice sites goes to zero (like the fraction of sites in the infinite cluster of percolation theory) [15]. Right at $p = p_c$ the number of damaged sites increases with some power (<1) of the number of lattice sites, and this power gives a fractal dimension.

More quantitatively, we check how long it takes for the damage to reach the edges of an L*L square lattice, if the initial spin flip happened in the center of this lattice, and we count how many sites are damaged at the moment the damage hits the lattice edge. If we denote this touching time by τ and the amount of actual damage at the moment of touching by M, we can define two asymptotic fractal dimensions D and d' by

$$M \sim L^D$$
$$\tau \sim L^D$$

for lattice length L going to infinity. For $p < p_c$ these critical exponents are undefined, for $p > p_c$ they are trivial ($D' = 1$, $D = d$ in d dimensions). The numerical values at $p = p_c$ are given in the table for the square and simple cubic lattice ($d = 2$ and 3), with an accuracy of about ten percent [16]. On a triangular lattice, $p_c \simeq 0.16$, fractal dimensions close to those on the square lattice were found [10], in agreement with the universality principle of critical phenomena. If instead of flipping initially one spin, we create the damage by changing the rule which the center spin obeys, again roughly the same exponents were observed on the square lattice [17]. Quite different results are obtained if with a finite probability the spins do not obey the Kauffman rule due to "noise" [18]. The hypercubic p_c for $d = 4$ is near 0.08 [27].

	$d = 2$	$d = 3$
D	1.5	1.8
D'	1.6	2.2

Fractal dimensions for mass (D) and time (D') on square ($p_c = 0.29$) and simple cubic ($p_c = 0.12$) lattices at $p = p_c$.

Even for $p > p_c$, the damage can remain limited or die out with some probability, for example because we flipped initially a spin which accidentally has no influence on his neighbors. Damage spreads only if he initial "error" happens on an infinite network of sites susceptible to damage. Numerically the transition to chaos seems to agree on the square lattice with the point were the set of spins which no longer flip with time stops percolating through the lattice, and where the spins which still flip now and then start percolating; at this percolation threshold about half of the spins never flip [19], except perhaps initially. Thus roughly one needs an infinite network of flipping spins for the damage to spread throughout the lattice. (For 4 dimensions the situation is more complicated [27]).

For *"neural"* networks, a transition between spreading and nonspreading damage was also found numerically [26] but I am not aware of numerical studies for the fractal dimensions D and D'. The dynamics of the Kauffman model with *infinite* range of interactions seems to be a special case of these neural networks (K.E. Kürten, priv. comm.).

In *Q2R automata* it first looks as if one finds also a transition to chaos at conditions corresponding to the Curie point of the Ising model ($p = 8\%$ of the spins up, the others down in a random initial configuration on the square lattice). For higher "temperatures" the damage spreads, and for lower p it does not. Closer inspection with many hours of simulation on an IBM 3090 showed, however [12], that even for $p < p_c$ damage still may spread, even after it stayed constant for hundreds or thousands of time steps. Thus the Curie point does not represent a sharp phase transition between spreading and nonspreading damage. Square and simple cubic lattice seem to be similar in this aspect.

More interesting is the situation in the *Ising model*. Damage spreads for temperatures above the Curie temperature but not for much lower temperatures [12,13]. Thus qualitatively the Ising model confirms what was found in Kauffman and Q2R models: The more disordered the system is, the easier damage can spread. (Unionization of workers is based on a similar principle). The data were not accurate enough to find a fractal dimension D smaller than d.

In contrast to the Kauffman model, a spreading damage eventually will affect all spins of the lattice at least once during the time development, and thus we can measure the time τ' needed to damage each spin at least once. This time goes to infinity if the temperature goes to infinity, since for very high temperatures the interactions between the spins are much weaker than the thermal fluctuations.

The time τ' also goes to infinity if we cool the system down to the spreading temperature below which damage no longer spreads. On the square lattice this spreading temperature agrees numerically with the Curie temperature but on the simple cubic lattice the spreading temperature is about 5% lower, as Fig. 1 shows [13].

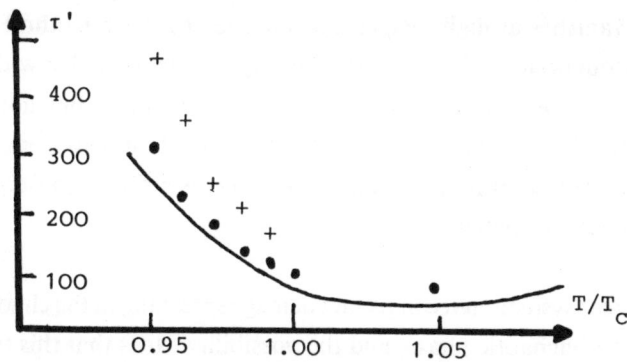

Fig. 1: Average time τ' for damage to hit all sites in L*L*L simple cubic lattice [13]. L = 120 (+), 60 (dots), and 40 (line).

This difference between spreading and Curie temperature reminds us of the 4% difference [20] between the percolation temperature, where the up spins have their percolation threshold (when most spins are down), and the Curie temperature, where there are as many up as down spins. This percolation temperature and the spreading temperature even agree within numerical accuracy, similarly to the above mentioned agreement of percolation and chaos in the Kauffman model [9,19].

Interestingly the situation is quite different if we change in the Ising case from the Metropolis algorithm to the "heat bath" method [14]. In the first case, a spin is flipped with a certain probability, in the latter case a spin is pointed up with a corresponding probability. Both methods give, on average, the same results if one simulates one lattice only. For damage spreading, on the other hand, we simulate two lattices using the same random numbers, and then a difference occurs. If one goes through the tedious details of the comparison of the two methods, one finds them equivalent *on average* because a random number z between 0 and 1 is smaller than p with the same probability as it is larger than 1-p. But for a given spin in a given configuration and for a given random number z (as for damage spreading studies) the two conditions $z<p$ and $z>1-p$ are very different. As a result, at high temperatures the damage vanishes at high temperatures in the heat bath method whereas it spreads throughout the lattice in the Metropolis method. Below the Curie temperature, the heat bath damage vanishes in most cases but not always. Presumably the damage spreading temperature for the heat bath method agrees with the Curie temperature [21], and the damage is a direct consequence of the spontaneous magnetization.

Thus different systems agree in having damage spreading in the chaotic phase but not in the nonchaotic phase, and the possibility exists that this transition is a percolation transition. However, just as with the periods, also in the damage question the Q2R automata do not fit into the picture. And for Ising model simulations two different simulation methods give nearly opposite results.

MULTIFRACTALS

Fractal dimensions like D and D' may be related through other critical exponents such that from two or three exponents all other exponents can be calculated. Such relations are known as scaling laws and work well for second-order phase transitions. Instead, it may also happen that one has an infinite hierarchy of *independent* exponents for the various quantities. Then one calls these systems multifractals, turbulence being one example [22]. One may even have both fractal and multifractal behavior in one problem: Percolation cluster *numbers* are determined by two exponents, whereas the *structure* of percolation clusters (as measured through moments of the electrical conductivity) is multifractal [23]. For the models discussed here, multifractality was investigated for the Kauffman model.

The periods of the Kauffman model do not seem to cause multifractality [10], but the damage spreading does [24]. Let us count how often in the development of a Kauffman lattice each site is damaged. We ignore the sites which are never damaged, and look at moments M_q of the probability to get a site damaged. Again, damage means that we simulate two nearly identical Kauffman models with initially only the center spin different in the two lattices.

In n_i counts how often in a damage spreading simulation the lattice site i was damaged, then $p_i = n_i / \Sigma_i n_i$ is some sort of damage probability. Its moments are defined through

$$M_q = \Sigma_i \, p_i^q$$

and vary for long times t in large systems as

$$M_q \sim t^{\varphi(q)}$$

where the sum runs over all sites damaged at least once. Here q is any real number, not necessarily 1, 2, 3. The zero'th moment (q = 0) gives the total damage up to the observation time, the first moment M_1 is always unity. It these moments would obey the usual two-exponent scaling known from many critical phenomena, then two exponents $\varphi(q)$ for two different q values would determine all other exponents $\varphi(q)$. (Since $\varphi(1) = 0$, only one other q value would be needed). If on the other hand these damage probabilitiees p_i are multifractal, the exponents $\varphi(q)$ of the corresponding moments M_q would form a set of infinitely many apparently unrelated exponents, as observed for example for the growth of diffusion-limited aggregates [25].

Simulations of square Kauffman lattices showed both types of behavior [24], as our Fig. 2 indicates: For p = 0.5, the most chaotic case, all $\varphi(q)$ plotted versus q followed numerically a straight line, as required by two-exponent scaling. (According to Euclid, two points determine a straight line). At the transition to chaos, $p = p_c = 0.29$, they do not for the lattice sizes and times observed and instead follow some smooth curve, as required by multifractality. Thus the growth of the damage cloud is multifractal at p_c and normal at p = 0.5. It would be nice to look for corresponding answers in neural networks or Ising models.

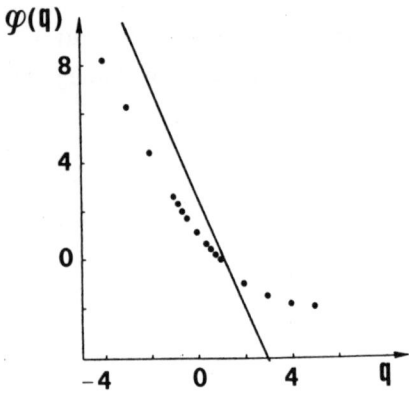

Fig. 2: Variation of moment exponent $\varphi(q)$ with q on square Kauffman model [24]. The straight line belongs to p = 0.5, the curve to $p = p_c = 0.29$.

SUMMARY

We have seen some similarity between different models: The variation of the limit cycle period with system size was one, the spreading of damage another. However, there are problems with Q2R; along with slightly different algorithms for the Ising model give drastically different results is shattering our hope that "everything" will be as simple as scaling laws for critical phenomena seem today: The transition to chaos may be more complex.

The author is indebted to H.O. Carmesin, A. Coniglio, L. de Arcangelis, B. Derrida, H.J. Herrmann, J. Kertész, N. Jan, H.E. Stanley, and G. Weisbuch for their collaboration in some of the research presented here, and numerous others for discussion and information.

REFERENCES

1. G. Vichniac, Physica D **10**, 96 (1984)
   Y. Pomeau, J. Phys. A **17**, L 415 (1984
   H.J. Herrmann, J. Statist. Phys. **45**, 145 (1986)

2. J.G. Zabolitzky and H.J. Herrmann, J. Comp. Phys., in press

3. W.M. Lang and D. Stauffer, J. Phys. A **20**, 5413 (1987)

4. S.A. Kauffman, J. Theor. Biol. **22**. 437 (1969); Physica D **10**, 145 (1984); p. 339 in: *Disordered Systems and Biological Organization*, edited by E. Bienenstock, F. Fogelman-Soulié and G. Weisbuch, Springer Verlag, Heidelberg 1986
   D. Stauffer, Phil. Mag. B **55**, in press
   B. Derrida, Phil. Mag. B **55**, in press

5. K. Binder, editor, *Applications of the Monte Carlo Method in Statistical Physics*, Topics in Applied Physics **36**, Springer Verlag, Heidelberg, 2nd edition 1987

6. M. Schulte, W. Stiefelhagen and E.S. Demme, J. Phys. A **20**, L 1023 (1987)

7. J. Deckert, S. Wansleben and J.G. Zabolitzky, Phys. Rev D **10**, 683 (1987)

8. H.J. Herrmann, H.O. Carmesin and D. Stauffer, J. Phys. A **20**, 4939 (1987)

9. G. Weisbuch and D. Stauffer, J. Physique **48**, 11 (1987)

10. L. de Arcangelis and D. Stauffer, J. Physique **48**, 1881 (1987)

11. S. Isola, R. Livi and S. Russo, Europhys. Lett. **3**, 407 (1987)

12. H.E. Stanley, D. Stauffer, J. Kertész and H.J. Herrmann, Phys. Rev. Letters **59**, 2326 (1987)

13. U.M. S. Costa, J. Phys. A **20**, L 583 (1987)

14. B. Derrida and G. Weisbuch, Europhys. Letters **4**, 657 (1987)

15. B. Derrida and D. Stauffer, Europhys. Letters **2**, 739 (1986)

16. L. de Arcangelis, J. Phys. A **20**, L 369 (1986)

17. M. Corsten and P. Poole, J. Statist. Phys, in press

18. P.M. Lam, J. Statist. Phys., in press
19. D. Stauffer, preprint for J. Theor. Biol.
20. H. Müller-Krumbhaar, Phys. Letters **50**, 27 (1974)
21. B. Derrida, private communication
22. B.B. Mandelbrot, J. Fluid Mech. **62**, 331 (1972)
    G. Paladin and A. Vulpiani, Phys. Reports **156**, 147 (1987)
23. L. de Arcangelis, lecture notes *"Disorder and Mixing"*, June 1987
    A. Coniglio, Physica **140** A, 51
24. A. Coniglio, D. Stauffer and N. Jan, J. Phys. A **20**, L 1103 (1987)
25. C. Amitrano, A. Coniglio and F. di Liberto, Phys. Rev. Letters **57**, 1016 (1986)
26. K.E. Kürten, IEEE Proceedings (San Diego Conference, June 1987)
27. A. Hansen, to be published.

GLOBAL AND LOCAL PERIODS IN KAUFFMAN CELLULAR AUTOMATA

by

Lucilla de Arcangelis

Service de Physique Théorique
Institut de Recherche Fondamentale, CEA
CEN-Saclay
91191 Gif-sur-Yvette Cedex, France

Abstract

We present a quantitative study by Monte Carlo simulation of the local and global periods for the two dimensional Kauffman model on the square lattice. A scaling law is found for the median period at the transition to chaos, whereas the moments of the local period distribution appear not to be critical.

Random networks of Boolean automata have been proposed by Kauffman [1] as a simple model for the interactions between biological genes leading to cell differentiation. The N Boolean variables, or spins, are placed at the sites of a lattice of size L and they can either be equal to one or zero. The time evolution of the system is determined by a set of Boolean functions depending, for each spin, on the K neighbours of that site on the lattice, and choosen at random among the 2^{2^K} possible Boolean functions of K variables. More generally, one can assign a probability $p \leq 0.5$ for the outcome of these functions to be unity.

Depending on p, these networks have a transition to chaos: two distinct phases can be observed, a chaotic and a frozen one. These phases can, for example, be characterized by saying that an initial damage introduced in the system remains confined or eventually dies out in the frozen phase, whereas it spreads throughout the whole system in the chaotic phase. References [2] and [3] review recent analytical and numerical work on Kauffman models.

Once the random Booloean functions are choosen for each spin, the dynamics of the Kauffman model is completely deterministic. Since there exist only 2^N different configurations for an ensemble of N spins, finite systems are forced into limit cycles. That is, no matter which configuration the system starts from, after a time $\leq 2^N$ it will return in one of the previously visited configurations; the limit cycle is defined as the set of all the *different* configurations that happen before a double visit occurs and the elapsed time for such event, τ, is called the ("global") period of such limit cycle.

Previous work on the square lattice with nearest neighbour interactions (K = 4), pointed out that the period of these limit cycles, averaged over different initial configurations and sets of rules, seems to increase exponentially with the number of sites L^2 in the chaotic phase and weaker ($\sqrt{L}$) in the frozen phase [5]. We present here a more detailed numerical study of the global period at the transition to chaos, together with the first attempt to quantify the concept of "local" periods and their distribution [6].

We initialize the spins on the square lattice at random; then for each site and at fixed p, we determine the value of the function for each of the possible $2^K = 16$ neighbours configuration, by means of a random number drawing. The transition to a chaotic regime can then be observed by continously varying the parameter p and we take p_c = .28 as value for the threshold on the K = 4 square lattice [3]. We let then the system evolve according to the fixed rules and we store in memory the whole history of configurations in order to detect the period of the cycles. More technical details are given in Ref. [6].

At $p = p_c$, the global periods have large fluctuations depending on the net and initial conditions; moreover, as the size L increases, we are also able to determine the period of fewer and fewer configurations within the limits of our observation time. Therefore we find that a more reliable quantity than the average global period is the "median" determined in such a way half of the observed periods are smaller and half are larger than that period.

This median period is found to vary (Fig.1) at the onset of chaos as $\sim \exp(L/2.66)$. This law differs from the variation with $\exp(\text{const. } L^2)$ observed for above p_c [5], and the proportionality to $\sqrt{L}$ found in the frozen phase below p_c [3]. Figure 1 also shows that, if we average over those periods which were found within the finite observation time, then both $\langle \tau \rangle$ and $\langle \ln \tau \rangle$ will for large periods have systematic errors due to finite observation times, and the curves flatten for larger lattice sizes, whereas the median period can always be determined as long as at least half of the periods were found.

We also study the local period T of one site, which gives the time after which this site repeats the same pattern of bits in its orientation [5]. The global period of the whole lattice will be the lowest common multiple of all the local periods in the lattice and thus in general much larger.

We focus our attention on the local period distribution n(T), representing the number of sites having a period T, and its moments defined as

$$M_q = \langle T^q \rangle = \left\langle \sum_T n(T) \, T^q \right\rangle \bigg/ \left\langle \sum_T n(T) \right\rangle$$

where we neglect in the average those sites which have period too long to be determined within our observation time. We ask the question weather these moments are critical at the onset of chaos, that is if they diverge as $p \to p_c$. Since the positive moments fluctuate strongly, we investigate mostly the behaviour of the negative moments as function of p. We find that in the chaotic phase most sites have period far longer than can be observed, and the negative moments are dominated by those sites which have period T = 1 (stable core), that is those spins which never move. Therefore all the moments go to unity for $p \to .5$ and we find no sign of divergence near p_c. If we remove from our average the stable core sites, the negative moments will go sharply to zero in the chaotic phase but the behaviour near the critical point is unchanged. A similar result is found if T is replaced by the natural logarithm of the period in the definition of the moments; in this case, though, all the M_q's will vanish as $p \to .5$.

This lack of criticality could be caused by our necessarily finite observation times (the maximum period that we can observe is 4500 time steps): the moments result then dominated by the many sites with small periods instead of the few sites with a very large one. On the other hand, the average maximum local period observed in each configuration, T_{max}, has indeed a maximum near p_c, but this maximum is found not to diverge and to change only slightly as the observation time is increased by an order of magnitude.

Another possible cause for the moment to remain finite is the effect of the finite size L of the system analysed. We therefore investigate if the moments M_q as function of L at $p = p_c$ do increase to infinity with increasing L. Figure 2 shows our results for some of the positive and negative moments of the logarithm of the period. The moments do not scale with system size and by extrapolating the value of each moment to infinite observation times, the curves appear to approach a constant value as $L \to \infty$.

In conclusion, we found that at $p = p_c$ the logarithm of the median global period varies linearly with L, whereas it varies as L^2 above p_c and as ln L below p_c.

Furthermore, the moments of the distribution of local periods do not show any critical behaviour at the onset of the chaotic phase, neither as function of p nor with a finite size scaling analysis at $p = p_c$.

We would like to thank our collaborator in this research, D. Stauffer.

REFERENCES

[1] Kauffman S.A., J. Theor. Biol. 22, 437 (1969); Physica D 10, 145 (1984); p.339 in "Disordered systems and Biological Organization", edited by E. Bienenstock, F. Fogelman-Soulié and G. Weisbuch, Springer Verlag, Heidelberg 1986.

[2] Derrida B., Phil. Mag. B 55, in press.

[3] Stauffer D., Phil. Mag. B 55, in press; and these proceedings.

[4] Derrida B. and Stauffer D., Europhys. Letters 2, 739 (1986).

[5] Weisbuch G. and Stauffer D., J. de Physique 48, 11 (1987).

[6] de Arcangelis L. and Stauffer D., J. de Physique 48, 1881 (1987).

FIGURE CAPTIONS

Figure 1: Semilog plot of the median global period (dots) and the average global period (squares) versus L. In the insert, the average natural logarithm of the global period (crosses) is shown in a log-log plot versus L.

Figure 2: Log-log plot of the q-th root of the moments of the logarithm of the local periods T, versus L for several values of q and observation time equal to 4500 time steps (dots). The crosses show the values for q = 1 and 4 for L = 10 and 25 with observation time 450.

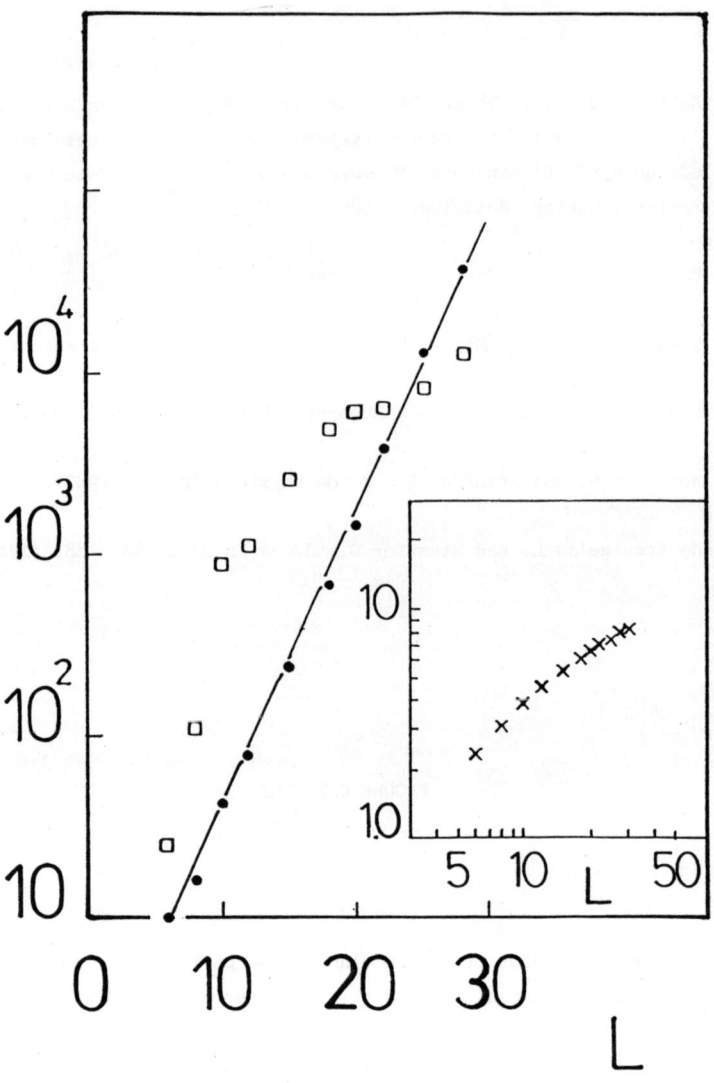

Figure 1

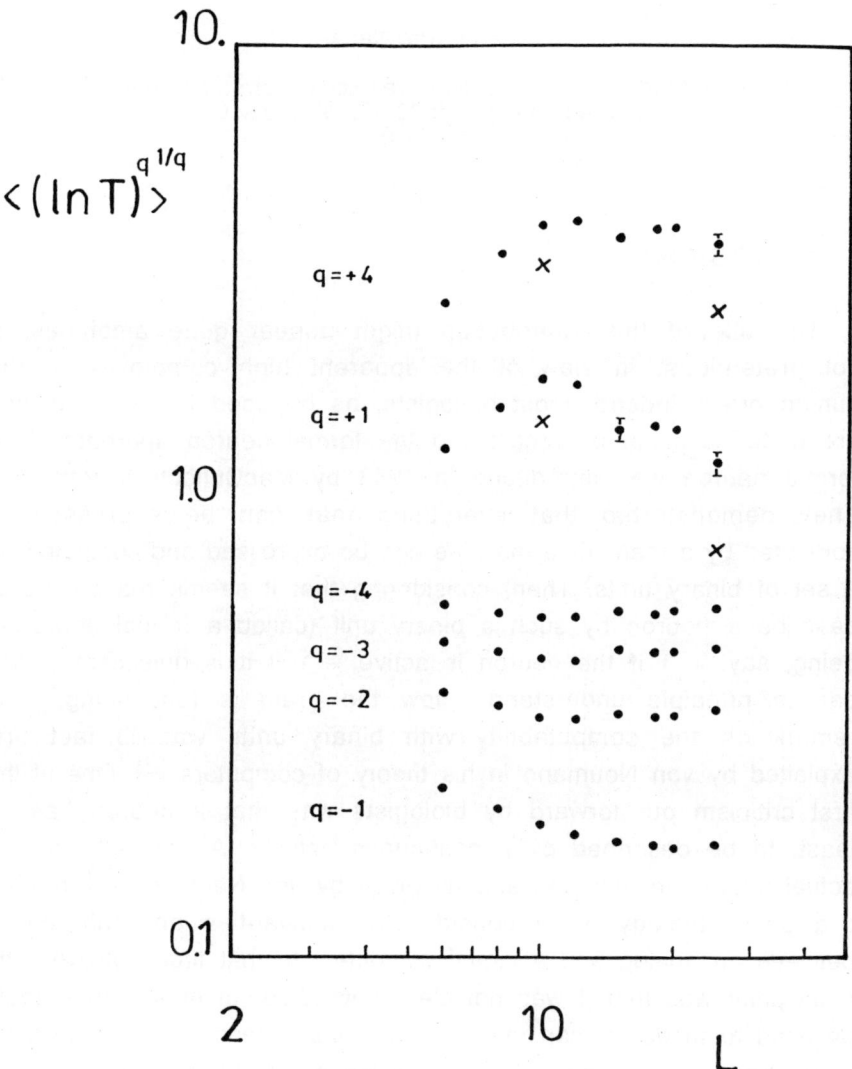

Figure 2

NEURAL NETWORKS : A PATH FROM NEUROBIOLOGY TO PSYCHOLOGY ?

Jean-Pierre Nadal

Groupe de Physique des Solides de l'Ecole Normale Supérieure
24 rue Lhomond, 75231 Paris Cedex 05
FRANCE

1. AN OLD STORY

The title of this contribution might appear quite ambitious, if not pretentious, in view of the apparent high complexity of the human brain. Indeed, most biologists, as opposed to psychologists, are quite sceptical as regards to the formal neuron approach. The formal neuron was introduced in 1943 by MacCulloch & Pitts [1]. They demonstrated that everything that can be expressed or computed by a man or a machine can be expressed and computed by a set of binary units. Then, considering that it seems reasonable to describe a neuron by such a binary unit (called a formal neuron) - being, say, + 1 if the neuron is active, - 1 if it is quiescent -, one can in principle understand how the brain is functioning. This remark on the computability with binary units was in fact first exploited by von Neumann in his theory of computers [2]. One of the first criticism put forward by biologists was that a neuron has, at least, to be described by a continuous activity. A detailed and still actual answer to this was already given by von Neumann [2] in 1948 : a good analogy is to consider the similarities and differences between an analog and a digital computer. At that time however, the main point was that it was not clear how a set of binary units could be used to model human memory. The basic ideas, upon which most actual research on memory relies, were given by the neurobiologist Hebb, in 1949 [3]. His project was to launch a bridge between the neurobiological level and the psychological level. Since antiquity [4] it is emphazised that learning occurs through associations. We can retrieve what we have learned when we are given some related (associated) information : human memory is associative, and content addressable. The opposition with the memory of a classical computer is clear. As another example, we can find a word in a dictionary because we know its address, given by the alphabetical order. Human memory operate in the opposite way : given the definition of a word, it will in general retrieve the word. How is this possible, and by the way, how do we know this alphabetical order ? We learned it as a child by associating each letter with the next one, forming a temporal sequence that we remember for ever. Sequences are particular associations, clearly of great importance in the organization of long term memory.

At the neurobiological level, one can make the following observations : each neuron is connected to, typically, several thousand other neurons. At least locally, and in the parts of the cortex which might be responsible for a memory function, these connexions seem to be distributed at random. The basic idea of Hebb is then the following : an event will be coded by the activity of a set of neurons. Learning occurs by associations : "... any two cells or systems of cells that are repeatedly active at the same time will tend to become 'associated' so that activity in one facilitates activity in the other" [3]. Thus memorization occurs by a modification in the way two neurons interact, that is in the property of the synapses between these neurons. This has been formalized in a simple way, as explained in the following. One first remark is that each synapse will contain a little bit of information on many memorized activities : the memory is distributed. A second and related remark is that retrieval relies on a cooperative behavior of many neurons. For these reasons, the memory is fault tolerant : a local error will not affect the global behavior. And recent experimental studies [5] have shown that, indeed, learning induces modifications in the properties of synapses.

In the following I will present simple models, studied within the framework of statistical physics. The models are probably the simplest possible formalization of Hebb's ideas. They do not intend to modelize in a precise way some specific part of the brain, but the similarity of their behavior with some behavior of human memory allows to take them as relevant metaphors, which should help us to ask new questions. I will not at all try to make a review on neural modeling (see for example [6,7,8,9]). Other approaches are presented in this workshop by F. Fogelman-Soulié, E. Bienenstock, P. Carnevali and S. Patarnello, and other specific results on models closely related to the ones studied here are given by M. Opper, K.E. Kürten and R. Serra.

2. COGNITIVE ASSOCIATIVE MEMORIES

Different versions of the C.A.M. model I will consider have been introduced and studied by many authors [8,10,11]. The most elegant version however was given by Hopfield in 1982 [10]. Two points were made clear : first, retrieval of a stored pattern is viewed as the convergence of a dynamical system towards an attractor - the set of activity coding this pattern. This notion of memories as attractors renders natural the possibility of retrieving a complete pattern starting with a noisy or incomplete pattern. Second, the formalism made explicit the link with statistical physics, so that it was possible to apply the methods of this domain.

So let us consider a set of N formal neurons, S_i, i = 1, N. S_i is a spin like variable, being + 1 (neuron firing) or - 1 (neuron quiescent). The dynamics of this network is governed by the synaptic efficacies J_{ij} which characterize the property of the synapse from neuron j to neuron i. In view of the high connectivity in the cortex, a simple choice is to connect each neuron to every other neuron : $J_{ij} \neq 0$ for all (i $\neq$ j). Each neuron compute its "local field" $h_i(t)$:

$$h_i(t) = \sum_j J_{ij} S_j(t) \tag{1}$$

and

$$S_i(t+\Delta t) = \begin{cases} +1 & \text{with probability } 1/[1+\exp-2\beta h_i(t)] \\ -1 & \text{with probability } 1/[1+\exp+2\beta h_i(t)] \end{cases} \quad (2)$$

where $\beta = 1/T$ is a temperature-like parameter modeling the intrinsic noise of the system. For random sequential updating, the dynamics corresponds to the usual Monte Carlo algorithm. Taking the J_{ij} symmetric - an unrealistic assumption ! - one can define an energy E :

$$E = -\frac{1}{2}\sum_{i\neq j} J_{ij} S_i S_j \quad (3)$$

and we end up with a spin like system. Now we would like to choose the J_{ij} such that a set ($\{\xi_i^\mu, i = 1, N\}$, $\mu = 1,p$) of p patterns to be learned are fixed points of the dynamics. The standard scheme [10] follows the "generalized Hebb rule" :

$$i \neq j \quad J_{ij} = \frac{1}{N}\sum_{\mu=1}^{p} \xi_i^\mu \xi_j^\mu \quad (4)$$

If the ξ_i^μ are randomly chosen to be + 1 or - 1, J_{ij} can be either positive (the synapse is then excitatory) or negative (the synapse is inhibitory) : indeed we know that many (meta)stable states would not be obtained if all the J_{ij} were of the same sign. Eq (4) expresses the idea of Hebb : if i and j are in the same state in pattern μ, the contribution $\xi_i^\mu \xi_j^\mu$ in (3) contributes to increase J_{ij} (= J_{ji}). (3) can be rewritten

$$E = -\frac{1}{2}\frac{1}{N}\sum_\mu (\sum_i \xi_i^\mu S_i)^2 + \text{constant}$$

Thus if the patterns are orthogonal, the ground states of the system are the patterns ξ^μ. Now if we choose the ξ^μ at random, they are only statistically orthogonal. In this case, the thermodynamics of the model can be solved [7]. The main results are the following :
- if the number of pattern p is smaller than a critical value $p_c = \alpha_c(T)N$, with $\alpha_c \sim 0.14$ at T = 0, then the network operates as an associative memory. Starting with an initial configuration not

too far from a stored pattern ξ^μ, the network converges toward an attractor S^μ which is very close to the stored pattern : if we call m_μ the overlapp of S^μ on ξ^μ :

$$m_\mu = \langle \frac{1}{N} \sum_i S_i^\mu \xi_i^\mu \rangle \tag{5}$$

m_μ is at least equal to 0.97 (at zero temperature).
- if $\alpha = p/N$ is greater than α_c, no pattern at all can be retrieved : the network forgets everything !

Many studies have been made to explore variants of this model - e.g. in order to store correlated patterns [7,12], sequences [13]. In particular it has been shown that random dissymmetrization of the synapses, or random dilution of the network, do not affect qualitatively the properties of this system [7]. In fact, the dynamics of a highly diluted and asymmetric model [14] can be solved exactly. The main results are that one keeps the notion of memories as attractors, and that the capacity is proportional to the connectivity C - that is the typical number of neurons to which any neuron is connected :

$$p_c = \alpha_c C . \tag{6}$$

It is easy to understand why it is so : the information concerning the p patterns of N "bits" is stored in NC synaptic efficacies, so that we can expect pN ~ NC. This seems to indicate that the maximal possible capacity of the network (for some optimized choice of the J_{ij}) is $\alpha_{max} = 1$. Surprisingly, it is twice more [15,16] :

$$\alpha_{max} = 2 \tag{7}$$

Furthermore, this is the optimal capacity when one requires *exact* retrieval ($m_\mu = 1$ for every μ). An other interesting aspect is that, in order to reach (7), one has to use some iterative learning algorithm [16,17]. That is, to learn more, the network has to practice !

Now I would like to consider the deterioration of the memory due to overloading in the Hopfield model. This will lead to simple models of short term memory.

3. PALIMPSESTS

"Qu'est-ce-que le cerveau humain, sinon un palimpseste immense et naturel ?" (What is the human brain, but a natural and huge palimpsest ?), Baudelaire, Les paradis artificiels.

In a Hebbian learning scheme such as (4), learning a new pattern $\{\xi_i, i = 1,N\}$, means adding to J_{ij} a quantity

$$\Delta J_{ij} = \sigma \xi_i \xi_j \tag{8}$$

The new pattern is indeed stabilized if for all i

$$\sum_j \xi_i (J_{ij} + \Delta J_{ij}) \xi_j \geq 0 \tag{9}$$

For patterns chosen independently at random, for large N the right hand side is a random Gaussian variable :

$$\sum_j \xi_i (J_{ij} + \Delta J_{ij}) \xi_j = \sigma + z_i \sqrt{N \overline{J_{ij}^2}} \tag{10}$$

(where z_i is a normalized Gaussian variable). Thus memorization occurs only if σ is large enough compared to the noise level due to the previously stored patterns, and we have a criterion for learning [18] :

$$\sigma \geq \varepsilon_c \sqrt{N \overline{J_{ij}^2}} \tag{11}$$

where ε_c is some unknown numerical factor. For the Hopfield scheme, $\sigma = 1/N$ is independent of the patterns. Thus after p patterns have been learned, $N J_{ij}^2 = \sigma^2 (p/N) = \alpha \sigma^2$. Thus, (11) reads

$$\alpha \leq \frac{1}{\varepsilon_c^2} \tag{12}$$

Hence when α becomes greater than $\alpha_c = 1/\varepsilon_c^2$, all the patterns are simultaneously forgotten. (11) gives immediately a way to get rid of this catastrophy : one has to control the relative weight of the acquisition amplitude σ to the noise level J_{ij}^2. This is achieved in two basic schemes : in the first one, called "learning within bounds" [10,18,19], the synaptic efficacies are constraint within bounds, a biological constraint in fact. The iterative learning scheme is still (8) with a constant value of σ, provided the synaptic efficacy remains within the bounds. The second scheme, called "marginalist scheme" [18,20], is defined by

$$J_{ij}(p+1) = \lambda [J_{ij}(p) + \sigma \xi_i^{p+1} \xi_j^{p+1}] \qquad (13)$$

with the proper scaling

$$\sigma = \frac{\varepsilon}{N} , \quad \lambda = \exp{-\varepsilon^2/2N} \qquad (14)$$

After an infinite number of patterns have been learned, and if we now number the patterns with remote ancestry, (13) gives

$$J_{ij} = \sum_{\mu \geq 1} \lambda^\mu \xi_i^\mu \xi_j^\mu \qquad (15)$$

Analytical and numerical studies of these models [18-21] show that they share the same qualitative behavior : the most recently learned patterns are well memorized; their number is still proportional to the connectivity C, $p_c = \alpha_c C$, but α_c is here much smaller than in the Hopfield scheme. Moreover, all the basic phenomena observed on human short term memory [22,23] are mimicked by these models. Two main points have to be stressed. First it is well known that the capacity of human short term memory is "7 plus or minus 2" [23], whatever the material which is learned (numbers, syllables...). The fact that the capacity is proportional to the connectivity, which is known to be fairly uniform in the cortex, would explain this universality. This small number, 7, as opposed to the large value of the connectivity (~ 10 000), would come from the prefactor α_c, which is very small, due to the constraints on the system : learning has to be fast, thus the learning scheme cannot be elaborate; α_c is again diminished by the constraint on the plasticity of the synapses - e.g. the synaptic efficacies are bounded -, preventing overloading. Second, these models are formalizations of the hypothesis of forgetting by interference : it is the learning of new patterns which provokes the forgetting of older ones. A long debate, still unclosed, in experimental psychology [23] is on finding whether forgetting in short term memory is due to such an active masking effect, or to a passive relaxationnal time decay (of the synaptic efficacies). There are experiments in favor of both hypothesis, but for short times interference theory is more plausible. We can comment on this debate in view of our models. First, the biological constraint on the plasticity of the synapses is sufficient to provide a short term memory effect : there is no need here for a relaxational time decay. Let us suppose however that we have some iterative learning scheme given by (8), together with

$$J_{ij}(t+\Delta t) = \lambda J_{ij}(t)$$

with here

$$\lambda = e^{-\Delta t/\tau}$$

In a typical experiment, where items are learned at a constant rate, the expression of the J_{ij} would still be given by an expression similar to (15). This has the following consequences : i) a non zero capacity is possible only if $\tau/\Delta t$ scales with the connectivity. ii) even if the weights of each learned pattern comes from this relaxational effect, the mechanism responsible for forgetting is still the superposition of patterns. iii) now if no more patterns are learned during some times, the overall value of J_{ij} decreases. This is strictly equivalent to keep the J_{ij} unchanged and to increase the temperature : $T(t + \Delta t) = T(t) e^{\Delta t/\tau}$. Thus patterns are progressively forgotten, since α_c decreases with the temperature T, until this effective temperature reaches the critical value where $\alpha_c = 0$. Here comes the only useful property I can see in having a time decay : if one starts again a learning task, the noise level due to previously stored patterns is very small . Then there is a transient time during which the capacity can be very large. The moment when the capacity starts to decrease towards the stationary value (of about 7 items) is known as the "proactive interference effect" [22,23]. Note that any mechanism, able to set to zero (or to a small value) the synaptic efficacies before a learning task begins, will produce this effect, again due to the superposition of patterns.

CONCLUSION

Although the models described above are quite simple as regards to the biological complexity, it is encouraging that their behavior is very similar to the one of human short term memory. This, however, is not really surprising : we are looking at the most elementary, mechanical, aspects of human memory, finding basic - but then universal - properties due to interference.

ACKNOWLEDGEMENTS

My contribution to the field exposed here is the result of collaborations with G. Toulouse, J.P. Changeux, S. Dehaene, M. Mézard and B. Derrida. Discussions with K. O'Regan on the psychological aspects are gratefully acknowledged. I thank W. Krauth for a critical reading of the manuscript.

REFERENCES

[1]   Mac Culloch, W.W. and Pitts, W., Bull. Math. Biophys. $\underline{5}$, 115 (1943).
[2]   von Neumann, J., in Collected Works, vol. V, (Pergamon Press 1963), pp 288-328; "The computer and the brain", (Yale Univ. Press, 1958).
[3]   Hebb, D.O., "The organization of behavior" (Wiley, New-York, 1949).
[4]   Warren, "A history of Association Psychology", 1967.
[5]   Di Prisco, G.V., Progr. in Neurobiology $\underline{22}$, 89 (1984); Thompson, R.F., Science $\underline{233}$, 941 (1986).
[6]   Cowan J.D. and Sharp, D.H., preprint 1987; Tank D.W. and Hopfield, J.J., Scientific American $\underline{257}$, 62 (1987); Bienenstock, E.,

Fogelman, F. and Weisbuch, G., "Disordered Systems and Biological Organization", (Springer-Verlag, Berlin 1986).
[7] Amit, D.J., in Heidelberg Colloquium on Glassy Dynamics, von Hemmen, J.L. and Morgenstern, I., eds., (Springer-Verlag, Berlin 1987), p.430.
[8] Kohonen, T., "Self Organization and Associative Memory", (Springer-Verlag, Berlin 1984).
[9] Rumelhart, D.E., Hinton G.E. and Williams, R.J., Nature 323, 533 (1986).
[10] Hopfield, J.J., Proc. Natl. Acad. Sci. USA 79, 2554 (1982).
[11] See references in Shinomoto, S., Biol. System 57, 197 (1987) and in Cowan and Sharp [6].
[12] Parga, M. and Virosoro, M.A., J. Physique (Paris) 47, 1857 (1986); Feigelman, M.V. and Ioffe, L.B., Int. J. of Mod. Phys. B1, 51 (1987); Gutfreund, H., preprint 1987; Cortes, C., Krogh, A. and Hertz, J.A., J. Phys. A, submitted.
[13] Nadal, J.P., proceedings of "Measures of Complexity", Rome 1987 and references therein.
[14] Derrida, B., Gardner, E. and Zippelius, A., Europhys. Lett. 4, 167 (1987).
[15] Venkatesh, S., Proceedings of the Conference on Neural Networks for Comptuing (Snowbird, 1986); T.M. Cover, IEEE transactions EC14, 326 (1985).
[16] Gardner, E., Europhys. lett. 4, 481 (1987); J. Phys. A21, 257 (1988).
[17] Diederich, S. and Opper, M., Phys. Rev. Lett. 58, 949 (1987); Pöppel, G. and Krey, U., Europhys. Lett. 4, 979 (1987); Kleinfeld, D. and Pendergraft, D.B., Biophys. 51, 47 (1987); Krauth, W. and Mézard, M., J. Phys. A20, L745 (1987).
[18] Nadal, J.P., Toulouse, G., Changeux, J.P. and Dehaene, S., Europhys. Lett. 1, 535 (1986).
[19] Parisi, G., J. Phys. A19, L617 (1986); Vedenov, A.A. and Levchenko, E.B., JETP Lett. 41, 402 (1985).
[20] Mézard, M., Nadal, J.P., Toulouse, J. Physique (Paris) 47, 1457 (1986).

[21] Derrida, B. and Nadal, J.P., to appear in J. Stat. phys., Dec. 1987; Geszti, T. and Pazmandi, F., to be published; Gordon, M., to appear in J. Physique (Paris).

[22] Nadal, J.P., Toulouse, G., Mézard, M., Changeux, J.P. and Dehaene, S., in "Computer Simulation in Brain Science", R.M.J. Cotterill Ed., (Cambridge Univ. press 1987).

[23] Massaro, D.W., "Experimental Psychology and Information Processing", (Rand McNally College Publ. Co., Chicago, 1975).

STATISTICAL MECHANICS OF LEARNING IN NEURAL NETWORK MODELS

Manfred Opper
Institut für Festkörperforschung der Kernforschungsanlage Jülich
Sigurd Diederich and Joachim K. Anlauf
Institut für Theoretische Physik der Justus-Liebig-Universität Giessen

ABSTRACT

The performance of two learning algorithms is studied for the storing of random patterns in neural networks. For large networks and random patterns exact results for storing capacities and learning times are obtained.

1. INTRODUCTION

In recent years neural networks as models of associative memory have become increasingly popular in physics. Methods of statistical physics have proven to be very useful in understanding characteristic features of the pattern retrieval in large networks[1]. In this paper we apply similar methods to the network's learning phase, where a set of patterns is stored by adaptive algorithms.

We consider a simple neural network consisting of N interconnected neurons with two possible states $S=\pm 1$. Patterns are retrieved by

$$S_i(t+1) = \text{sign}(\sum_{j \neq i} J_{ij} S_j(t)) . \qquad (1)$$

The weights J_{ij} are the strengths of the synapses which connect the neurons. To serve as a memory for a set of P "spin"-patterns $\underline{S}^\nu = (S_1^\nu, \ldots S_N^\nu)$ the couplings J_{ij} must be adjusted such that the $\underline{S}^\nu$ are locally stable fix points of the dynamics (1), i.e. a set of inequalities

$$S_i^\nu h_i^\nu \geq c_i > 0 , \quad h_i^\nu = \sum_{j \neq i} J_{ij} S_j^\nu \qquad (2)$$

must be obeyed for each neuron i and pattern ν. There have been several attempts to design learning algorithms, which adjust the synaptic couplings iteratively. Effective learning rules should optimize the stability of the patterns, so that patterns can be recognized even from distorted (noisy) input versions. It has been argued that using large normalized thresholds $\Delta_i = c_i / (\sum_{j \neq i} J_{ij}^2)^{1/2}$ should give satisfactory results[2].

In the following we discuss the performance of two learning rules in the limit of large networks N, P→∞, α=P/N fixed.

2. ADALINE-LEARNING RULE

The simplest way to satisfy (2) is to require, that all internal fields h_i^ν are of equal modulus (which may be set equal to 1) after learning, i.e

$$s_i^\nu h_i^\nu = 1 \text{ for all } i, \nu. \qquad (3)$$

The solutions of these equations (if they exist) coincide with the minimum of the "energy" density

$$H(t) = (NP)^{-1} \sum_i \sum_\mu (1 - s_i^\nu h_i^\nu(t))^2. \qquad (4)$$

Minimizing H might give approximate solutions even if an exact solution does not exist. An algorithm which can find this minimum is the Adaline learning rule:

$$J_{ij}(t+1) = J_{ij}(t) + \frac{\gamma}{N} \sum_\nu (1 - s_i^\nu h_i^\nu(t)) s_i^\nu s_j^\nu, \quad J_{ij}(0) = 0 \qquad (5)$$

γ is a relaxation parameter, which must be adjusted to guarantee convergence. One can show, that the algorithm yields the solution of (3) with minimum norm $\|J_i\| = (\sum_{j \neq i} J_{ij}^2)^{1/2}$, and thus maximal stability Δ_i.

To find out how many patterns can be stabilized and how fast the learning process operates one must specify the types of patterns to be stored. To study at least some "typical" behaviour we can use statistical ensembles of patterns. The simplest choice is $s_i^\nu = \pm 1$ with equal probability, independently for each ν and j.

By definition, the value of H(t) depends on the realization of the patterns. For large networks $N\to\infty$ however, fluctuations become negligible ($\simeq 1/\sqrt{N}$). H(t) is a "self averaging" quantity, i.e. its values coincide with its average value $\bar{H}(t)$ with probability 1. Assuming that the algorithm converges one can find the average over random patterns for the minimized $H=H(t=\infty)$

$$\bar{H} = \lim_{\beta\to\infty} (N\beta)^{-1} \overline{\ell nZ}, \text{ where } Z = \int D[J] e^{-\beta NH} \delta[\|J\|-1/\Delta] \tag{6}$$

the limit $\beta\to\infty$ ensures that H is a minimum for given stability Δ. To average over the randomness we use the "Replika-trick" $\ell nZ = \ell im_{n\to 0}(Z^n-1)/n$. For $\alpha<1$ the resulting mean field equations are found to have solutions as long as $\Delta<\sqrt{(1-\alpha)/\alpha}=\Delta_{max}$ and the energy equals $\bar{H}=0$, i.e. for P<N random patterns eqs.(3) are fulfilled exactly. For $\alpha>1$ one can show, that an increasing fraction of bits for each pattern become destabilized.

Using random matrix theory we have also calculated $\bar{H}(t)$ for finite times with the result[3]

$$\overline{H(t)} = (2\pi\alpha\gamma)^{-1} \int_{\lambda_1}^{\lambda_2} \frac{\lambda 2t}{1-\lambda} \sqrt{(\lambda-\lambda_1)(\lambda_2-\lambda)} \, d\lambda + \theta(\alpha-1))\cdot(1-\alpha)/\alpha$$

$$\lambda_{1,2} = 1-\gamma(1\pm\sqrt{\alpha})^2 \tag{7}$$

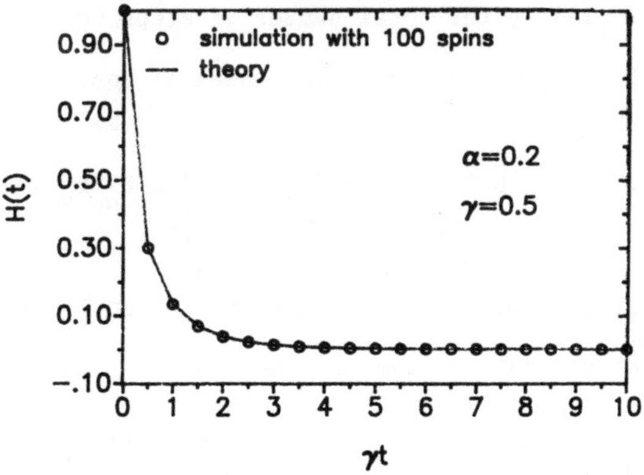

Fig. 1: Decay of error (7).

in the limit $N\to\infty$. For $\alpha<1$ and γ small enough, the error will relax to zero (Fig.1).

3. PERCEPTRON LEARNING RULE

This rule has the nice property to converge if a solution to (2) exists[4,5,6]. Its storage capacity has thus the maximal values[6] possible for neural networks. A variant which reaches optimal stability asymptotically, has been recently introduced by Krauth and Mezard[2]. The algorithm operates in parallel for each neuron i. An elementary time step consists of a change $\delta J_{ij}(t)$

$$\delta J_{ij}(t) = N^{-1} S_i^{\nu(i,t)} S_j^{\nu(i,t)} \qquad (8)$$

where $\nu(i,t)$ is the pattern which is stored worst (minimal $S_i^\nu h_i^\nu$) at time t. Subsequently all fields are updated and the procedure is repeated for the next time step t+1. The algorithm stops for a neuron, when all patterns satisfy (2). Maximal stability is obtained for $c_i\to\infty$. Our aim is to find the speed of the learning process in this limit. The basic idea is as follows: After termination of the algorithm the couplings have the form

$$J_{ij} = N^{-1} \sum_\nu t_\nu(i) \cdot S_i^\nu S_j^\nu \qquad (9)$$

where t_ν denotes the number of time steps, pattern ν has led to a change δJ_{ij}. Clearly the total number of steps for neuron i equals $T=\sum t_\nu$. With $x_\nu = t_\nu/c$, (omitting the index i) eqs.(2) and optimality of Δ become

$$\sum_\nu B_{\mu\nu} x_\nu \geq 1, \quad B_{\mu\nu} = N^{-1} \sum_{j\neq i} S_i^\mu S_j^\mu S_i^\nu S_j^\nu \quad \text{and} \quad \sum_{\mu,\nu} x_\mu B_{\mu\nu} x_\nu = N\Delta^{-2} = \text{Minimum!} \qquad (10)$$

The minimization problem (10) can be treated analogous to eq.(6). As the result of the pattern average one finds[7]

$$\tau = \lim_{\substack{N\to\infty \\ c\to\infty}} \frac{\bar{T}}{N\cdot c} = \alpha \bar{x}_\mu = \Delta^{-2}, \text{ where } \alpha \cdot \int_{-\Delta}^{\infty} \frac{dt}{\sqrt{2\pi}} e^{-t^2/2}(t+\Delta)^2 = 1 \qquad (11)$$

τ diverges quadratically for $\alpha \to 2$ (maximal storage capacity). A comparison with simulations is given in Fig.2.

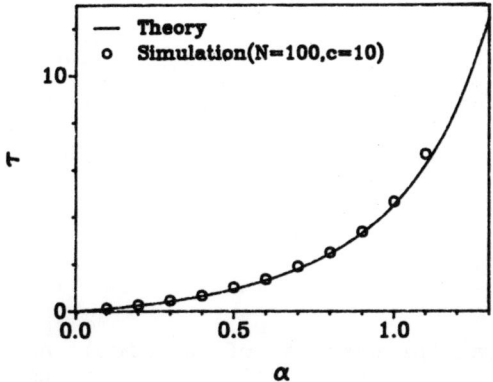

Fig. 2: Number of learning steps (11) for Perceptron algorithm (8).

We would like to thank Prof. W. Kinzel for many inspiring discussions.

REFERENCES

1) Amit D.J., Gutfreund H. and Sompolinsky H., Phys. Rev. Lett. 55, 428 (1985).
2) Krauth W., Mezard M., J. Phys. A: Math. Gen. 20, L745 (1987).
3) Opper M., Diederich S. and Anlauf J.K., in preparation.
4) Block H.D., Rev. Mod. Phys. 34, 123 (1962).
5) Diederich S. and Opper M., Phys. Rev. Lett. 58, 949 (1987).
6) Gardner E., J. Phys. A: Math. Gen. 21, 257 (1988).
7) Opper M., submitted to Phys. Rev. Lett..

"Training" Quasirandom Neural Networks

K.E. Kürten

Institut für Theoretische Physik

Zülpicher Str. 77, 5000 Köln, West Germany

Abstract

Dynamical properties of a model neural network consisting of binary decision elements interacting asymmetrically at synaptic junctions with activity-dependent synaptic efficiencies are investigated statistically. Computer simulations show that various experience-dependent plasticity algorithms designed to closely fit neurophysiological experiments applied to networks with initially quasirandom connectivity lead to networks showing exclusively ordered behavior in contrast to the virgin networks where chaotic activity has been observed.

The first decisive step towards a "dynamical" neural network model with activity-dependent synaptic interactions was Hebb's classical postulate that the efficacy of an excitatory synapse increases when the two neurons it links fire simultaneously [1]. In fact, experiments carried out by Levy et al.[2] show that temporal correlations between pre- and post-synaptic firings strongly enhance the *excitatory* synaptic strength and that, if the pre-synaptic cell is silent and the post-synaptic element is active, the excitatory synaptic efficiency decreases considerably. These findings have been fully confirmed by electrophysiological conditioning experiments on the cortex of the cat by Rauschecker and Singer[3], who observed that synaptic efficiency even decreases slowly in time independent of pre-synaptic activity whenever the post-synaptic cell is silent. In summary, they came to the conclusion that self-organizing and learning effects might be based on the same neuronal mechanism, which can be formulated in terms of a few simple rules formally closely related to those postulated by Hebb.[1]

Actually much less is known about the plasticity of

inhibitory synapses. In this report Hebb's postulate is extended to inhibitory couplings by a rule symmetrical to that obeyed by excitatory synapses as has been suggested earlier, though there is no experimental evidence so far.

The model network consists of a set of N interacting threshold elements called neurons, which can take two possible states $\sigma_i = +1$ or $\sigma_i = -1$. Each neuron is supposed to have the same number of K inputs, chosen differently at random among all other neurons of the network. In the spirit of the pioneering work of Mc Culloch and Pitts [4] the time is quantized in units of the delay time τ for signal transmission from one neuron to another and the time evolution of the system is given by

$$\sigma_i(t+\tau) = \text{sign}(\sum_j c_{ij}(t) \sigma_j(t)) \quad , i = 1,\ldots,N . \quad (1)$$

such that the state of each neuron at the next time step is determined by whether or not its total excitation exceeds or equals zero. Thus, the dynamics of the network is synchronous and fully deterministic. Since there are only 2^N distinct firing patterns, the time evolution of the system will eventually be periodic evolving to attractors consisting of limit cycles or fixed points. In analogy to the Kauffman model [5] it has convincingly been demonstrated by analytical results as well as by computer simulations that there exist two phases, a chaotic phase and a frozen phase [6,7], depending on the choice of the network parameters. In the ordered phase the average period increases with a power of the total number of cells, whereas in the chaotic phase the mean cycle length increases exponentially with N. At the critical point there is some indication that the increase is linear. [7]

In the following section we want to explore several theoretically possible training rules designed to demonstrate their effectiveness in gradually leading to networks showing more ordered behavior in contrast to the corresponding virgin quasirandom networks, which for the same net parameters exhibit strongly chaotic behavior. Following Rauschecker's and Singer's suggestions [3] the coupling coefficients are only modified if the postsynaptic neuron i is active. Thus, one possible ansatz for the dynamical time evolution for the modification of the coupling coefficients is

$$|c_{ij}(t+\tau)| = |c_{ij}(t)|(1 + \delta\Delta_s(\sigma_i(t),\sigma_j(t))) \quad s = 1,\ldots,4 \quad ,(2)$$

where the learning constant δ characterizes the rate of modification being of the order of 0.01 and the quantity Δ_s depends on the specific training rule.

$$\Delta_1 = \frac{\sigma_i(t)+1}{2} \frac{\sigma_j(t)+1}{2} \frac{\text{sgn}(c_{ij}(t))+1}{2} \qquad (2.1)$$

$$\Delta_2 = \frac{\sigma_i(t)+1}{2} \frac{\sigma_j(t)+1}{2} \text{sgn}(c_{ij}(t)) \qquad (2.2)$$

$$\Delta_3 = \frac{\sigma_i(t)+1}{2} \sigma_j(t) \frac{\text{sgn}(c_{ij}(t))+1}{2} \qquad (2.3)$$

$$\Delta_4 = \frac{\sigma_i(t)+1}{2} \sigma_j(t) \text{sgn}(c_{ij}(t)) \qquad (2.4)$$

Note that rule (2) is local in time and space. Since $c_{ij}(t+\tau)$ is proportional to $c_{ij}(t)$ the synaptic coefficients cannot change sign consistent with observations in biological systems. At the beginning of the training procedure the $c_{ij}(t=0)$ are chosen randomly according to a uniform distribution $\varrho(c_{ij})$ in the intervall $[-1,+1]$. In order to avoid that their magnitude rise to unrealistic values any further increase of their magnitude is ceased when it reaches unity. The connectivity parameter K has been chosen as K=3 throughout this study.

We first study the effects of the "genuine" Hebb rule on the dynamic properties, when only the strengths of excitatory links are modified. The second rule under consideration is a straightforward generalization to inhibitory couplings such that the inhibitory strengths are weakened whenever the pre- and post-synaptic cell fire simultaneously. The third rule takes into account Levy's, Rauschecker's and Singer's experimental findings.[2,3] This rule has extensively been studied theoretically by Peretto [8] in a general study of sixteen possible rules, where he found that with respect to stability this rule was by far the best the network could achieve. In a last step we extend the former one to inhibitory couplings.

Figure 1 represents histograms of the distribution $\varrho(c_{ij}(t>1200))$ averaged over 500 specimen nets after convergence of the coupling coefficients for the above training rules. With increasing training steps the peaked structure of $\varrho(c_{ij})$ gets more and more pronounced indicating that the rule forms a synchronously activated assembly consisting of excitatorily connected cooperating neurons. On the other hand, pathways which remain unstimulated, gradually decrease their synaptic strength to very small values. These effects have already been observed by Buhmann and Schulten[9], who found that the magnitude of synapses which contribute to a learned pattern are either very large if excitatory, or are very small, if inhibitory, whereas for synapses which do not contribute to a learned pattern they observed the opposite effect.

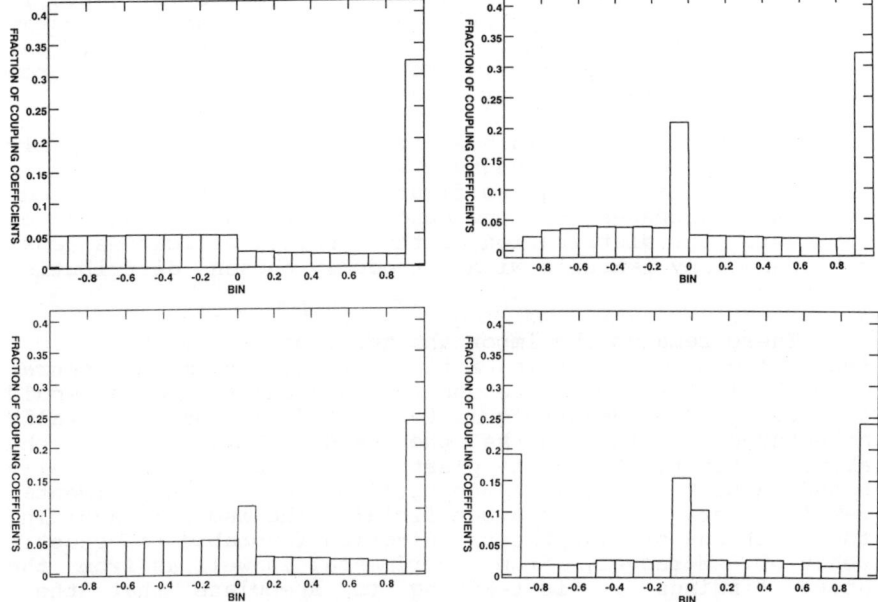

Fig.1: Histograms of the final distribution $\varrho(c_{ij}(T))$ for different training rules.

Figure 2 shows the average period as a function of the number of training steps for different learning algorithms for N = 20, averaged over 500 specimen nets. The average period falls off rapidly as the number of training steps increases, finally reaching its asymptotic value. Results for N=50 and N=100 are qualitatively similar.

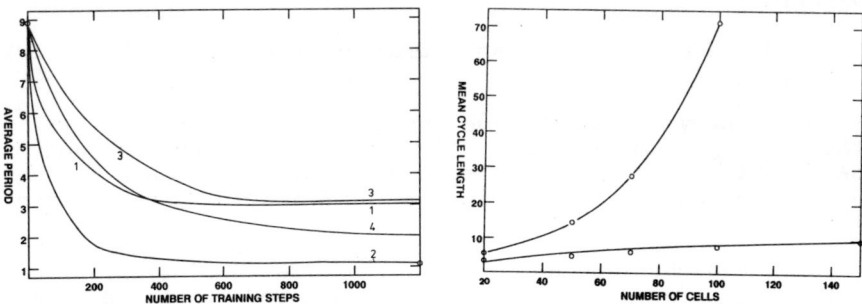

Fig.2: Mean cycle length as a function of T for different training rules. (N=20)

Fig.3: Mean cycle length as a function of N for 200 and 600 steps according to rule (2.3).

Figure 3 represents the average period as a function of the total number of cells for 200 and 600 training steps T according to rule (2.3). As already demonstrated in ref. [6] the increase of the mean cycle length is exponential for T=200, representing the chaotic phase, whereas for T=600 it increases with a small power of N reflecting that the system is in its ordered phase. Thus, there exists a critical number of training steps, where a phase transition from chaotic to ordered behavior occurs.

It should be stressed however that the system definitely shows quite different behavior, if one chooses simply a distribution according to figure 1. Since the "trained" network possesses a pronounced intrinsic correlated structure the distribution of the synaptic efficacies does not share many features with a randomly chosen distribution any more.

There remains the important question of the biological implications of chaotic activity found in model neural networks.[10] A general tendency toward chaotic modes would definitely be undesirable in terms of detailed temporally structured memories. On the other hand we have seen that the above activity-dependent plasticity schemes essentially based on Hebb's postulate supported by various experimental results turn out to be a beneficial mechanism for avoiding the patterns of synaptic organization favorable to chaotic behavior. Moreover, from experimental as well as from the above findings it is tempting to speculate that these Hebbian principles incorporated in our model might be quite universal and could find useful applications in various biological systems for a variety of different functions.

The author likes to thank R.Nemeth, G.Porenta, M.L.Ristig, G.Senger and D.Stauffer for stimulating discussions. Funding for this work was provided in part by the Deutsche Forschungsgemeinschaft under grant No. Ri 267/9.

References

[1] D.O. Hebb : The organization of behaviour, New York (Wiley) (1949)

[2] W.B. Levy : Associative changes at the synapse: LTP in the hippocampus. In: "Synaptic modification, neuron selectivity and nervous system organization." W.B. Levy, J.A. Anderson, S. Lehmkuhle, Lawrence Erlbaum Ass.,(London) 1985

[3] J.P. Rauschecker, W. Singer : The effects of early visual experience on cat's visual cortex and their possible explanation by Hebb synapses, J. Physiol. (London), 310,215-239,(1981)

4) W.S. Mc Culloch and W.H. Pitts : A logical calculus of ideas immanent in nervous activity. *Bull. Math. Biophys.*,5,115-133,(1943).

5) S.A. Kauffman : Emergent properties in random complex automata, Physica 10D, 146-156 (1984)

B. Derrida and D. Stauffer : Phase transitions in two-dimensional Kauffman cellular automata, Europhys. Lett.2,10 (1986)

6) K.E. Kürten : Phase transitions in quasirandom neural networks, Proceedings of "IEEE First Annual International Neural Network Conference", (San Diego, 1987)

7) K.E. Kürten : Critical phenomena in model neural networks, preprint 1987

8) P. Peretto : On learning rules and memory storage abilities of asymmetrical neural networks, preprint (1987)

9) J. Buhmann and K. Schulten : Associative recognition and storage in a model network of physiological neurons, Biol. Cybern. 54,319-355 (1986)

10) K.E. Kürten and J.W. Clark : Chaos in neural systems, Phys. Lett., 114A,(1986), 413

GENERALIZED HOPFIELD LEARNING RULES

R. Serra (*), G. Zanarini (#), F. Fasano (#)

(*) TEMA and ENIDATA, Via A.Moro 38 - 40127 Bologna (Italy)
(#) Physics Dept., Via Irnerio 46 - 40126 Bologna (Italy)

ABSTRACT

The deterministic Hopfield model is modified, in order to avoid the presence of the complementary patterns in the set of attractors. The phase space of the modified Hopfield rules is discussed. Since mixed states may also appear, a precise definition of the concept of exclusion of the complementary patterns is given, along with a sufficient condition for achieving it.

The well known Hopfield model [1] displays some interesting "cognitive" properties. Among the slightly different versions which have been proposed [1-3], we will refer to the following: the N degrees of freedom $X_1, \ldots X_N$ can take only two values (say 0 and 1), and they evolve deterministically in discrete time according to the following rule:

$$X_i(t+1) = 1 \quad \text{if} \quad G_i(X(t)) > \theta$$
$$X_i(t+1) = 0 \quad \text{if} \quad G_i(X(t)) \leq \theta \quad (1)$$

The "input function" of neuron i is defined as:

$$G_i(X) \equiv \sum_j T_{ij} X_j \quad (2)$$

Parallel iteration is chosen, i.e. all spins are updated at each time step. The case of finite N will be considered, thus departing from most published papers on this topic, which consider the limit $N \rightarrow \infty$.

The phase space portrait must be shaped in accordance with the cognitive task to be performed: this is generally assumed to be pattern recognition, but it should be observed that recognition and association of patterns are the basic ingredient of more general cognitive tasks, including rule based reasoning, inferencing and deduction.

Let us us suppose that we want to teach to the system M different patterns $A^1, \ldots A^M$, each of them being an N-dimensional vector. The Hopfield learning rule is then

$$T_{ij} = (1-\delta_{ij}) \sum_m (2A_i^m - 1)(2A_j^m - 1) \quad (3)$$

One would ideally desire that the M different patterns could be stored as different fixed points; the limitations to this goal have been thoroughly investigated, mainly in the infinite N limit [1,3,4]. Some results are known also regarding the finite case [5-7]; in particular, two different patterns always coexist as different fixed points, while this guarantee is lost if M ≥ 3 [6,7].

Let us give some useful definitions. The overlap function $Q(X,Y)$ ($\in [0,N]$) measures the number of places where both patterns present high values:

$$Q(X,Y) = \Sigma_k X_k Y_k \qquad (4)$$

The complementary of an arbitrary pattern X, X', is defined as:

$$X'_i = 1 - X_i \qquad i=1,\ldots,N \qquad (5)$$

We remark that the coupling matrix T (Eq.(3)) is invariant under the transformation $A^m \to A'^m$. In other words, learning a given pattern is the same as learning its complementary pattern. However, these complementary patterns are usually not meaningful, and they overload the set of attractors.

In view of the limitations of the storage capacity of Hopfield networks [1,3,4], which are particularily severe in the case of correlated patterns, it is natural to suggest the use of a hierarchical system. The input state to be recognized is sent in parallel to several Hopfield networks ("subnetworks"). Each subnetwork will relax to the attractor closest to the input state, thus giving its own "interpretation" of the state. The competition among the different interpretations will be handled at a higher hierarchical level, e.g. by some measures of closeness to the input, or perhaps also by combining this measure with further information of different kind.

However, the use of the original Hopfield rule in such an architecture would lead to problems, due to the presence of the complementary patterns, which fill the phase space with useless states. Some of them might even be closer to a distorted input than any "true" pattern, and they might win the competition for recognition. It would therefore be necessary to introduce a very demanding supplemetary control of the admissibility of the final states.

A way to avoid such overload of the higher level consists in modifying the learning rule (Eq.3) in such a way to maintain the attractive features of the Hopfield original rule, eliminating the complementary patterns. The general form of a symmetric synaptic matrix, quadratic in the learnt patterns, is

$$\delta T_{ij} = \alpha A_i A_j + \beta(1-A_i)(1-A_j) + \Gamma A_i(1-A_j) + \Gamma(1-A_i)A_j \qquad (6)$$

where δT is the variation of the coupling matrix when a new pattern A is learnt. The Hopfield original rule corresponds to the choice $\alpha = \beta > 0$, $\Gamma = -\alpha$. In the spirit of the Hebbian hypothesis, we are led to consider three other major classes of learning rules of the kind described by Eq.6:

Class 1 : $\alpha > 0$, β and $\Gamma < 0$.
Class 2 : $\alpha > 0$, $\beta = \Gamma = 0$.
Class 3 : $\alpha > 0$, $\Gamma < 0$, $\beta = 0$ (introduced by Toulouse et al. [8])

These three rules display different behaviours, which have been analyzed elsewhere [6,7]. It has been also shown that the stored patterns coexist as different attractors if they are not too similar to each other; otherwise they give rise to mixed attractors. Typically, in the two-pattern case, the mixed attractors are the AND and the OR of the learned patterns, in the language of boolean algebra. The AND (OR) of two patterns is the state whose elements are equal to 1 where the corresponding elements of both (at least one of) the two originary patterns are equal to 1, zero elsewhere.

These mixed states, however, can still be used for recognition, if the two similar stored patterns "have the same meaning", i.e. if they correspond to slightly different versions of what can be considered the same thing, from the viewpoint of the higher processing level. In this case the attractors, which are some combinations of the stored patterns, would also mean the same as their "parent" patterns.

The presence of mixed states requires to clarify the meaning of the idea of eliminating the complementary patterns from the phase space. Beyond the obvious requirement that they be no fixed points, we must secure that final states are logical combinations of the stored patterns (like the AND and OR discussed above), with no contributions from the complementaries.

So, let us first define the concept of positive combination (pc) of a given set of patterns $P \equiv \{A^1 ..., A^M\}$. A state $X = (X_1, ..., X_N)$ is a pc of the given set of patterns P iff it can be written as a logical sum (OR) of logical products (AND) of patterns in P.

We will restrict our considerations to learning rules such that the most general form of the input function is the following :

$$G_i(X) = F\{A_i^1, ..., A_i^M, Q(A^1, X)...Q(A^M, X), X_i, Q(X,X)\} \qquad (8)$$

It can be proven that, given an arbitrary state $X(t)$, its successor $X(t+1)$ is a pc of the learned patterns $A^1...A^M$, provided that, for every pair of indices i,m, and for every X with components $\in \{0,1\}$:

$$F(*, A_i^k = 0) \leq F(*, A_i^k = 1) \tag{9}$$

where the symbol * means all the other variables, which should be equal on both sides of Eq.9. If F is differentiable the previous condition amounts to the requirement that the partial derivatives of F with respect to all the A_i^k's should be non negative. For a proof of this result, see [7]. Class 1 and 2 rules obey the condition 9, while class 3 rules do not: indeed, the descendants of complementary patterns are included in their set of attractors

We also remark that consideration of asymmetric coupling matrices is important, both for the simulation of biological neural networks - where asymmetry is the rule - and for artificial cognitive systems. In particular, the following asymmetric learning rule, which obeys the no complementary patterns condition, has interesting information processing features.

$$T_{ij} = \Sigma_m (2A_i^m - 1) A_j^m \tag{10}$$

The basic idea behind rule (10) is that, if neuron j is quiescent, no modification in the synapse from it to other neurons takes place: only active neurons influence learning, which takes place according to the usual hebbian style.

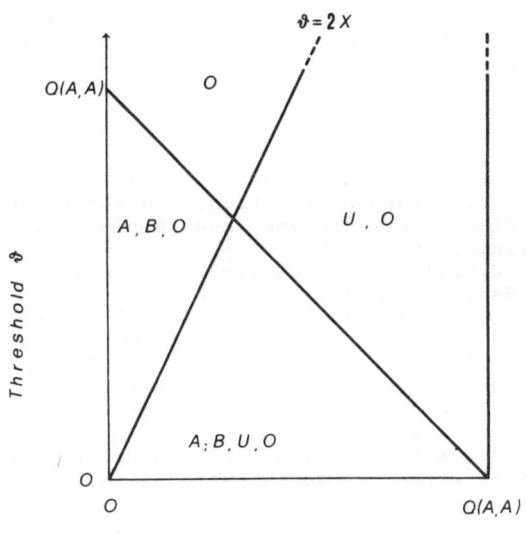

FIG.1 PHASE PLOT, ASYMMETRIC RULE

In the one pattern case, A is the only nontrivial attractor, and its basin is composed by all initial states X such that $Q(X,A) > \theta$. The phase plot for the two pattern case is given in fig.1.

This rule is both complementary-free and endowed with good storage capacity, in the case where the patterns have a fraction of high values smaller than that of low values. A detailed analysis is given in [7].

Another interesting feature of this asymmetric rule is its way to handle the case of correlated patterns. In particular, it is capable of categorizing the set of input patterns in a sensible way. This capability is exemplified by the analysis of an example, regarding the identification of environments (e.g., bathroom, kitchen, town, etc.) starting from information about what they contain. It is a simplified version of a task studied by Rumelhart et al. with a Boltzmann machine [9]; it is shown in ref. [7] that the simpler and faster algorithm which implements the deterministic rule (10) achieves good results.

From a training set of typical examples of different environments, the asymmetric rule succeeds not only in identifying the correct environment starting from incomplete input, but it also extracts the relevant information in the case of partially contradictory inputs. For example, if the information is that there is a bathtub and a refrigerator, the network will relax neither to "bathroom" nor to "kitchen", but it is able to identify the input state as "room", and it evokes the features expected for all rooms (door, floor, ceiling, etc.). For further details, see ref. [7].

REFERENCES

(1) J.J.Hopfield, Proc. Natl. Acad. Sci. USA $\underline{79}$, 2554 (1982).
(2) F.Fogelman-S., Lyapunov functions and their use in automata networks, in Disordered Systems and Biological Organization, Springer, Berlin, 1986.
(3) D.J.Amit, H.Gutfreund, H.Sompolinsky, Phys. Rev. A $\underline{32}$, 1007 (1985); Phys. Rev. Lett. $\underline{55}$, 1530 (1985).
(4) G.Weisbuch, F.Fogelman-S., J. Physique Lett. L623 (1985).
(5) F.Fogelman-S., in SIAM Journal of Comput., Marseille, 1984, in press.
(6) R.Serra, G.Zanarini, F.Fasano, Attractors, learning and recognition in generalized Hopfield networks, in Proceed. Cognitiva 87, 459, Vol.1, May 87.
(7) R.Serra, G.Zanarini, F.Fasano, Cooperative Phenomena and Artificial Intelligence, in Journ. Molec. Liquids, special issue, 1987, in press.
(8) G.Toulouse, S.Dehaene, J.Changeux, Proc. Natl. Acad. Sci. USA $\underline{83}$, 1695 (1986).
(9) J.L. Mcclelland, D.E.Rumelhart and the PDP Research Group, Parallel Distribuited Processing, MIT press, Cambridge, 1986.

BOOLEAN NETWORKS WHICH LEARN TO COMPUTE

Stefano Patarnello and Paolo Carnevali
IBM ECSEC
Via Giorgione 159
00147 Rome
Italy

ABSTRACT

Through a training procedure based on simulated annealing, Boolean networks can 'learn' to perform specific tasks. As an example, a network implementing a binary adder has been obtained after a training procedure based on a small number of examples of binary addition, thus showing a generalization capability. Depending on problem complexity, network size, and number of examples used in the training, different learning regimes occur. For small networks an exact analysis of the statistical mechanics of the system shows that learning takes place as a phase transition. The 'simplicity' of a problem can be related to its entropy. Simple problems are those that are thermodynamically favored.

The study of the collective behavior of systems of 'formal neurons' which are designed to store a number of patterns ('associative memories') or to perform a task has recently gained increasing interest in physics and engineering applications as well as in biological science [1].

As far as models with biological motivations are concerned, many efforts have clarified, with numerical and analytical methods, the behavior of Hopfield's model [2,3]. Systems with asymmetric 'synapses' which appear to be a more realistic model, have also been proposed [4]. The study of the storage capacity of such systems has taken advantage of methods typical of statistical mechanics, in particular by exploiting the connection between learning systems and spin glasses.

Coming to practical applications in engineering (see [5] and references therein), applications in many areas, including speech synthesis [6], vision [7], and artificial intelligence [8] have been proposed. In these cases less attention has been paid to the general properties of the models, while research has concentrated on the actual capabilities of the systems for specific values of the parameters involved.

In our model [9] we consider networks of N_G boolean gates with two inputs. Each gate implements one of the 16 possible Boolean functions of two variables. Each of its inputs can be connected to another gate in the circuit or to one of the N_I input bits. The last N_O gates produce at their output the N_O desired output bits. To rule out the possibility of feedback we number the gates from 1 to N_G and we do not allow a gate to take input from an higher numbered gate. On the other hand, we ignore fan-out problems allowing each gate to be the input of an arbitrary number of gates. When the gate types and the connections are fixed, the network calculates the N_O output bits as some Boolean function of the N_I input bits.

If we want the network to 'learn' to implement a particular function, we use the following training procedure. We randomly choose N_E examples of values of the input bits, for which corresponding values of the output bits are known. Then, we try to optimize the circuit in order to minimize the average discrepancy, for these N_E examples, between the correct answer and the one calculated by the circuit. This optimization is performed by simulated annealing [10]: the network is considered as a physical system whose microscopical degrees of freedom are the gate types and the connections. With simulated annealing one then slowly cools down the system until it reaches a zero temperature state, which minimizes the energy. In our case the energy of the system is defined as the average error for the N_E samples.

$$E \equiv \sum_{l=1}^{L} E_l \equiv \sum_{l=1}^{L} \frac{1}{N_E} \sum_{k=1}^{N_E} (E_{lk} - A_{lk})^2 .$$

Here E_{lk} is the exact result from the l-th bit in the k-th example, while A_{lk} is the output for the same bit and example as calculated by the circuit. Therefore A_{lk} is a function of the configuration of the network. Thus, E is the average number of wrong bits for the examples used in the training. For a random network, for example one picked at high temperatures in the annealing procedure, $E_l \sim 1/2$.

As an example, we have considered the problem of addition between two binary integers. We have considered 8-bit operands, so that $N_I = 16$, and ignored overflow (as in standard binary addition), so that $N_O = 8$. In principle the performance evaluation of the system is straightforward: given the optimal circuit obtained after the learning procedure, one checks its correctness over the exhaustive set of the operations, in the specific case all possible additions of 2 L-bit integers, of which there are $N_o = 2^L \cdot 2^L$. This can be afforded for the set of experiments which will be described here, for which $L = 8$ and $N_o = 65536$. Thus another figure of merit is introduced:

$$P \equiv \sum_{l=1}^{L} P_l \equiv \sum_{l=1}^{L} \frac{1}{N_o} \sum_{k=1}^{N_o} (E_{lk} - A_{lk})^2 \ .$$

This quantity is defined in the same way as E, but the average is done over all possible operations, rather than just over the examples used in the training. We stress that P is only used *after* the training procedure as a tool for performance evaluation.

Roughly speaking, the quantities E and P are all is needed to understand the behavior of the network: low values of E mean that it has been capable at least to 'memorize' the examples shown to it during the training. If P is small as well, then the system has been able to generalize properly since it is able to calculate the correct result for operations it has never been exposed to. Therefore one expects the existence of these two regimes (discrimination and generalization) between which possibly a state of 'confusion' takes place.

A network of 160 gates has been able to organize itself in a completely correct binary adder after a training procedure with $N_E = 224$ only, out of the 65536 possible binary additions of two 8-bit numbers. This means that the system has been able to recognize the rule that was to be used to generate the output, thus generalizing to construct the correct result of any addition not contained in the 224 used during the training. This means that only a fraction .003 of the total samples is necessary to generalize. It is a priori not clear whether or not training could be improved introducing correlations among examples shown, i. e. implementing a sort of 'didactic' teaching.

More generally, we can draw a qualitative picture of the learning processes as they occur in the different cases. As previously mentioned, these are essentially of two kinds. One is lookup-table like: namely, when the system is poorly trained (low N_E), it simply builds a representation of the examples shown, which has nothing to do with any general rule for the operation. Therefore this regime is characterized by values of E near to 0 and values of P near to that of a 'random' circuit, which gives the correct result for each bit with probability 1/2. Therefore $P \sim 1/2 \cdot L = 4$ in this look-up table regime. Providing the system with more and more examples, it will find it hard to follow this brute-force strategy, unless its capability is infinite (the somewhat trivial case $N_G \sim O(N_o)$). Therefore E will increase from 0 as a function of N_E, and P will practically stay constant. As the number of examples used in the training becomes critically high, the onset of the 'generalization regime' occurs provided that the number of gates is large enough, and P will decrease toward 0. This is the region of parameters in which genuine learning takes place.

The specific features for different regimes are somewhat hidden in the 'global' parameters E and P, due to the fact that memorization and learning for each bit start to occur for different N_G and N_E , and are all weakly coupled among each other. Typically the two least significant bits are always correctly processed, and one can roughly say that, as complexity grows when considering more significant binary digits (because of the potentially high number of carry propagations needed), learning 'harder' bits is in a way equivalent to work with less gates. To get a clearer insight in the whole process it is better to focus the attention on the behavior of central bits (to minimize 'border' effects) plotting the quantities E_t and P_t introduced in previous formulae. Figs. 1a, 1b, and 1c are obtained for N_G fixed respectively at 20, 40, and 160. One can recognize the following distinct behaviors:

a) At low N_G (Fig. 1a) only look-up table behaviour occurs. Storing of examples is perfect until $N_E \sim \overline{N}_E = .4N_G$, which estimates the capacity of the system. It is remarkable that after this value is reached the system does *not* enter a confusion state. In other words this maximum number of 'patterns' is preserved, and simply no more examples are kept. As a consequence, for $N_E > \overline{N}_E$ one has

$$E_I \sim 1/2\,(1 - \frac{\overline{N}_E}{N_E}).$$

In the look-up table region $P_I = 1/2$ for all N_E.

b) For intermediate N_G there is a cross-over to partial generalization. This is clearly shown in Fig. 1b where P_I shows a decrease from $P_I = 1/2$ to a 'residual' value still greater than 0.

c) Finally for large N_G (say $N_G > \overline{N}_G$) the system is able to switch from a perfect storing regime ($E_I = 0$, $P_I = 1/2$) to a complete generalization ($E_I = 0$, $P_I = 0$). For N_G very large we expect this transition to be abrupt, i. e. there is not an intermediate regime where partial generalization takes place. To put it in another way, we conjecture that in this limit there is a critical number of examples N_E^c such that correspondingly the systems switches from perfect storing to complete generalization.

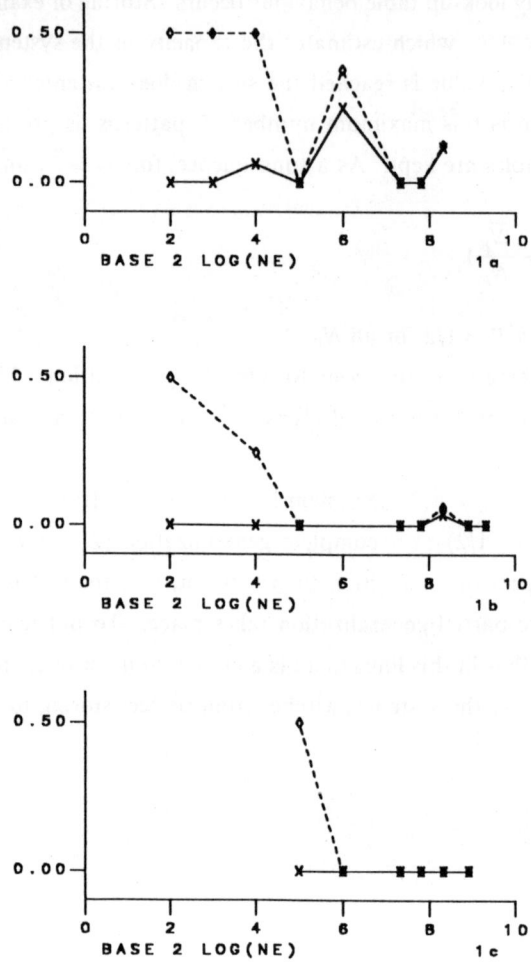

Fig. 1. Behavior of E_l (solid lines) and P_l (dashed lines) as a function of N_E, for various values of N_G (see text).

To summarize this first part, the learning behavior of the system is dependent on its size (N_G), on the complexity of the problem, and on the number of examples used in the training (N_E). For N_G and N_E large enough for the problem to be solvable, generalization and learning take place as described above. If N_G is decreased, the system is no longer able to generalize. For small N_E and for any N_G the system is not able to generalize, but may be able to 'memorize' the N_E examples and construct a circuit that gives the correct answer at least in those N_E cases, or in a significant fraction of them.

Given an explicit example in which the training has led to a network configuration which implements the problem correctly, we want now to address the most puzzling question: how is it that such system is able to perform correctly over *all* possible cases, when given information only on a partial set of examples? In other words, where does generalization come from?

For small enough networks one can study in detail all the properties of the system through a complete enumeration of all possible circuits. As an example, we will refer in the following to a network with $N_G = 4$, $N_I = 4$ and $N_O = 1$. Thus, one can calculate the thermodynamical properties of the system, as well as, for any rule, the average learning probability as a function of N_E and N_G. This analysis entirely confirms the picture sketched above containing the different learning behaviors. In addition, a direct calculation of the specific heat as a function of temperature clearly shows the existence, for most rules, of a peak which, in the limit of large systems, would transform in a singularity characteristic of a phase transition. The intensity of this peak is higher for more 'difficult' rules. Thus, learning clearly appears to be a process of ordering that takes place, when temperature is lowered, in a phase transition. We have been able to recognize a hierarchical structure for the feasible rules, with some degree of ultrametricity.

The analysis based on complete enumeration also clearly indicates that the simplicity of a rule is related to its entropy: simple rules are those that have a large entropy, which means that can be realized in many different ways. As a matter of fact, this kind of approach allowed us to compute *exactly* the learning probability for a given problem, as a function of the number of examples N_E used in the training [11]. This quantity measures the probability that, performing the training with N_E examples, the network will organize in a configuration which implements correctly the problem for *all* possible inputs. In the following we report results on some particular problems.

Let's start by studying the training on a very simple problem, consisting of producing a value of 0 at the output bit regardless of the values of the input bits. In Fig. 2, curve a, we plot the probability of learning as a function of N_E. The curve is for a network with $N_G = 4$. The curve rises quite fast, and reaches 50% already for $N_E = 2$, thus showing that for that N_E the training has 50% probability of resulting in a *perfect* network, i. e., one that produces always 0 at its output, even for the $16 - 2 = 14$ input configurations not used in the training. This already shows clearly the generalization capabilities of the system we are considering. This fast rise of the learning curve is related to the fact that there are very many circuits that always produce zero at their output. In fact 14% of all possible networks with $N_G = 4$ implements the '0 function'.

Now let's consider a more difficult problem, consisting of reproducing at the output bit the value of a specified input bit. The corresponding learning probability is plotted in Fig. 2, curve b, (again the curve is valid for $N_G = 4$). Generalization still occurs, but now we need $N_E = 4$ to get 50% chances of finding a perfect network. At the same time only a fraction $\sim 3.5\%$ of the total number of configurations of the network solve this problem.

We then turn to the even more difficult problem of producing at the output of the network the AND of 3 of the 4 input bits. This problem is solved by a much smaller number of circuits (.047% of the total number). From the plot of the corresponding learning probability (Fig. 2, curve c) one can see that generalization almost does not occur at all, and N_E quite close to 16 (which amounts to give complete information describing the problem to be solved) is needed for the learning probability to be reasonably different from zero ($N_E = 11$ for 50% learning probability). It is clear at this point that the occurrence of generalization and learning of a problem is directly related to the fact that that problem is implemented by many different networks and that this provides also a definition (architectural-dependent) for the complexity of a given problem.

In conclusion, the model we have defined has shown clearly a self-organization capability, when trained on a problem. Moreover, we have been able to provide in this context a clear characterization of generalization processes. We believe that this latter issue could provide some useful hints for other classes of learning machines, as well as for the understanding of learning in biological systems.

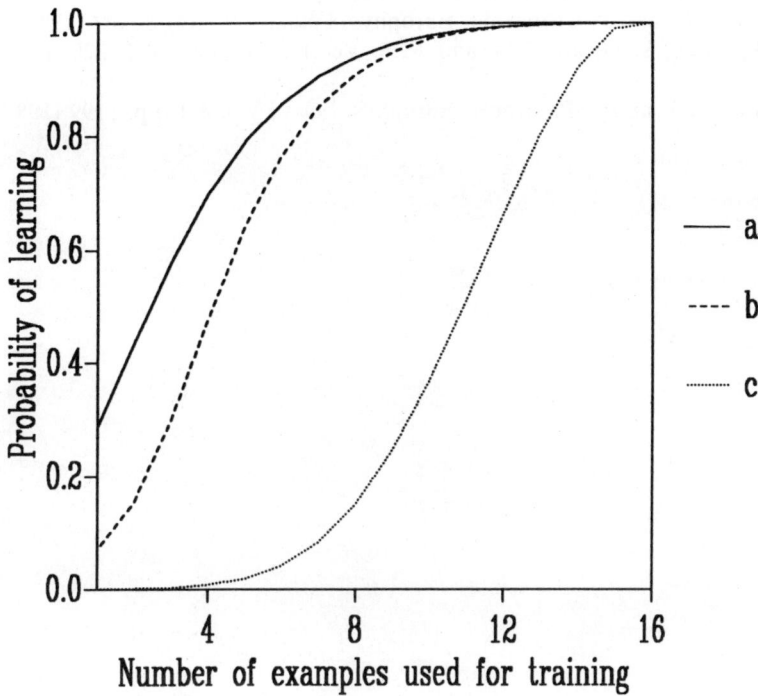

Fig. 2. Learning probability as a function of N_E for three problems.

REFERENCES

1. Hopfield, J.J: Proc. Nat. Acad. Sci. USA, Vol. 79 p. 2254 (1982)
2. Amit, D.J., Gutfreund, H. and H. Sompolinsky: Phys. Rev. A, Vol. 32 p. 1007 (1985)
3. Amit, D.J., Gutfreund, H. and H. Sompolinsky: Phys. Rev. Lett.: Vol. 55 p. 1530 (1985)
4. Parisi, G.: Jour. Phys. A (Math. Gen.), Vol. 19 p. L675 (1986)

5. Personnaz, L., Guyon, I. and G. Dreyfus: in Disordered Systems and Biological Organization, (Eds. E. Bienenstock et al.), Springer & Verlag (1986)
6. Sejnowsky, T.J. and C. Rosenberg: John Hopkins University Tech. Rep., Vol. 86/01 (1986)
7. Hinton, G.E., and T.J. Sejnowsky: Proc. IEEE Comp. Soc. Conference on Computer Vision and Pattern Recognition, p. 488 (1983)
8. Cruz, C.A., Hanson, W.A. and J.Y. Tam: in Neural Networks for Computing, Am. Inst. of Phys. Proc., Vol. 151 (1986)
9. Patarnello, S. and P. Carnevali: Europhys. Letts., Vol. 4(4) p. 503 (1987)
10. Kirkpatrick, S. Gelatt, S.D. and M.P. Vecchi: Science, Vol. 220, p. 671 (1983)
11. Carnevali, P. and S. Patarnello Europhys. Letts., Vol. 4(10) p. 1199 (1987)

OPTIMIZATION PROBLEMS AND STATISTICAL MECHANICS

J. Bernasconi
Brown Boveri Research Center
CH-5405 Baden, Switzerland

ABSTRACT

The problem of finding low autocorrelation binary sequences is used to demonstrate that valuable information about the nature of an optimization problem can be obtained by studying its statistical mechanics. We analyze different simulating annealing strategies and optimize the cooling schedule by adapting it to the thermodynamic behavior of the problem. This enables us to find sequences of length $N \gtrsim 100$ with very high autocorrelation merit factors. We further speculate that a complexity catastrophy is responsible for the inability of stochastic search procedures to discover near-optimal solutions at increasingly larger N-values. Finally, we also discuss some preliminary results concerning the statistical mechanics of the so-called football pool problem.

1. INTRODUCTION

The complications that arise in combinatorial optimization problems are closely related to those encountered in the statistical physics of disordered an frustrated systems. This analogy, in which the cost function of the optimization problem is viewed as the energy function of a physical system, has recently led to the development of a novel type of optimization strategies, generally known as "Simulated Annealing Strategies".[1-7] Corresponding algorithms simulate the evolution of a physical system in configuration space during a slow cooling process.

Simulated annealing methods have successfully been applied to a number of complex optimization problems.[1-10] It is not obvious, however, that a simulated annealing strategy should always perform better than a conventional local improvement strategy. Local improvement procedures invariably terminate in the nearest local minimum, but they converge fast and can thus be applied to a large number of initial configurations. Simulated annealing algorithms, on the other hand, accept transitions which lead to an increase in the cost function (with a probability that decreases with decreasing "temperature"), so that the system can climb out of bad local minima. Simulated annealing, however, is generally a rather slow procedure, and to make effective use of its advantages, the annealing schedule has to be designed very carefully.[1]

Valuable information about the nature of an optimization problem can be obtained by studying its "thermodynamic" properties, and various analytical and numerical techniques have therefore been applied to analyze the statistical mechanics of some prototype combinatorial optimization problems.[2-4,10-16]

In the present paper, we shall explicitly demonstrate that the efficiency of an annealing strategy can be improved considerably by adapting it to the thermodynamic behavior of a given optimization problem. We shall mainly be concerned with the specific problem of finding long binary sequences with small off-peak autocorrelations[10,17-21] which play an important role in several communication engineering applications (synchronization, radar and sonar ranging, etc).[21] The construction of such sequences has turned out to be a very hard mathematical problem, and the question about the maximum possible autocorrelation merit factor for long sequences is still open.[10,17-20] In a recent paper[10], we have analyzed the statistical mechanics of low autocorrelation binary sequences in some detail, and the most important conclusions from this study are summarized in section 3 below. We then analyze different simulated annealing strategies and show how the thermodynamic properties of the problem can guide us in optimizing the cooling schedule. This will allow us to find sequences of length $N \gtrsim 100$ with merit factors that are considerably higher than those found with conventional optimization methods.

Finally, we briefly discuss some preliminary results concerning the statistical mechanics of the so-called football pool problem[22-25] for which new upper bounds have recently been obtained by simulated annealing methods.[9,26]

2. SIMULATED ANNEALING STRATEGIES FOR COMBINATORIAL OPTIMIZATION PROBLEMS

We consider optimization problems which consist in minimizing a given cost function $E(\underline{S})$, where $\underline{S} = \{S_1,\ldots,S_N\}$ denotes a configuration of the system. In the case of combinatorial optimization, the number of possible configurations is finite, but increases very rapidly with N (e.g. as 2^N or as N!). Due to so-called frustration effects, complex optimization problems moreover possess a very large number of local minima which considerably increase the difficulty of finding an optimal or near-optimal solution.

If $E(\underline{S})$ is viewed as an energy function, our optimization problem becomes equivalent to the problem of finding a low-energy configuration of the corresponding physical system. Statistical mechanics now tells us that the probability of finding the system in configuration $\underline{S}$ at temperature T is given by the Boltzmann distribution

$$P_T(\underline{S}) = \frac{1}{Z} e^{-E(\underline{S})/T} \quad . \tag{1}$$

Z is the partition function,

$$Z = \sum_{\underline{S}} e^{-E(\underline{S})/T} \quad , \tag{2}$$

from which we can calculate thermodynamic averages such as the average energy at temperature T,

$$\langle E \rangle = - \frac{d\ln Z}{d(1/T)} \quad , \tag{3}$$

or the specific heat,

$$C = \frac{d\langle E \rangle}{dT} = \frac{1}{T^2} (\langle E^2 \rangle - \langle E \rangle^2) \quad . \quad (4)$$

From Eqs. (1) and (2), we see that $P_T(\underline{S})$ becomes nonnegligible only for configurations with lower and lower energy values as T decreases towards zero. We may therefore hope to find a configuration with very low energy, i.e. a near-optimal solution of our problem, by simulating the temporal evolution of the system in a slow cooling process ("<u>Simulated Annealing</u>").

Metropolis et al [27] have introduced a simple algorithm to carry out such a simulation. In each step of this algorithm, the current configuration $\underline{S}$ of the system is changed randomly to create a trial configuration $\underline{S}'$, and the energy difference

$$\Delta E = E(\underline{S}') - E(\underline{S})$$

is calculated. The new configuration $\underline{S}'$ is then accepted with probability $p(\Delta E)$, where

$$p(\Delta E) = \begin{cases} 1 & \text{if} \quad \Delta E \leq 0 \\ e^{-\Delta E/T} & \text{if} \quad \Delta E > 0 \end{cases} \quad . \quad (5)$$

With this choice of $p(\Delta E)$, the system evolves according to the equilibrium distribution $P_T(\underline{S})$, if the procedure is iterated long enough at a fixed temperature T and if the random configuration changes $\underline{S} \rightarrow \underline{S}'$ satisfy the following two conditions. Firstly, we must be able to reach any configuration by a succession of allowed changes, and secondly, the changes have to fulfill detailed balance.

According to Eq.(5), configuration changes with $\Delta E > 0$ are possible at any nonzero temperature. Simulated annealing strategies, which consist in combining the Metropolis algorithm with a suitably chosen cooling schedule, thus need not get stuck in a local minimum, in contrast to conventional local improvement methods which correspond to an infinitely rapid cooling ("quenching") of the system.

The efficiency of an annealing strategy may depend crucially on the set of allowed configuration changes and, in particular, on the choice of an appropriate cooling schedule. Valuable information to guide us in the development of an efficient algorithm can be obtained by analyzing the statistical mechanics of the optimization problem to be solved [1-4,10-16]. Predictions for the groundstate energy, e.g., can be used to test the performance of an algorithm, and the temperature dependence of the specific heat can give useful hints for the design of an effective cooling schedule. Information about the structure of low-energy configurations, finally, may eventually lead to the development of new or improved optimization strategies.

3. LOW AUTOCORRELATION BINARY SEQUENCES

In this section, we consider the problem of finding binary sequences of length N,

$$\underline{S} = \{S_1, S_2, \ldots, S_N\} \quad , \quad S_i = \pm 1 \quad ,$$

whose off-peak autocorrelations,

$$R_k = \sum_{i=1}^{N-k} S_i S_{i+k} \quad , \quad k = 1, 2, \ldots, N-1 \quad , \tag{6}$$

are small. To be precise, we measure the "goodness" of a sequence in terms of Golay's [17,18] merit factor F,

$$F = N^2 / (2 \sum_{k=1}^{N-1} R_k^2) \quad , \tag{7}$$

so that our optimization problem can be stated as

$$E = \sum_{k=1}^{N-1} R_k^2 = \sum_{k=1}^{N-1} \sum_{i=1}^{N-k} \sum_{j=1}^{N-k} S_i S_{i+k} S_j S_{j+k} = \text{minimum} . \quad (8)$$

The problem of finding such low autocorrelation binary sequences has a long history (cf. references 10, 17-21), but the question about the maximally attainable merit factor for long sequences is still open. The most extensive studies have concentrated on skew-symmetric sequences of odd length $N = 2n-1$. These sequences satisfy

$$S_{n+\ell} = (-1)^\ell S_{n-\ell} \quad , \quad \ell = 1,\ldots,n-1 \quad , \quad (9)$$

from which it follows that all R_k with k odd vanish,

$$R_{2\ell-1} = 0 \quad , \quad \ell = 1,\ldots,n-1 \quad . \quad (10)$$

Skew-symmetry reduces the size of the problem by a factor of 2, and Golay[17] has made an exhaustive search to find all optimal skew-symmetric sequences for lengths up to $N = 59$. By using what he calls an "ergodicity postulate", Golay[17] also derives an estimate for the maximum merit factor of very long sequences (see below), and the conjectured asymptotic value of $F = 12.32$ is not inconsistent with an extrapolation of the known optima below $N = 59$ (see Ref. 10). On the other hand, no F-values substantially larger than $F \approx 6$ have actually been found for long sequences (say $N \gtrsim 100$)[18-20], where an exhaustive search is no longer possible. This has caused Jensen and Høholdt[20] to conjecture that the true asymptotic value of the maximum merit factor is 6, rather than 12.32.

To obtain some insight into this apparent discrepancy, we have recently made an attempt[10] to analyze the statistical mechanics of the problem as well as the structure of the corresponding energy landscape. Based on Golay's "ergodicity hypothesis"[17], which assumes that in Eq.(8) the R_k^2 can be treated as independent random variables, we have first derived a simple approximate expression for the partition function Z. The corresponding result for skew-symmetric sequences, in the limit as $N \to \infty$, becomes

$$-\frac{4}{N} \ln Z = \frac{1+2\beta}{2\beta} \ln(1+2\beta) - (1+2\ln 2) \quad , \tag{11}$$

with

$$\beta \equiv 2N/T \quad , \tag{12}$$

and we note that the only nontrivial change with respect to the expression for unrestricted sequences is the different normalization of the inverse temperature ($\beta = 2N/T$ instead of $\beta = N/T$).

Using Eq.(11), we can derive approximate expressions for thermodynamic averages such as the average energy or the specific heat, and in Ref. 10 we have compared these predictions with results obtained by a simulated annealing procedure. We have found that the ergodicity hypothesis leads to a reasonably accurate description of the thermodynamic behavior at temperatures above about $\beta^{-1} = 0.25$. At lower temperatures, however, it breaks down completely, and we see indications of a phase transition at $\beta_c^{-1} \approx 0.20\text{-}0.25$ (see below).

Using Eq.(11), the trivial inequality

$$Z > e^{-E_{min}/T} \tag{13}$$

leads to

$$F_{max} = \frac{N^2}{2E_{min}} \leq 12.3248 \quad . \tag{14}$$

This estimate is identical with Golay's conjecture [17] for the maximum merit factor of long ($N \to \infty$) binary sequences. The merit factors we have found with our simulated annealing procedure, on the other hand, seemed to approach a value below $F = 5$ as $N \to \infty$.

We do not think, however, that this should be taken as evidence against an asymptotic value of $F_{max} \approx 12$. Our results on the structure of the energy landscape rather indicate that we observe the manifestation of a type of complexity catastrophy in the sense of Kauffman [28]. The local minima appear to be completely uncorrelated [10], and the distribution of their merit factors becomes more and more peaked as N increases. For very large N, a stochastic search procedure will therefore not be able to find one of the extremely rare F-values above that peak whose position seems to approach a value between $F = 4$ and $F = 5$.

The annealing procedure we used in Ref. 10, on the other hand, was rather crude, so that in principle we should at least be able to find much better sequences in an intermediate range of lengths (say $N \gtrsim 100$, which is actually a range of interest for practical applications). We have therefore attempted to optimize the cooling schedule of our annealing strategy by taking into account the thermodynamic behavior of the problem.

The specific heat, depicted in Fig. 1, points to the existence of a phase transition at $T_c/2N \approx 0.25$, and this is confirmed by the behavior of the acceptance probability (the probability that a configuration change in the Metropolis algorithm is accepted) which drops to zero at precisely this temperature. This implies that we just waste our computer time if we cool below this temperature. In addition, we should spend most of our time at temperatures around $T_c/2N \approx 0.3$, where the fluctuations in energy are maximal. Similar conclusions concerning the choice of a good cooling schedule have been drawn by Morgenstern and Würtz [16].

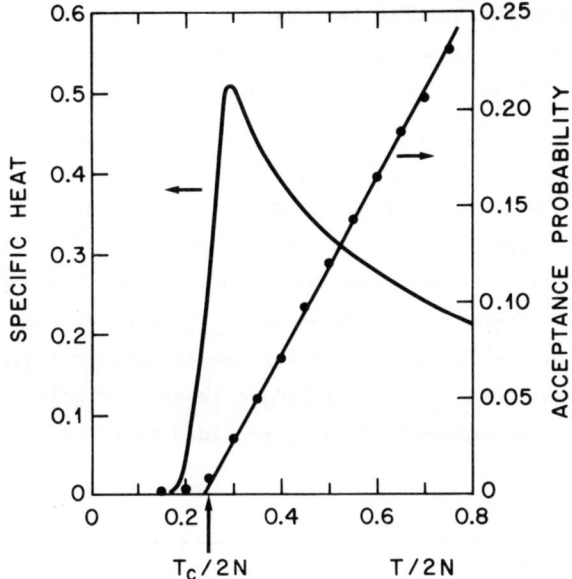

Fig. 1. Specific heat (normalized as $4C/N$) and acceptance probability vs. $T/2N$ for skew-symmetric sequences. The data refer to simulations for sequences of length $N = 103$. If scaled as shown, however, the results are independent of N.

In most applications of simulated annealing, the temperature is decreased exponentially,

$$T_k = \lambda^k T_o \quad , \quad \lambda < 1 \quad . \tag{15}$$

With such a schedule, however, the time spent in a constant temperature interval ΔT only increases as T^{-1}. A much more appropriate choice for our situation seems to be a logarithmic temperature decrease,

$$T_k = \frac{C_o}{\ln(1+k)} \quad , \quad k = 1, 2, \ldots, k_{max} \quad , \tag{16}$$

where C_o fixes the initial temperature. Here the number of temperature steps in an interval ΔT increases much faster [$\sim T^{-2} \exp(C_o/T)$], so that most of the time is spent in a narrow region just above the final temperature $T_{min} = C_o/\ln(1+k_{max})$.

The results of our corresponding simulations for skew-symmetric sequences of length N = 103 are shown in Fig. 2. The data refer to an average over 10 annealing runs, and we used "single spin flip" configuration changes. At each of the k_{max} temperatures, we made one sweep through the first half of the skew-symmetric sequence. We see that the logarithmic cooling schedule leads to significantly better results than an exponential one, provided that we choose C_o such

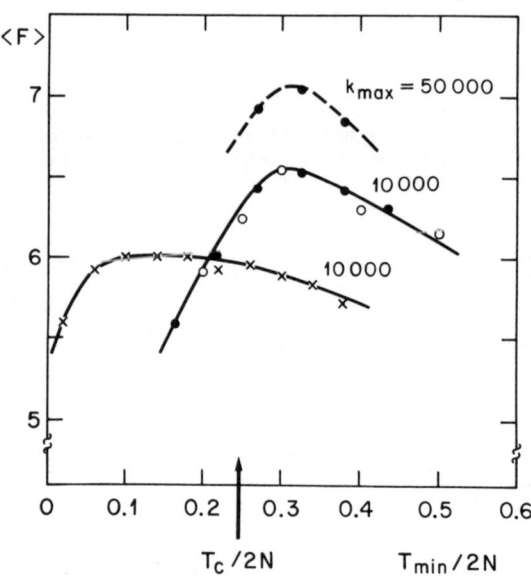

Fig. 2. Average optimal merit factor <F> for different cooling schedules vs. $T_{min}/2N$, where T_{min} is the final temperature in the annealing process. The full circles refer to the logarithmic cooling schedule of Eq. (16), the crosses to an exponential temperature decrease, Eq. (15), and the open circles to a simulation at $T = T_{min}$ = const.

that $T_{min}/2N$ corresponds to the temperature where the specific heat has its pronounced peak. We also observe that the results of a simulation at $T = T_{min}$ = const. are indistinguishable from those obtained with a logarithmic temperature decrease. It thus seems that in our problem the dominating requirement for a good cooling schedule is simply to spend most of the searching time in the temperature region of the specific heat peak.

Fig. 3 shows that an optimally adapted annealing strategy is considerably more efficient than a conventional local improvement method. If we look for F-values larger than 7, e.g., the gain is more than an order of magnitude in computing time. With our optimized cooling

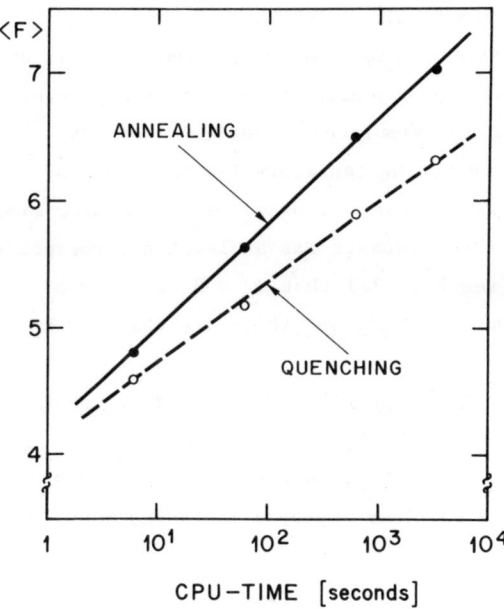

Fig. 3. Average optimal merit factor <F> vs. CPU-time (on a VAX 8600). Comparison of the simulated annealing method (logarithmic cooling schedule) with a local improvement strategy ("quenching").

schedule, we can now easily find sequences of length $N \gtrsim 100$ that have merit factors between 7 and 8, [29] and the best value we have found so far is F = 9.56 for a skew-symmetric sequence of length N = 103. Only three sequences are known that have a still higher merit factor, and they are all rather short [17,19]: the Barker sequences of lengths 11 and 13 (with F = 12.10 and 14.08, respectively), and one of length 27 with F = 9.85.

While our results give additional support to Golay's conjecture, they do of course not imply that the complexity catastrophy can be overcome by an improved annealing strategy. The onset of the catastrophy is merely pushed to somewhat larger N-values. This, however, may be important for some engineering applications.

4. SOME REMARKS ON THE FOOTBALL POOL PROBLEM

The so-called football pool problem [22-25] is another hard combinatorial optimization problem for which new results have recently been obtained by simulated annealing methods [9,26]. The problem consists in determining the minimal number, $N = \sigma_n$, of forecasts for the outcome (win, loose, or draw) of n football games such that at least one of the forecasts has at least n-1 correct results. It has been proved (see Ref. 22) that $\sigma_2 = 3$, $\sigma_3 = 5$, $\sigma_4 = 9$, $\sigma_5 = 27$, and $\sigma_n = 3^{n-k}$ if $n = \frac{1}{2}(3^k-1)$, and these are the only known exact results.

For n = 6, the best upper bound obtained by combinatorial arguments was $\sigma_6 \leq 79$ (Ref. 23). Using a simulated annealing algorithm, however, Wille [9] recently found a solution with N = 74 forecasts which implies $\sigma_6 \leq 74$ and thus improves the upper bound considerably.

To analyze the statistical mechanics of the problem, we define the energy E associated with a fixed set of N forecasts as the number of possible outcomes which differ from all the forecasts in more than one result. It follows that $E_{min} = 0$ for $N \geq \sigma_n$, and that $E_{min} > 0$

for $N < \sigma_n$. At zero temperature, the system thus exhibits a phase transition at $N = \sigma_n$, and we have wondered whether we could detect indications of this phase transition already at temperatures where it is still feasible to study the thermodynamic behavior by numerical simulations.

If $N > \sigma_n$, the system is observed to undergo a freezing transition at a temperature around $T = 0.3$ to 0.4 (see also Ref. 9). We therefore analyzed the N-dependence of $<E>$ and of $\Delta E = (<E^2>-<E>^2)^{1/2}$ just above this freezing temperature, and our preliminary results can be summarized as follows.

For $n = 5$, the $<E>$ vs. N curve exhibits a pronounced change in slope at $N = 27$, i.e. at the known exact value for σ_5. For $n = 6$, the change in the $<E>$ vs. N behavior appears to be less abrupt, but the corresponding fluctuations ΔE are strongly enhanced between about $N = 73$ and $N = 80$. We might thus be tempted to conjecture that $\sigma_6 = 73$, but with our limited data we can realistically do no better than confine σ_6 to an interval of $69 \lesssim \sigma_6 \leq 73$.

Using a simulated annealing strategy which spends most of its time close to the freezing temperature of $T_F \approx 0.3$, we have indeed found several solutions with $N = 73$, but no solution with $N < 73$. Our best result for $N = 72$, however, corresponds to an E-value of only $E = 2$.

Solutions with $N = 73$ have recently also been found independently by Wille, Aarts, and van Laarhoven [26] who, in addition, have used their simulated annealing method to improve the existing upper bounds for σ_7 and σ_8.

5. CONCLUSION

We have considered two hard combinatorial optimization problems, the problem of finding long low autocorrelation binary sequences, and the football pool problem. In both cases, a study of the corresponding statistical mechanics gives useful insights into the difficulties of finding good solutions, and thus provides us with valuable information concerning the design of effective optimization strategies. In particular, we have demonstrated that the efficiency of a simulated annealing procedure can be improved considerably by adapting the cooling schedule carefully to the thermodynamic behavior of the problem.

REFERENCES

1. For an overview see P.J.M. van Laarhoven and E.H.L. Aarts, "Simulated Annealing: Theory and Applications" (Reidel, Dordrecht), 1987.

2. S. Kirkpatrick, C.D. Gelatt Jr., and M.P. Vecchi, Science 220, 671 (1983).

3. S. Kirkpatrick, J. Stat. Phys. 34, 975 (1984).

4. E. Bonomi and J.-L. Lutton, SIAM Reviw 26, 551 (1984).

5. R.E. Burkhard and F. Rendl, European J. Operational Res. 17, 169 (1984).

6. P. Siarry and G. Dreyfus, J. Physique Lett. 45, L-39 (1984).

7. V. Černy, J. Optimization Theor. Appl. 45, 41 (1985).

8. A.A. El Gamal, L.A. Hemachandra, I. Shperling, and V.K. Wei, IEEE Trans. Inform. Theory IT-33, 116 (1987).

9. L.T. Wille, J. Combin. Theory, Ser. A 45, 171 (1987).

10. J. Bernasconi, J. Physique 48, 559 (1987).

11. S. Kirkpatrick and G. Toulouse, J. Physique 46, 1277 (1985).

12. M. Mézard and G. Parisi, J. Physique Lett. 46, L-771 (1985); J. Physique 47, 1285 (1986).

13. Y. Fu and P.W. Anderson, J. Phys. A 19, 1605 (1986); G. Baskaran, Y. Fu, and P.W. Anderson, J. Stat. Phys. 45, 1 (1986).

14. M.A. Moore, Phys. Rev. Lett. 58, 1703 (1987); R. Ettelaie and M.A. Moore, J. Physique Lett. 46, L-893 (1985); J. Physique 48, 1255 (1987).

15. S. Rees and R.C. Ball, J. Phys. A 20, 1239 (1987).

16. I. Morgenstern and D. Würtz, Z. Phys. B 67, 397 (1987).

17. M.J.E. Golay, IEEE Trans. Inform. Theory IT-28, 543 (1982); IT-23, 43 (1977).

18. M.J.E. Golay, IEEE Trans. Inform. Theory IT-29, 934 (1983).

19. G.F.M. Beenker, T.A.C.M. Claasen, and P.W.C. Hermens, Philips J. Res. 40, 289 (1985).

20. H.E. Jensen and T. Høholdt (Preprint, 1987).

21. R.H. Petit, Microwave J. 10, 63 (1967).

22. H.J.L. Kamps and J.H. van Lint, J. Combin. Theory 3, 315 (1967).

23. E.W. Weber, J. Combin. Theory, Ser. A 35, 106 (1983).

24. H. Fernandez and E. Rechtschaffen, J. Combin. Theory, Ser. A 35, 109 (1983).

25. A. Blokhuis and C.W.H. Lam, J. Combin. Theory, Ser. A 36, 240 (1984).

26. L.T. Wille, E.H.L. Aarts, and P.J.M. van Laarhoven (Preprint, 1987).

27. N. Metropolis, A. Rosenbluth, M. Rosenbluth, A. Teller, and E. Teller, J. Chem. Phys. 21, 1087 (1953).

28. S. Kauffman, this volume.

29. Using simulated evolution, Q. Wang [Biol. Cybern. 57, 95 (1987)] has recently also found two sequences of lengths $N = 103$ and 105, respectively, that have merit factors larger than 7.

Cellular Automata

CELLULAR AUTOMATA AND COMPLEX SYSTEMS

Gérard Y. Vichniac

Plasma Fusion Center,
Massachusetts Institute of Technology, Cambridge, MA 02139, USA

and

BBN Communications Corporation, Cambridge, MA 02140, USA

1 Introduction

1.1 Definitions

A cellular automaton is a fully discrete dynamical system: the dynamical variables take their values in a finite set, they are arranged on a lattice, and they evolve in discrete time[6, 9, 19, 5, 49, 41]. The overall dynamics of an automaton with values 0 or 1 on N sites (or "cells") is given by the iterations of a *global* mapping F:

$$F : \{0,1\}^N \mapsto \{0,1\}^N. \tag{1}$$

Cellular automata are furthermore defined to be local (or short-ranged) and homogeneous. This means that the global F can be described in a compact way in terms of a *local* transition rule f

$$f : \{0,1\}^n \mapsto \{0,1\} \tag{2}$$

which maps the occupancies at time t of a *neighborhood* (of size $n \ll N$) around any cell i to the next value x_i^{t+1} of that cell. In the simple case of the one-dimensional (1D) peripheral automata, $n = 3$ and the new value x_i^{t+1} at a site i is determined from its current value and that of the adjacent neighbors:

$$x_i^{t+1} = f(x_{i-1}^t, x_i^t, x_{i+1}^t). \tag{3}$$

All sites update their state in parallel: the arguments of f are unupdated values, taken at the same time t. The system is uniform in time and homogeneous in space: the same rule f applies at all sites i and at all time steps t. For example, the state at step $t+2$ of site i's right neighbor is

$$x_{i+1}^{t+2} = f(x_i^{t+1}, x_{i+1}^{t+1}, x_{i+2}^{t+1}), \tag{4}$$

or, in terms of the original values at time t:

$$x_{i+1}^{t+2} = f(f(x_{i-1}^t, x_i^t, x_{i+1}^t), f(x_i^t, x_{i+1}^t, x_{i+2}^t), f(x_{i+1}^t, x_{i+2}^t, x_{i+3}^t)). \tag{5}$$

1.2 Complexity and Computational Irreducibility

While a single iteration of F is rarely instructive, a large number of them often leads to genuine surprises. Cellular automata are particularly adapted to study the *emergence* of complexity that can occur for many choices of f. Indeed, when complexity exists in these systems, it arises out of the uniform iteration of simple interactions between the simplest possible elements (bits). Moreover, this complexity growth can be followed step by step and in full detail. The inner structure of cellular automata is fully known since these mathematical objects are our own creation.

An evolution can be characterized as truly complex when believed to be *computationally irreducible*, i.e., when there is *no shortcut* to the knowledge of a state x_j^{t+m} of an arbitrary site j in the future. The computation of x_{i+1}^{t+2} with (5) involves four evaluations the rule f. In general, for peripheral cellular automata, it costs m^2 evaluations of f to compute the value of an arbitrary site at time $t + m$ from the knowledge of the array at time t. Irreducibility means that this cost reflects the minimal computational effort to predicting an arbitrary value x_j^{t+m}.

It is further often claimed that for complex rules, the *only* way to arrive at an arbitrary value in the future is the repeated applications of the definition, as in (5); no closed mathematical expression can yield x_j^{t+m} if it does not invoke m^2 explicit evaluations of the rule. We shall see in the next section that this claim is erroneous.

2 Taming Complexity with Boolean Derivatives

Let us define the *Boolean derivative* F' of F as the $N \times N$ Jacobian matrix with elements

$$F'_{ij} = \frac{\partial x_i^{t+1}}{\partial x_j^t}. \tag{6}$$

Here x_i^{t+1} is expressed with a local rule f as in (3). The partial derivative in this expression is in turn defined as

$$\frac{\partial f}{\partial x_j} = f(x_1, \ldots, \overline{x}_j, \ldots x_n) \oplus f(x_1, \ldots, x_j, \ldots x_n), \tag{7}$$

where $\oplus$ is the Exclusive OR (XOR) Boolean operation, and $\bar{x}_j = x_j \oplus 1$ is the binary complement of x_j. (For more on the derivatives of Boolean functions, see the monographs by Thayse[38] and Robert[32]).

In words, the two last equations define a matrix element F'_{ij} as 1 if varying (or "flipping") site j at time t affects site i at time $t+1$, and as 0 otherwise. With proper index labeling, the Jacobian matrix is n-diagonal for neighborhoods of size n, an expression of the local nature of cellular automata.

2.1 Propagation of Information

Let us now be concerned with the propagation of information over several time steps. Figure 1 shows cell i at time t and below it, neighboring cells under its influence with a rule of type (3) during two successive steps. As information travels at the maximal rate of one cell per step (the "speed of light"), these cells are said to lie within the "future light-cone" of (i, t). We now ask under what conditions site $i + 2$ will be affected by a complementation at site i. In 1D peripheral automata, site $i + 2$ will "know" that site i has flipped if and only if ($\iff$) site $i + 1$ is affected in the first place and ($\wedge$) can influence site $i + 2$. In terms of Boolean derivatives:

$$\frac{\partial x_{i+2}^{t+2}}{\partial x_i^t} = 1 \iff (\frac{\partial x_{i+2}^{t+2}}{\partial x_{i+1}^{t+1}} = 1 \wedge \frac{\partial x_{i+1}^{t+1}}{\partial x_i^t} = 1). \tag{8}$$

Expressing $\wedge$ with the Boolean multiplication ($\cdot$):

$$\frac{\partial x_{i+2}^{t+2}}{\partial x_i^t} = \frac{\partial x_{i+2}^{t+2}}{\partial x_{i+1}^{t+1}} \cdot \frac{\partial x_{i+1}^{t+1}}{\partial x_i^t}. \tag{9}$$

The last equation is nothing but the chain rule for the derivative of a composite function. In the case of Figure 1, this rule extends readily for Boolean derivatives.

The chain rule is not valid, though, when information propagates along multiple paths, as for $\partial x_{i+1}^{t+2}/\partial x_i^t$ (Figure 2). The influence of x_i^t upon x_{i+1}^{t+2} can be mediated at time step $t + 1$ through sites i, $i + 1$, or ($\vee$) both:

$$\frac{\partial x_{i+1}^{t+2}}{\partial x_i^t} = \frac{\partial x_{i+1}^{t+2}}{\partial x_{i+1}^{t+1}} \cdot \frac{\partial x_i^{t+1}}{\partial x_i^t} \cdot \overline{\frac{\partial x_{i+1}^{t+1}}{\partial x_i^t}} \vee \frac{\partial x_{i+1}^{t+2}}{\partial x_{i+1}^{t+1}} \cdot \frac{\partial x_{i+1}^{t+1}}{\partial x_i^t} \cdot \overline{\frac{\partial x_i^{t+1}}{\partial x_i^t}} \vee D_{i,i+1} \cdot \frac{\partial x_i^{t+1}}{\partial x_i^t} \cdot \frac{\partial x_{i+1}^{t+1}}{\partial x_i^t}, \tag{10}$$

where $D_{i,i+1}$ is 0 or 1 whether f varies or not under simultaneous flips of x_i^{t+1} and x_{i+1}^{t+1}:

$$D_{i,i+1} = f(\ldots \bar{x}_i \ldots \bar{x}_{i+1} \ldots) \oplus f(\ldots x_i \ldots x_{i+1} \ldots). \tag{11}$$

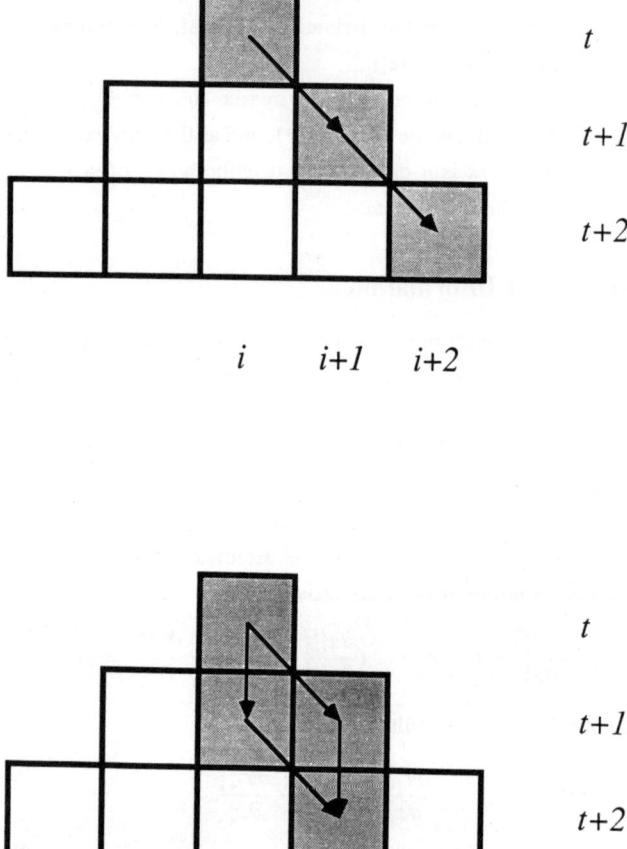

Figure 1: Future light-cone of cell (i, t), future points downward. Top: the influence of x_i^t upon x_{i+2}^{t+2} is mediated at time $t+1$ by cell $i+1$. Bottom: the influence of x_i^t upon x_{i+1}^{t+2} is mediated at $t+1$ by cells i, $i+1$, or both. Shaded cells lie within both the future light cone of x_i^t and the past light-cone of x_{i+2}^{t+2} (top), x_{i+1}^{t+2} (bottom).

$D_{i,i+1}$ can be expressed in terms of second derivatives $\partial^2 f/\partial x_i \partial x_j \equiv \partial/\partial x_i(\partial f/\partial x_j)$ as

$$D_{i,i+1} = \frac{\partial x_{i+1}^{t+2}}{\partial x_i^{t+1}} \oplus \frac{\partial x_{i+1}^{t+2}}{\partial x_{i+1}^{t+1}} \oplus \frac{\partial^2 x_{i+1}^{t+2}}{\partial x_{i+1}^{t+1} \partial x_i^{t+1}}. \tag{12}$$

Notice that the monomials in (10) represent mutually exclusive predicates, and thus the disjunctions have the same value whether computed as exclusive or not. The $\vee$'s can thus be replaced by $\oplus$'s, yielding the simplification:

$$\frac{\partial x_{i+1}^{t+2}}{\partial x_i^t} = \frac{\partial x_{i+1}^{t+2}}{\partial x_i^{t+1}} \cdot \frac{\partial x_i^{t+1}}{\partial x_i^t} \oplus \frac{\partial x_{i+1}^{t+2}}{\partial x_{i+1}^{t+1}} \cdot \frac{\partial x_{i+1}^{t+1}}{\partial x_i^t} \oplus \frac{\partial^2 x_{i+1}^{t+2}}{\partial x_{i+1}^{t+1} \partial x_i^{t+1}} \cdot \frac{\partial x_i^{t+1}}{\partial x_i^t} \cdot \frac{\partial x_{i+1}^{t+1}}{\partial x_i^t}. \tag{13}$$

This formula expresses the dependence as an exclusive sum over over all possible paths of inheritence, with possible destructive interferences, because of the $\oplus$'s (which cannot, as in (10), be replaced by $\vee$'s).

In a similar way, $\partial x_i^{t+2}/\partial x_i^t$ is expressed in terms of third derivatives as well second and first derivatives.

2.2 Complexity out of the Simplest Seed

Complex initial conditions are quite valuable for studying *self-organization* with cellular automata—and, for that matter, with other fully discrete dynamical systems, such as Kauffman nets[23]. However, complex seeds are not the most useful for investigating *complexity growth*. To take the most extreme case, *random* configuration encapsulate an infinite amount of computation—cf. Kolmogorov and Chatin's notion of algorithmic complexity[8]—and thus do not leave any room for complexity growth. The study of that growth is most instructive when complexity is not there to start with, i.e., when the initial condition itself is simple.

It turns out that a wealth of rules are capable of very rich and complex evolutions even out of the simplest initial configuration, made of a single 1 out of a background of 0's. Wolfram has shown for example that the 1D peripheral "rule 30" generates out of this seed pseudo-random sequences of very high quality[48].

This 1-seed represents a unit source in the "quiescent state," and its evolution is thus a kind of response function or discrete Green function. As a consequence, an arbitrary value x_j^{t+m} is simply given by the derivative $\partial x_j^{t+m}/\partial x_i^t$. For example,

$$x_{i+1}^{t+2} = \frac{\partial x_{i+1}^{t+2}}{\partial x_i^{t+1}} \cdot \frac{\partial x_i^{t+1}}{\partial x_i^t} \oplus \frac{\partial x_{i+1}^{t+2}}{\partial x_{i+1}^{t+1}} \cdot \frac{\partial x_{i+1}^{t+1}}{\partial x_i^t} \oplus \frac{\partial^2 x_{i+1}^{t+2}}{\partial x_{i+1}^{t+1} \partial x_i^{t+1}} \cdot \frac{\partial x_i^{t+1}}{\partial x_i^t} \cdot \frac{\partial x_{i+1}^{t+1}}{\partial x_i^t}. \tag{14}$$

It is important to notice that in this formula all the derivatives are to be evaluated at the unperturbed (i.e., the quiescent) configuration, once and for all. Their presence do *not* involve any table look-up: the derivatives are simply constants. The generic value x_j^{t+m} is given by a generalization of (14), involving a sum over all "causal" paths connecting (i, m) and $(j, t + m)$. The expansion in terms of Boolean derivatives is (unlike Taylor series) always finite: derivatives of order higher than the neighborhood size n vanish identically[45].

Does the existence, for a generic value x_j^{t+m}, of a *finite* expansion involving $\oplus$'s and $\wedge$'s of *constants* defeat a chief motivation for studying cellular automata, namely, their ability to generate complexity out of simple sources? Not necessarily. For truly complex rules, the computation of x_j^{t+m} via a generalization of (14) should incur a computational effort of the order of that required for performing m iterations of f, as in (5). (A quantitative comparison is not readily obtained, since each evaluation of f is achieved in practice by a table look-up while formulæ of type (14) involve Boolean operations.) At any rate, expression (14) is arguably more pleasant and lucid than the blind iteration (5). It makes explicit the inheritence from the source at (i, t), and for some f's it can simplify and reveal previously unsuspected structure. Since irreducibility of computational cost is a necessary condition of true complexity, the Boolean derivative offers a test for complexity.

3 Universality vs. Simulating Complex Systems

3.1 Universality

Cellular automata owe their origin to their ability of representing some aspects of the very most complex systems: life itself. Von Neumann was led to inventing cellular automata in an program to "extract the logical structure" of life, where life is defined as the ability of survival and self-replication[46]. "Life" is also the name of the best known cellular automaton[13], a two-dimensional system capable of generating remarkably organized patterns out of a disordered initial condition (the "primeval soup"), and thus circumventing the Second Law of thermodynamics (see, e.g., ref.[43]). Commenting on "Life," Roger Penrose wrote: *It is hard to resist the tempting argument that the game offers a model for our universe—itself presumably governed by rules of the utmost simplicity, yet exhibiting the richness and complexity we observe around us, especially in living things*[29].

The belief that cellular automata can represent phenenoma of arbitrary complexity can be justified by a hard mathematical fact. "Life," among other rules, has been shown to be capable of universal computation: given the proper initial configuration, an evolution can simulate the operations of any general-purpose computer. This is a powerful result. However, it achieves the simulation of a *sequential* universal computer. In view of the recent progress in parallel architecture, invoking an array of finite-state machines working in parallel for the sake of simulating sequential computation is definitely a backward step. Unless it solves the formidable problem of efficient programming massively parallel computers, the quest for universal computation with cellular automata does not answer a question asked as early as 1961: *What kinds of operations can iterative networks perform, how can they be analyzed, and how can an iterative array be designed to do a specific job?*[20].

3.2 Non-numerical Simulation of Complex Systems

This question has recently found an unexpected answer. The answer actually sidesteps the problem of parallel universal computation, and yet it aims at exploiting of parallel computing potential of present and future hardware. The idea is that cellular automata are naturally adapted to the modeling of distributed natural phenomena, in particular those occurring in homogeneous physical systems with short-ranged microscopic interactions[42, 39, 19]. Actually, these binary objects have been found capable of representing in a non-numerical way paradigms of continuum mathematics (e.g., differential equations and wave motion). To be sure, these simulations are a far cry from universal computation, but they turn out to be precisely the task on which current supercomputers spend most of their cycles: the numerical solution of partial differential equations.

It must be emphasized that the cellular automaton method differs sharply from standard use of computer for treating differential equations. Being fully discrete, cellular automata let digital computers do only what they do best, i.e., iterative Boolean decisions rather than necessarily imprecise floating-point arithmetic. Cellular automata models lend themselves to exact simulation by "finitistic" means (as von Neumann called them[14]); they allow the economy of numerical analysis, an inevitable and often opaque screen between continuous models and their standard discrete simulations. Moreover, when freed from the requirement of universal computation capability, cellular automata modeling lends itself gracefully to parallel hardware implementation[28, 41].

The most spectacular successess of the cellular automaton approach have been the result of the intense activity in the direction pioneered in 1976 by Hardy, de Pazzis, and Pomeau[18]: the fully discrete modeling of fluid flows[12, 28, 7]. As Chopard, Droz, and Lallemand have given at this meeting detailed reports on these developements (see their contributions in this volume), hydrodynamics will not be discussed here.

4 Cellular Automata as Bridges between Physics and Computation

Cellular automata constitute a rich cross-fertilizing bridge between physics and computation, they reveal new connections between Boolean algebra and continuous physical and mathematical systems. Boolean derivatives are but an instance of such connection. Their definition (6) is justified[45] by the following "good" properties of F' that one expects from a derivative:

- F' vanishes if the rule f is constant;
- $F' = F$ if F is linear, i.e., f is the Exclusive OR of some neighbors. Hence F' extracts the linear part of a general nonlinear global mapping F;
- The Jacobian $\det|F'|$ does not vanish if the rule is invertible (inverse function theorem on cellular automata).

It should be emphasized that the interplay between the discretum and the continuum that cellular automata provide is quite different than that of numerical analysis. First, as indicated above, cellular automata do not attempt a frontal solution to differential equations but rather a simulation of the phenomena the differential equation aims at describing in the first place. Secondly, in many investigations with cellular automata, one does not take limits "when the time step and the mesh size go to zero." Discreteness is unashamedly taken as a constitutive part of the approach. We show in the rest of this section some examples of using cellular automata as bridges between physics and computation.

4.1 Reversible Computation

Physics imposes irreducible limits on the ultimate potential of computing mechanisms. Rapid progress in digital hardware makes these limits increasingly felt. The finiteness of the speed of light is one of such limits. As indicated above, the notion of propagation of information at finite speed is quite explicit in cellular automata. These dynamical systems offer a natural framework that encompasses both parallel computing and and the physical constraint of causality.

A second, and more subtle limit[24] is the necessary dissipation of high-grade energy into heat during standard computing, i.e., in the course of a computation using at the lowest level standard Boolean gates such as $\oplus, \wedge,$ and $\vee$. These gates are non-invertible (they have two inputs but a single output) and thus any computation based on them must submit to the Second Law of thermodynamics; logical non-invertibility eventually translates into physical irreversibility. But universal computation can be achieved without non-invertible elements[3, 11] and therefore be virtually nondissipative. What is more, reversible compuation can be formally cast as a quantum Hamiltonian acting on the input data and returning, when iterated, the result of the computation[2, 10]. Here again, cellular automata offer reversible models of computation that incorporates physical constraints[26, 27]. Admittedly, these reversible automata suggest neither the schematics of an actual nondissipative computer nor a solution to the formidable problem, mentioned above, of efficient parallel programming. But they are quite interesting in providing an intriguing bridge between physics and computation, thanks to their *paradoxical property of behaving like a computer for special initial conditions, but like a traditional statistical-mechanical model (e.g., a gas or Ising model) for generic initial conditions*[4].

4.2 Interface Motion

Very simple cellular automata rules give rise to dynamics analogous that of interfaces between two phases. A class of such rules (called in [43] "*twisted* majority rules") simulates with great accuracy the Allen-Cahn equation[1]:

$$v_r(P) = -\alpha \, \kappa(P), \tag{15}$$

where $v_r(P)$ and $\kappa(P)$ are the normal velocity and the curvature at point P of the interface, and α is the coefficient of surface tension. These rules can be easily described. A site assumes at time $t+1$ the value (or "phase") which is most prevalent in the neighborhood, but with the following twist: the majority is followed only when strong, a marginal majority will not be followed, the marginal minority will be assumed instead. This twist is instrumental in providing a kind of "frustration" that prevents a settling of the interface (which occurs with simple majority rules[42]), forcing it to evolve endlessly into smoother and smoother shapes. Figure 2, prepared using the CAM fast simulator[40, 41], shows how the twisted rule M46789[43] achieves the proportionality between $v_r(P)$ and $\kappa(P)$, filling fjords and eroding peninsulæ. Despite their short-ranged action, the twisted rules are thus capable of accurate measurements of radii of curvature of any magnitude. A shortsighted rule f cannot, of course, know the radius of curvature, an extended quantity. It is in fact the iterates f^m that do arrive at this quantity

272

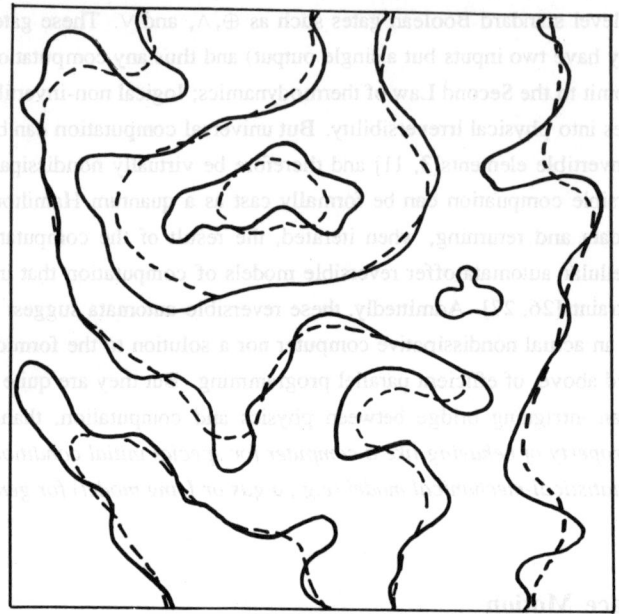

Figure 2: Self-ordering out of a random initial condition under the twisted majority rule M46789. The interface between the two phases "0" and "1" is shown after 300 steps (solid line) and 600 steps (broken line).

thanks to "Life"-like vehicles that continually travel along the interface, carrying the information necessary for computing the radius of curvature. Other pictures of M46789 are given in [41].

Other types of rules have recently be shown to be useful in describing interface motion and surface tension[7, 34, 33]

4.3 Q2R: Spin-Glass Behavior, Limits to Full Parallelism and the Four-Color Theorem

Q2R is a 2D invertible rule remarkable on several grounds. It was first studied for its ergodicity-breaking dynamics and spin-glass-like capability of storing a macroscopic amount of patterns[42]. Q2R's reversible dynamics conserves the Ising energy of the lattice[30], and provides the fastest algorithm to date for the simulation of the Ising model[21]. As Stauffer reported here, recent careful computer studies have shown that the Q2R admits exponentially (in N) many trajectories on the energy "surface"[22] and long spin-glass-like transients in damage spreading[37]. These works confirm that the Q2R dynamics is indeed not *uniquely* ergodic, i.e., the energy surface is not spanned by a single trajectory (see, e.g., Halmos[17]). Q2R has furthermore been shown to exhibit an exceptional amount of ultrametricity[36], another property it shares with spin-glasses. The lack of unique ergodicity, however, does not necessarily forbid averages of thermodynamically relevant Ising quantities from reaching their correct Gibbs-Boltzmann values in the large N limit[31]. Q2R is readily extended to 3D lattices[25] and to the Potts model[31].

Cellular automata simulation of spin models leads to an unexpected, albeit simple, application of the four-color theorem[31]. No cellular automaton rule, probabilistic or not, can yield correct Ising averages in a single square or cubic lattice unless the lattice is first split into two sublattices such that the spins in each sublattice are alternatively updated (possibly in parallel) and kept fixed[42]. Likewise, in the 2D triangular lattice, where each site has six nearest neighbors, a splitting into a least three sublattice is required if no adjacent sites are to be updated at the same time. Now, in the random 2D lattice with nearest-neighbor interaction, the highest allowed degree of parallelism is implemented with exactly four passes per global sweeping. This results directly from the four-color theorem[35] as applied on the map where the "countries" are the Wigner-Seitz (or Voronoi) cells of the random lattice.

4.4 Directed Percolation and the Halting Problem

In a probabilistic 1D peripheral cellular automaton where f is taken to be the XOR (resp. the OR) of x_{i-1} and x_{i+1} with probability p^{XOR} (resp. $p^{OR} = 1 - p^{XOR}$), localized initial seeds grow without bounds (at high p^{XOR}) or shrink and eventually vanish (at low p^{XOR})[44]. In this directed percolation system (a simplification of a model by Domany and Kinzel), a localized seed remains localized indefinitely at a critical concentration $p^{XOR} = p_c$ of XOR rules, very much as in Wolfram's[47] "class 4" deterministic rules.

If the phenomenon is a second order transition, it exhibits critical slowing down producing transients with unbounded lifetimes. The fate of an initial line of data evolving under the critical concentration is essentialy unpredictable. This is similar to the classic halting problem of the theory of computation, according to which no algorithm can decide whether a given pair of program and initial data will halt or not.

5 Open Problems

5.1 The Inverse Problem

We saw that some cellular automata constitute fully discrete (and hence exactly simulatable) models that can harness the power of distributed processing elements for the prediction of distributed natural phenomena, in a way that that bypasses both numerical analysis and the requirement of computational universality. The difficult question is now solving the *inverse problem:* given a differential equation, find a systematic way of identifying the cellular automata rules that simulate, in the macroscopic limit, solutions of that equation.

5.2 A Discrete Noether's Theorem

A way to achieve some progress in the previous problem is to search for more connections, such as lattice-gas hydrodynamics and Boolean derivatives, between digital logic and continuous physical and mathematical systems. A vexing gap between fully discrete and continuous mathematics is the apparent lack of a *discrete Noether's theorem,* i.e., a systematic way to derive conserved quantities from the knowledge of symmetries in discrete systems. Invariants of cellular automata dynamics must be obtained, it seems, one by one and with great effort[30, 15]. The sweeping generality and beauty of Noether's theorem appear to apply to continuous systems only. Perhaps this suggests that space and time in nature are continuous after all.

This work was supported in part by the Dean of the School of Engineering, Massachusetts Institute of Technology.

References

[1] S. M. Allen and J. W. Cahn, *Acta Metall.* 27 (1979) 1085.

[2] P. Benioff, *J. Stat. Phys* 22 (1980) 563; *J. Stat. Phys* 29 (1982) 515; and in [16].

[3] C. H. Bennett, *IBM J. Res. Develop.* 6 (1973) 525.

[4] C. H. Bennett, *Found. Phys* 16 (1986) 585.

[5] E. Bienenstock, F. Fogelman, and G. Weisbuch, eds. *Disordered Systems and Biological Organization*, Les Houches Winter School 1985, Proceedings (Springer-Verlag, 1986).

[6] A. W. Burks, *Essays on Cellular Automata* (University of Illinois Press, 1970).

[7] *Complex Systems* 1 (1987) No. 4 pp. 545–851, (Proc. Workshop on Large Nonlinear Systems, Santa Fe, Oct. 1986, G. Doolen, ed.)

[8] G. Chaitin, *Information, Randomness and Incompleteness: Papers on Algorithmic Information Theory* (World Scientific, 1987).

[9] D. Farmer, T. Toffoli, and S. Wolfram, eds. *Cellular Automata* (North-Holland, 1984).

[10] R. P. Feynman, *Opt. News* 11 (1985).

[11] E. Fredkin and T. Toffoli, *Int. J. Theor. Phys.* 21 (1982) 219.

[12] U. Frisch, B. Hasslacher, and Y. Pomeau, *Phys. Rev. Lett.* 56 (1986)

[13] M. Gardner, *Scientific American* 223:4 (October 1970) 120; 224:2 (February 1971) 112; M. Gardner, *Wheels, Life and other Mathematical Amusements* (Freeman, 1982); E. R. Berlekamp, J. H. Conway, and R. K. Guy, *Winning Ways*, vol 2. (Academic Press, 1982) Chap. 25; W. Poundstone, *The Recursive Universe* (W. Morrow, 1985).

[14] H. M. Goldstine and J. von Neumann, *unpublished* (1946), reprinted in *J. von Neumann's Complete Works*, Vol. 5 (Pergamon, 1961), pp. 1–32.

[15] E. Goles and G. Y. Vichniac, *J. Phys.* A19 (1986) L961.

[16] D. Greenberger, ed. *New Techniques and Ideas in Quantum Measurement Theory* (N. Y. Acad. Sc., 1986).

[17] P. Halmos, *Lectures in Ergodic Theory* (Chelsea, 1956).

[18] J. Hardy, O. de Pazzis, and Y. Pomeau, *Phys. Rev.* A13 (1976) 1949.

[19] B. Hayes, *Scientific American*, 250:3 (March 1984) 12.

[20] F. C. Hennie III, *Iterative Arrays of Logical Circuits* (MIT Press, 1961).

[21] H. J. Herrmann, *J. Stat. Phys.* 45 (1986) 145; J. G. Zabolitzky and H. J. Herrmann, *J. Comp. Phys.* to appear.

[22] H.J. Herrmann, H. D. Carmesin, and D. Stauffer, *J. Phys.* A20 (1987).

[23] S. A. Kauffman, this volume; see also [5] and *Physica* 10D (1984) 145, reprinted in [9].

[24] R. Landauer, in *Signal Processing*, S. Haykin, ed. (Prentice-Hall, 1988).

[25] W. M. Lang and D. Stauffer, *J. Phys.* A20 (1987); R. C. Desai and D. Stauffer, *J. Phys.* A21 (1988) L59.

[26] N. Margolus, *Physica* 10D (1984) 81, reprinted in [9].

[27] N. Margolus, in [16].

[28] N. Margolus, T. Toffoli, and G. Vichniac, *Phys. Rev. Lett.* 56 (1986) 1694.

[29] R. Penrose, "Lessons From the Game of Life," Review of W. Poundstone's book[13], *The New York Times*, March 17, 1985, VII, 34:2.

[30] Y. Pomeau, *J. Phys.* A17 (1984) L415.

[31] Y. Pomeau and G. Vichniac, *J. Phys. A (Comments)*, 1988, to appear.

[32] F. Robert, *Discrete Iterations: a Metric Study* (Springer-Verlag, 1986).

[33] D. H. Rothman and J. M. Keller, "Immiscible cellular-automaton Fluids," *DEAPS-MIT preprint*.

[34] A. Rucklidge and S. Zaleski, "A Microcanonical Model for Interface Formation," *Dept. Math. MIT preprint*.

[35] T. L. Saaty and P. C. Kainen, *The Four Colour Problem*, (Dover, 1986).

[36] G. Sorkin, private communication.

[37] H. E. Stanley, D. Stauffer, J. Kertesz, and H. Herrmann, *Phys. Rev. Lett.* 59 (1987) 2326.

[38] A. Thayse, *Boolean Calculus of Differences* (Springer-Verlag, 1981).

[39] T. Toffoli, *Physica* 10D (1984) 117, reprinted in [9].

[40] T. Toffoli, *Physica* 10D (1984) 195, reprinted in [9].

[41] T. Toffoli and N. Margolus, *Cellular Automata Machines: A New Environment for Modeling* (MIT Press, 1987).

[42] G. Y. Vichniac, *Physica* 10D (1984) 96, reprinted in [9].

[43] G. Y. Vichniac, in [5].

[44] G. Y. Vichniac, P. Tamayo, and H. Hartman, *J. Stat. Phys.* 45 (1986) 875.

[45] G. Y. Vichniac, *Complex Systems*, to appear.

[46] J. von Neumann, *Theory of Self-Reproducing Automata* (edited and completed by A. W. Burks), (University of Illinois Press, 1966).

[47] S. Wolfram, *Physica* 10D (1984) 1, reprinted in [9].

[48] S. Wolfram, *Adv. Applied Math.* 7 (1986) 123, reprinted in [49].

[49] S. Wolfram, ed. *Theory and Applications of Cellular Automata* (World Scientific, 1986).

FLUID DYNAMICS WITH LATTICE GASES

D. d'Humières and P. Lallemand

Laboratoire de Physique de l'Ecole Normale Supérieure
24 Rue Lhomond
75231 Paris Cedex 05
FRANCE

J.P. Boon, D.Dab and A. Noullez

Faculté des Sciences, CP 231
Université Libre de Bruxelles
B1050 Bruxelles
BELGIQUE

ABSTRACT

In this paper we present some results concerning the use of a particularly simple method to simulate fluid flows. As initially proposed by Frisch, Hasslacher and Pomeau, the fluid is modelled as a 2-D triangular lattice gas. We first describe the initial model and give basic results concerning its thermodynamic and transport properties. We then extend the model to treat diffusion phenomena. Using various collision rules, we obtain fairly large variations in the effective diffusion coefficient for a binary mixture. We also present a theoretical analysis for lattice gas diffusion.

Traditionally fluid motions are simulated starting from a continuum type description in which the relevant variables are related to conserved quantities : local density, local momentum flux and local energy density for the simple case of a classical monatomic fluid, with the possible addition of further quantities when the system is more complex : concentrations in fluid mixtures, local anisotropy in the case of a nematic fluid, etc....

These quantities are known to satisfy macroscopic equations of motion like the continuity equation and the Navier–Stokes equation. The theoretical justification for those equations arises from a microscopic analysis where the conservation laws of physics in the elementary behaviour of its components play an essential role. These macroscopic equations include nonlinear terms due to convection by large scale motion of the fluid. These nonlinear terms are responsible for the virtually infinite variety of fluid motions but at the same time preclude the possibility to obtain closed form solutions. So considerable effort has been devoted to design and apply computer techniques to solve numerically the macroscopic equations of fluid dynamics.

These methods very often use some discretization of the fields associated to the density, momentum and energy of the fluid. This discretization can be performed in several ways by projecting the fields on extended functions (spectral technique), on local functions (finite difference technique) or by approximating them locally with simple interpolation schemes (finite elements technique). In some cases the fluid is described as an assembly of "blobs" that can be thought of as representing fluid elements (particle in cell type approximation) or the motion is described in terms of a number of fictitious eddies (vortex methods). In spite of a large variety of techniques, these methods ignore the basic nature of the fluid as composed of individual particles.

Here we shall discuss a technique that can be thought of as a simplified approach to describe the fluid at the microscopic level and may be best introduced by referring to molecular dynamics. In molecular dynamics simulations a finite number N of particles is considered and their motion is computed by solving $3N$ equations of motion

$$\ddot{\mathbf{r}}_i = \mathbf{F}_i(\{\mathbf{r}_j\}), \qquad (1)$$

where $\mathbf{r}_i$ is the position of particle i at time t and $\mathbf{F}_i$ is the force acting on that particle due to all the other particles. Even for simple systems with central two-body forces the amount of computer work necessary to solve (1) is very large and therefore imposes severe limitations to the number of particles N. Consequently this method is not well suited for large scale flow simulations although recently interesting results have been obtained by Rapaport and Clementi [1], and by Kestemont and Mareschal [2].

It is known from Statistical Mechanics that the actual form of the interaction potential is essential to establish the exact equation of state, or to compute the values of the transport coefficients. However the equations describing the large scale behaviour of the fluid have the same general form independently of the details of the microscopic interactions. Furthermore the solutions of these equations can be cast in reduced form (using non–dimensional quantities) so that different fluids exhibit the same behaviour when the relevant reduced quantities are identical in different situations; this is the philosophy of Reynolds *similarity principle* which is applied in practice for instance to test the quality of an airplane model in a hydrodynamic tunnel or to measure the noise produced by turbulence around a submarine in a wind tunnel.

Accounting for the fact that the microscopic details are unimportant to the macroscopic behaviour, we can construct a simplified version of molecular dynamics to simulate flows provided we determine either theoretically or experimentally the following quantities : equation of state, speed of sound, transport coefficients, in order to set the values of the Mach and Reynolds numbers for a given situation.

The basics for such a program were established by Pomeau et al. in the 70's [3] and in a more satisfactory way by Frisch, Hasslacher and Pomeau (FHP) in 1985 [4]. The idea is to replace real particles by point particles which are restricted to move along the links of a lattice with a speed chosen among a limited set of values. The particles interact as hard spheres (of zero volume) when they encounter at the vertices of the lattice. For those encounters to occur it is assumed that the motions of all the particles are synchronized so that their time evolution can be decomposed in a succession of two–step sequences. During the *displacement step* each particle moves from one vertex (usually called a node) to a neighbouring one according to its velocity. During the *collision step* particles interact so as to produce a redistribution of their velocities since they are assumed to be indistinguishable. As a further simplification to the model, we introduce an exclusion principle so that no two particles with the same velocity can occupy the same link.

In spite of these formidable simplifications, the lattice gas behaves like a real fluid when a coarse–grain analysis is performed.

We shall first outline the statistical analysis of the model system, then we shall give results concerning its thermodynamic properties and show some examples of fluid flows. A more elaborate model will be presented to describe diffusion phenomena along with a theoretical analysis and results of numerical simulations.

1 Statistical analysis

Consider the two–dimensional lattice shown in Fig. 1, obtained by tiling the plane with equilateral triangles of size l. In the simplest case, particles can move from one node (or site) to any of its six neighbours; so the velocity set is given by $\{\mathbf{v}_1, \ldots, \mathbf{v}_6\}$ as shown in Fig. 1. If the time duration of the propagation step is equal to τ then the velocity scale is $v = l/\tau$. In most cases new units are chosen such that $l = 1$ and $\tau = 1$. Using our assumption that there is either 0 or 1 particle per cell of phase space, we can describe the complete state of the system by assigning values to 6 occupation numbers at each node of the lattice. Let $n_i(t, \mathbf{r}_j)$ be the number of particles (equal to 0 or 1) at location $\mathbf{r}_j$ and time t for the direction i. The complete system is then defined by the set $s : \{n_i(t, \mathbf{r}_j)\}$

for $i \in \{1,\ldots,6\}$ and $j \in$ the set of accessible lattice sites. For reasons that will appear later it is convenient to introduce a 7th possibility for the velocity that corresponds to $v_0 = 0$, that is to a particle at rest on the node.

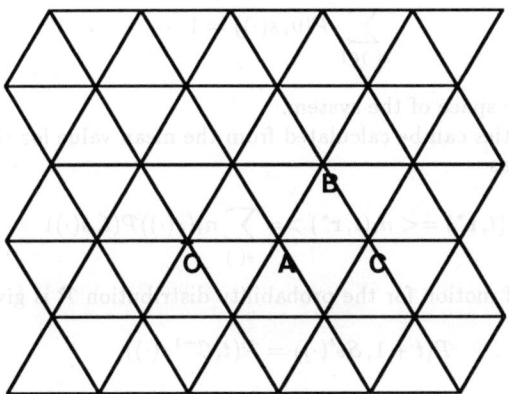

Fig. 1 Hexagonal lattice proposed by Frisch, Hasslacher and Pomeau [4] to support a gas of particles hopping from one vertex to another.

Following the rules presented in the introduction for the dynamics of the system, the microscopic equations of motion can be written in terms of two operators : the streaming operator $\mathcal{S}$ and the collision operator $\mathcal{C}$. These operators are defined by the following equations :

$$\mathcal{S} : n_i(\mathbf{r}^*) \mapsto n_i(\mathbf{r}^* - \mathbf{v}_i) \qquad (2)$$

$$\mathcal{C} : n_i(\mathbf{r}^*) \mapsto n_i(\mathbf{r}^*) + \Delta(n(\mathbf{r}^*)) \qquad (3)$$

where $\Delta(n)$ is the collision function that relates the state s of each node before collision to the state s' after collision. The explicit expression of the collision operator Δ is obtained for specific collision rules which will be given below.

The above equations that follow straightforwardly from the definition of the model are particularly well suited for numerical solutions on a digital computer. One can assign one computer word per node of the lattice and write a simple program in which the update of the system is performed by successively implementing the displacement step (Eq. 2) and the collision step (Eq. 3). For the simple 6+1 velocity model described above, 7 bits of data per node are necessary, so that some packing of the information can be performed when using a standard 32 or 64 bit machine. This allows to speed up the displacement step. The performance of a machine is measured by the number of nodes that can be updated per second (Mus will stand for million updates per second).

Statistical analysis of the problem starts with ensemble averages over sets of initial configurations of the system, each configuration having an occurrence probability $\mathcal{P}(0, s(\cdot))$ and such that

$$\sum_{s(\cdot) \in \Gamma} \mathcal{P}(0, s(\cdot)) = 1 \qquad (4)$$

where Γ is the phase space of the system.

Physical quantities can be calculated from the mean value for the occupation numbers n_i defined as

$$N_i(t, \mathbf{r}^*) = < n_i(t, \mathbf{r}^*) > = \sum_{s(\cdot)} n_i(s(\cdot)) \mathcal{P}(t, s(\cdot)) \qquad (5)$$

The equation of motion for the probability distribution $\mathcal{P}$ is given by

$$\mathcal{P}(t+1, \mathcal{S}s'(\cdot)) = \mathcal{P}(t, \mathcal{C}^{-1}s(\cdot)) \qquad (6)$$

which is the analog of the standard Liouville equation for classical statistical mechanics.

When non deterministic collision rules are used (i.e. one pre–collision configuration yields several possible post–collision configurations), the preceding equation for the time evolution of $\mathcal{P}$ is replaced by

$$\mathcal{P}(t+1, \mathcal{S}s'(\cdot)) = \sum_{s(\cdot) \in \Gamma} \prod_{r* \in \mathcal{L}} A(s(\mathbf{r}^*) \mapsto s'(\mathbf{r}^*)) \mathcal{P}(t, s(\cdot)) \qquad (7)$$

where $\mathcal{L}$ is the set of nodes of the system and A the probability for a collision to change state s into state s'. A satisfies the sum rule

$$\sum_{s'} A(s \mapsto s') = 1 \qquad (8)$$

Macroscopic quantities such as the density and the particle flux of the system are defined by

$$\rho(t, \mathbf{r}^*) = \sum_i N_i(t, \mathbf{r}^*) \qquad (9)$$

$$\mathbf{j}(t, \mathbf{r}^*) = \sum_i \mathbf{v}_i N_i(t, \mathbf{r}^*) \qquad (10)$$

Analysis of equilibrium states shows that if the probability distribution function $\mathcal{P}$ is factorized in terms of single node distribution functions, then $\mathcal{P}$ has a simple form. As shown in appendix C of Ref.[5] when the set of collisions satisfy the semi–detailed balance condition

$$\sum_s A(s \mapsto s') = 1 \qquad \forall s' \tag{11}$$

then $\{\log(N_i) - \log(1 - N_i)\}$ is a collisional invariant. As a result the mean occupation number N_i has the form :

$$N_i = \frac{1}{1 + \exp(h + \mathbf{q} \cdot \mathbf{v}_i)} \tag{12}$$

where h is a real scalar and $\mathbf{q}$ a vector. This result can be interpreted as a Fermi-Dirac distribution function due to the Boolean nature of the particles.

For low values of the mean velocity $\mathbf{u}$, h and $\mathbf{q}$ can be expanded as a power series in terms of the velocity components u_α. For the original FHP model, or any model with b moving particles of the same mass per node, the expressions for the equilibrium values of N_i are given to second order in u_α by

$$N_i^{eq}(\rho, \mathbf{u}) = \frac{\rho}{b} + \frac{\rho D}{v^2 b} v_{i\alpha} u_\alpha + \rho G(\rho) Q_{i\alpha\beta} u_\alpha u_\beta + \mathcal{O}(u^3) \tag{13}$$

where

$$G(\rho) = \frac{D^2}{2v^4 b} \frac{b - 2\rho}{b - \rho} \tag{14}$$

$$Q_{i\alpha\beta} = v_{i\alpha} v_{i\beta} - \frac{v^2}{D} \delta_{\alpha\beta} \tag{15}$$

Here D is the dimensionality of space and the usual summation convention over greek indices has been used.

As mentioned earlier the dynamics of the system can be computed from a set of initial conditions and definite collision rules. Let us first consider a deterministic situation where the probabilities A are either 0 or 1. Then the dynamics is completely deterministic and can be shown to be reversible. The ergodic nature of the trajectories of the system in phase space has not been studied in a general way. There exists special initial conditions for which the system has a short period (1 for extremely "pathological" cases) but experience shows that such states have very little weight. Using a specialized machine built at Ecole Normale Supérieure (RAP1 : Réseau d'Automates Programmables, version 1) [6], which updates a field of 512×256 sites at a rate of 50 Hertz (corresponding to approximately 6.5 Mus), we observe that starting from a "pathological" state of period 1 and changing just 1 bit of the system leads, in a time short compared to the usual macroscopic times associated with the field, to states that are undistinguishable from true random states. The finite size of phase space which has 2^{bN} different states, for a lattice of N nodes of b bits each, produces a finite period for the system, but a rather large one compared to the macroscopic times of interest here.

If the model is to be of any interest, at least in order to model the dynamics of a gas, we cannot take arbitrarily any collision rule among the 2^{2^b} possible rules that relate the set of initial states $\{s\}$ to the set of final states $\{s'\}$. We must make sure that the basic conservation laws of physics are satisfied. A simple way to guarantee this requirement is to impose that mass, momentum and energy are conserved in individual collisions at each node. Obviously one could also consider non local events in which the overall quantities would be conserved on average over several nodes. This alternative involving non–local interactions offers a procedure to simulate more complex phenomena, like phase transitions, anisotropic fluids, even shear waves; to date, this field is practically unexplored. This approach would imply elaborate statistical analysis and would lack the simplicity of the initial FHP model that maps readily on cellular *automata machines* thanks to the local nature of the collision step.

2 Macrodynamics

The macroscopic behaviour of the system is similar to that of an ideal gas, as its equation of state is $P = \rho/2$. Thus the lattice gas is compressible and pressure waves propagate at the sound speed $c_s = \sqrt{1/2}$.

To obtain the macroscopic equations of motion the standard Chapman–Enskog expansion can be used for the present lattice gas, by performing an asymptotic expansion of $N_i(t, \mathbf{r})$ in the presence of perturbations of ρ and $\mathbf{u}$ with long wavelength (of the order of ϵ^{-1}). This procedure yields the following results :

$$\partial_t \rho + \partial_\alpha (\rho u_\alpha) = O \tag{16}$$

$$\partial_t (\rho u_\alpha) + \partial_\beta (\rho G(\rho) T_{\alpha\beta\gamma\delta} u_\gamma u_\delta + \frac{v^2}{D}\rho\delta_{\alpha\beta}) + \partial_\beta [(\psi(\rho) + \frac{D}{2v^2 b}T_{\alpha\beta\gamma\delta}\partial_\gamma(\rho u_\delta)]$$
$$= \mathcal{O}(\epsilon u^3) + \mathcal{O}(\epsilon^2 u^2) + \mathcal{O}(\epsilon^3 u) \tag{17}$$

These equations are closely related to the usual Navier–Stokes equations except for the presence in (17) of the factor $G(\rho)$ and of the tensor $T_{\alpha\beta\gamma\delta}$.

The tensor T is a consequence at the macroscopic level of the symmetries of the underlying lattice where the local dynamics takes place.

$$T_{\alpha\beta\gamma\delta} = \sum_i v_{i\alpha} v_{i\beta} Q_{i\gamma\delta} = \sum_i v_{i\alpha} v_{i\beta} (v_{i\gamma} v_{i\delta} - \frac{v^2}{D}\delta_{\gamma\delta}) \tag{18}$$

has the same symmetry as the elasticity tensor involved in the stress–strain relations. So one can refer to elasticity theory to determine whether a given lattice exhibits a macroscopic behaviour that is isotropic or not. For the two dimensional case, the lattice composed of equilateral triangles leads to isotropy whereas the square lattice model of Pomeau et al. does not. In three dimensions, there is no lattice that produces the required isotropy. There are however alternative solutions to the three dimensional problem : either a simple cubic lattice with an appropriate combination of particles with velocity 1 and $\sqrt{2}$ (multispeed model), or the projection into three dimensional space of a four dimensional face centered cubic lattice (FCHC model) [9].

When centers (i.e. particles at rest) are included, the equation of state becomes $P = 3\rho/7$ and consequently the velocity of sound reads $c_s = \sqrt{3/7}$. The Navier–Stokes equations have to be modified to include an additional term due to the presence of bulk viscosity

$$\partial_t(\rho u_\alpha)+\partial_\beta(G(\rho)\rho u_\alpha u_\beta) = -\partial_\alpha P(\rho,u^2)+\partial_\beta(\nu\partial_\beta(\rho u_\alpha))+\partial_\alpha((\frac{D-2}{D}\nu+\zeta)\mathrm{div}(\rho\mathbf{u})) \qquad (19)$$

with :

$$G(\rho) = \frac{D}{D+2}\frac{\rho}{\rho_m}\frac{1-2d}{1-d} \qquad (20)$$

$$P(\rho,u^2) = \frac{\rho_m}{D}v^2 - \rho G(\rho)\frac{c_s^2}{v^2}(1+\frac{D}{2}-\frac{v^2}{2c_s^2})u^2 \qquad (21)$$

where ρ and ρ_m are the total density and the density of moving particles respectively and d is the mean occupation number of the bits corresponding to moving particles. ν and ζ are the kinematic shear and bulk viscosities respectively.

To compute the actual values of the transport coefficients (a.o. in order to determine the Reynolds number of a flow), one can start from the microdynamical equations of motion of the system (Eqs 2 and 3), then switch to an average equation with the assumption (like in the Boltzmann approximation for real gases), that higher order distribution functions factorize into products of single particle distribution functions. Furthermore small fluctuations around equilibrium are considered so that a linearized version of the Boltzmann equation can be used for the lattice gases

$$\partial_t N_i + v_{i\alpha}\partial_\alpha N_i = \sum_j \mathcal{A}_{ij}N_j \qquad (22)$$

where $\mathcal{A}$ is the linearized collision operator, which can be written in terms of the set of collisions as

$$\mathcal{A}_{ij} = -\frac{1}{2}\sum_{s,s'}(s_i - s'_i)(s_j - s'_j)A(s \mapsto s')\prod_k d_k^{s_k}(1-d_k)^{1-s_k} \qquad (23)$$

The transport coefficients can be obtained from the eigenvalues of the matrix $\mathcal{A}$, and consequently their exact values depend on the collision probabilities $A(s \mapsto s')$.

3 Collision rules

For the FHP model with rest particles, there are 7 bits at each node and so 2^7 possible states. Thus the outcome of the collision step can be determined by simple inspection. For that purpose we classify the input states according to the values of the quantities

$$m = \sum_i n_i$$
$$j_x = \sum_i v_{ix} n_i \qquad (24)$$
$$j_y = \sum_i v_{iy} n_i$$

Operationally, the collisions are the transitions from an initial state to another state within the same class so that conservation of mass and of linear momentum is satisfied. For the FHP model with centers, it is found that 76 states belong to classes with more than 1 element and that 5 is the largest number of elements in a class; so there is a possible choice of at most 5 output states (in practice 4 because a transition with the output identical to the input is uneffocient). As an illustration, Fig. 2 shows the only possible 2 body collisions (with the relevant symmetries).

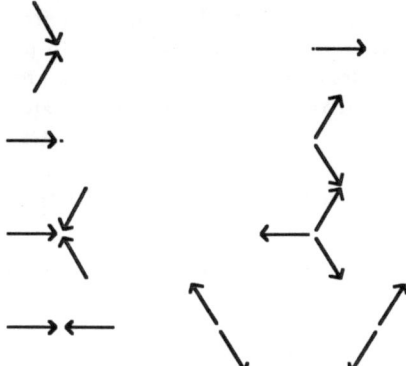

Fig. 2 Two body collisions that change the state of the gas.

A model is completely defined when the set of collisions is specified. For each set, the shear and bulk viscosities can then be calculated.

The so-called FHP–III model includes all 76 cases. The collision table is given below where the input and output states are in octal form (a state with only one particle at rest is coded 100). When there are two different possible outputs, their respective probabilities are 0.5.

input	5	11	12	13	15	21	22	23	24	25	26	27	31	32	33	35
output 1	102	22	104	105	26	140	44	45	110	52	112	113	150	54	66	154
output 2	102	44	104	46	112	140	11	142	110	52	15	146	64	124	55	132

input	37	42	44	45	46	50	51	52	53	54	55	56	57	62	64	65
output 1	156	101	11	142	13	120	62	25	145	124	33	126	127	121	31	162
output 2	156	101	22	23	105	120	121	25	123	32	66	115	127	51	150	151

input	66	67	72	73	75	76	101	102	104	105	110	111	112	113	115	120
output 1	55	153	131	165	172	135	42	5	12	46	24	122	15	146	56	50
output 2	33	153	164	165	172	135	42	5	12	13	24	144	26	27	126	50

input	121	122	123	124	125	126	127	131	132	133	135	140	142	144	145	146
output 1	51	144	53	32	152	115	57	164	35	166	76	21	23	111	123	27
output 2	62	111	145	54	152	56	57	72	154	155	76	21	45	122	53	113

input	150	151	152	153	154	155	156	162	164	165	166	172
output 1	64	65	125	67	132	133	37	151	72	73	155	75
output 2	31	162	125	67	35	166	37	65	131	73	133	75

This model leads to the following expressions for the kinematic viscosities

$$\nu = \frac{1}{28d(1-d)(1-8d(1-d)/7)} - \frac{1}{8} \tag{25}$$

$$\zeta = \frac{1}{98d(1-d)(1-2d(1-d))} - \frac{1}{28} \tag{26}$$

An alternative model (referred to hereafter as FHP–IV) is constructed with a different set of collisions as shown in the following table (using the same conventions)

input	5	11	12	13	15	21	22	23	24	25	26	31	32	33	37	42
output 1	102	22	104	46	26	140	44	45	110	52	15	64	54	55	156	101
output 2	102	44	104	46	26	140	11	45	110	52	15	64	54	66	156	101

input	44	45	46	50	51	52	54	55	57	62	64	66	67	73	75	76
output 1	11	23	13	120	62	25	32	66	127	51	31	33	153	165	172	135
output 2	22	23	13	120	62	25	32	33	127	51	31	55	153	165	172	135

input	101	102	104	110	111	113	115	120	122	123	125	126	127	131	132	133
output 1	42	5	12	24	122	146	126	50	144	145	152	115	57	164	154	155
output 2	42	5	12	24	144	146	126	50	111	145	152	115	57	164	154	166

input	135	140	144	145	146	151	152	153	154	155	156	162	164	165	166	172
output 1	76	21	111	123	113	162	125		67	132	166	37	151	131	73	133	75
output 2	76	21	122	123	113	162	125		67	132	133		37	151	131	73	155	75

This model leads to the following expressions for the kinematic viscosities

$$\nu = \frac{1}{28d(1-d)(1-6d(1-d)/7)} - \frac{1}{8} \tag{27}$$

$$\zeta = \frac{1}{98d(1-d)(1-3d(1-d))} - \frac{1}{28} \tag{28}$$

In this second model, the value of the shear viscosity is lower than for the FHP–III model, but the bulk viscosity is significantly larger. This is a drawback when the model is used to study sound waves, whose attenuation is proportional to $\nu + \zeta$ in two dimensions.

4 Lattice gas simulations

Over the last couple of years, a number of simulations have been performed to evaluate the "experimental" properties of the FHP hexagonal lattice gas. Transport coefficients have been determined by measuring the relaxation time of shear or sound waves [7]. They have been found to agree remarkably well with the theoretical predictions given above, which shows that the Boltzmann approximation is adequate for the whole density range. More recently a detailed analysis of Poiseuille flow in a two dimensional duct has been performed by Zanetti et al. [8]; they observe that the shear viscosity has a logarithmic dependence upon the size of the lattice as it is found for real two dimensional systems.

The lattice gas method has been used to simulate flows corresponding to actual experiments and to calculations performed by standard techniques to solve the Navier–Stokes equations for a real fluid; it has been found that up to Mach numbers of the order of 0.5, similarity laws hold with a good accuracy provided the velocity of the lattice gas flow is renormalized by the factor $G(\rho)$ (see Eq.(17)).

Techniques have been proposed by d'Humières, Lallemand and Frisch to extend the lattice gas technique to three dimensional flows [9]. As mentioned above, it is necessary to find a set of velocities such that the tensor $T_{\alpha\beta\gamma\delta}$ (Eq. 18) is isotropic. This is not possible for sets of velocities obtained by taking a three dimensional lattice and linking one lattice site to a number of its neighbours. A way out of this difficulty is to consider a four dimensional face–centered hypercubic lattice (FCHC) and taking the set of 24 vectors linking one lattice node to its nearest neighbours. In cartesian coordinates, this leads to the following velocities

$$(\pm 1, \pm 1, 0, 0), \quad (\pm 1, 0, \pm 1, 0), \quad (\pm 1, 0, 0, \pm 1)$$
$$(0, \pm 1, \pm 1, 0), \quad (0, \pm 1, 0, \pm 1), \quad (0, 0, \pm 1, \pm 1)$$

The shear viscosity of the FCHC model has been studied by Hénon [10] who obtained a general expression for ν in the framework of the Boltzmann approximation and proposed several ways to construct the collision table in order either to simplify the collision algorithm or to minimize the value of the shear viscosity. Recently, Rivet [11] has implemented (on a Cray–2 computer) the FCHC model together with efficient collision rules; he so determined "experimentally" the shear viscosity and performed simulations of three dimensional flows at Reynolds number close to that used in computations of viscous flows using the exact Navier–Stokes equations. The agreement confirms that the Boltzmann approximation is adequate meaning that the higher order distribution functions do factorize in products of single particle distribution functions. Note that the necessity for a very large computer is due to the fact that in 3–D 24 bits are needed to describe the state at each node of the lattice, which implies that the collision table has 2^{24} entries of 24 bits words. Here, we shall not elaborate on the results of these investigations but rather concentrate on an extension of the lattice gas technique to diffusion phenomena.

5 Lattice gas diffusion

In order to investigate diffusion phenomena, fluid particles should be "marked" to make them distinguishable. This can be realized very easily by adding a type bit to the moving particles. These type (or "color") bits propagate with the particles, and remain attached to their respective particles during collisions. The 7-bit per node models then become 14-bit models, with for each link the possibility to be in any of the states (expressed in binary)

00 : Empty Link
01 : Link carrying a A "blue" particle
10 : Link carrying a B "red" particle

Note that with two bits allocated to each link, we have the additional possibility for a third type of particle (state 11), which renders the model suitable to study reactive systems with three species. The exclusion principle now reads: there can be at most one particle of any color with a given speed sign on any link.

If we define the concentration of type s particles as

$$C_s(t,\mathbf{r}^*) = \frac{1}{\rho}\sum_i n_i(t,\mathbf{r}^*)\delta_{ss_i} - \frac{1}{2} \qquad (29)$$

it can be shown that the concentrations obey the macroscopic diffusion equation

$$\partial_t(\rho C_s) + \partial_\alpha(\rho C_s u_\alpha) = \partial_\alpha\bigl(D_s(\rho)\partial_\alpha(\rho C_s)\bigr) + \omega(C_s) \qquad (30)$$

where D_s is a diffusion coefficient and $\omega(C_s)$ is a source term that is present if type exchange (chemical or other) reactions are included.

For a two–species model, each node can be in one of $3^7 = 2187$ states, so the number of possible collision rules is very large. We again reduce the number of possibilities by imposing total number of particles and total momentum conservation during collisions. Chemical reactions can be modelled by allowing the number of particles of each type to vary in the collisions. Such models have been used in [12] to introduce interfaces between two species using majority–rule type reactions. Here we will restrict our discussion to purely diffusive collision rules in which the number of particles of each type is conserved during collisions. Moreover, the collision rules will be such that they are invariant through the exchange of the types of particles (the two species are physically identical) and such that they reduce to either FHP–III or FHP–IV when the type of the particles is ignored. In that case, the color plays the role of a passive tracer that is simply advected with the flow.

Even with such restricted collision rules, there are many configurations for which the particles can be redistributed in various ways in the resulting configurations (up to 35 for the 7–speed FHP models). If the selection between these output states is made at random, then we have a model for self–diffusion in a lattice gas. Another approach is to minimize mutual diffusion of the species in order to slow down their mixing, so as to make the model efficient to study flow dispersion. This can be realized by maximizing the momentum variation for each particle type during collisions while conserving total momentum. For instance, in the collision shown in Fig. 3a, only the first two configurations will be kept so as to minimize both the viscosity coefficient and the diffusion coefficient. Fig. 3b is an example of transparent collision in the non–colored case, but which contributes to the diffusion coefficient when color is introduced. We can also group together particles of the same type in collision processes, in order to minimize the concentration spread (Fig. 3c).

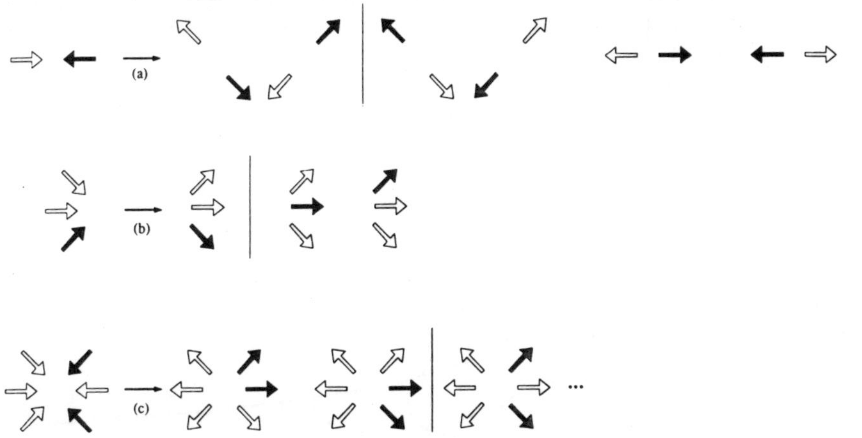

Fig. 3 Examples of elementary collisions for limited diffusion colored lattice gases.

The experimental realization of these lattice gases is relatively easy. The propagation step is the same as for simple lattice gases, and the collision step can be implemented using a collision table that gives the post-collision states for each possible pre-collision state (such a method is possible only in 2-D models; the size of the table becomes prohibitive in 3 dimensions (2^{48} 48-bit words)). The models that have been studied are a self-diffusion version of FHP-III [13,14,15] and a limited diffusion version of FHP-IV.

The diffusion coefficient D_s has been determined "experimentally" by studying the time relaxation of a one-dimensional concentration step and fitting the results to the corresponding standard solution of the diffusion equation. Very good agreement is obtained and Fig. 4 shows the "measured" diffusion coefficient as a function of density for both models. It is to be noted that, in the case of limited diffusion, the diffusion coefficient goes to zero for densities per link d approaching one, since the particles momentum is then reversed at nearly each step.

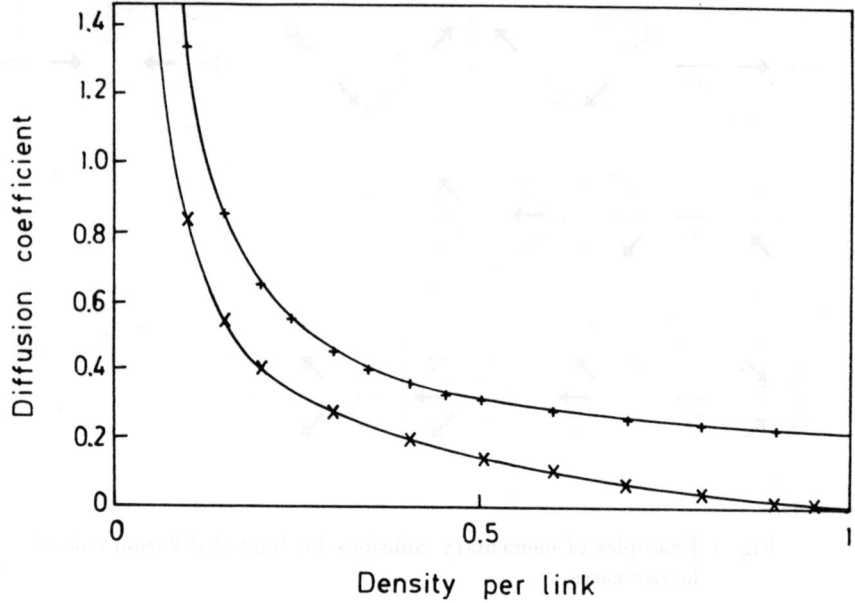

Fig. 4 Diffusion coefficient as a function of density for self-diffusion (+) and limited diffusion (×) models.

Another way to obtain the diffusion coefficient is to study the displacement of one "labeled" particle in the fluid as performed by Binder and d'Humières [16]. In this case, the mean-square displacement should be linear versus time, with a proportionality coefficient equal to $4D_s$ (in two dimensions), which yields a good test for diffusive behaviour. To minimize the effect of fluctuations, we split a large record of the succession of displacements for the labeled particle into many smaller records, each one being considered as the displacements of a different particle. This procedure should be correct provided the system behaves ergodically and the time separation between records is sufficiently large for all correlations to have vanished. Fig. 5a shows the mean–square displacement as a function of time for a density of 0.3, obtained with 128 records separated by 500 time units in a 65536 displacements record for the FHP–IV model. At low densities and short times, one can observe the free flight regime over distances corresponding to the mean free path (see Fig. 5b).

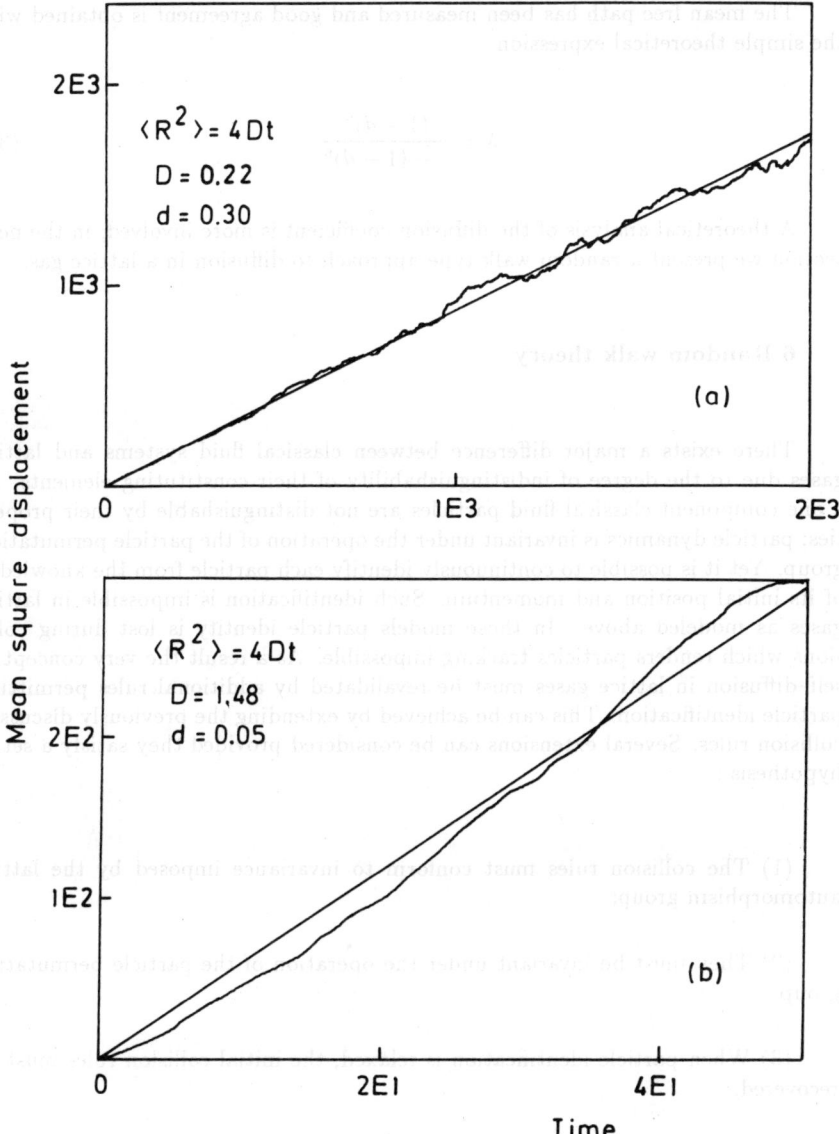

Fig. 5 Mean–Square Displacement as a function of time for densities 0.3 and 0.05

The mean free path has been measured and good agreement is obtained with the simple theoretical expression

$$\lambda = \frac{(1-d)^6}{1-(1-d)^6} \tag{31}$$

A theoretical analysis of the diffusion coefficient is more involved; in the next section we present a random walk type approach to diffusion in a lattice gas.

6 Random walk theory

There exists a major difference between classical fluid systems and lattice gases due to the degree of indistinguishability of their constituting elements. In a one component classical fluid particles are not distinguishable by their properties; particle dynamics is invariant under the operation of the particle permutation group. Yet it is possible to continuously identify each particle from the knowledge of its initial position and momentum. Such identification is impossible in lattice gases as modeled above. In those models particle identity is lost during collisions which renders particles tracking impossible. As a result the very concept of self–diffusion in lattice gases must be revalidated by additional rules permitting particle identification. This can be achieved by extending the previously discussed collision rules. Several extensions can be considered provided they satisfy a set of hypothesis :

(1) The collision rules must conform to invariance imposed by the lattice automorphism group;

(2) They must be invariant under the operation of the particle permutation group;

(3) When particle identification is relaxed, the initial collision rules must be recovered.

Hypothesis (1) imposes a natural invariance, whereas hypotheses (2) and (3) guarantee that with the extended rules, marked particles behave as passive tracers. In general, specific modifications of the collision rules induce specific self-diffusion behaviours. An example is shown in Fig. 6.

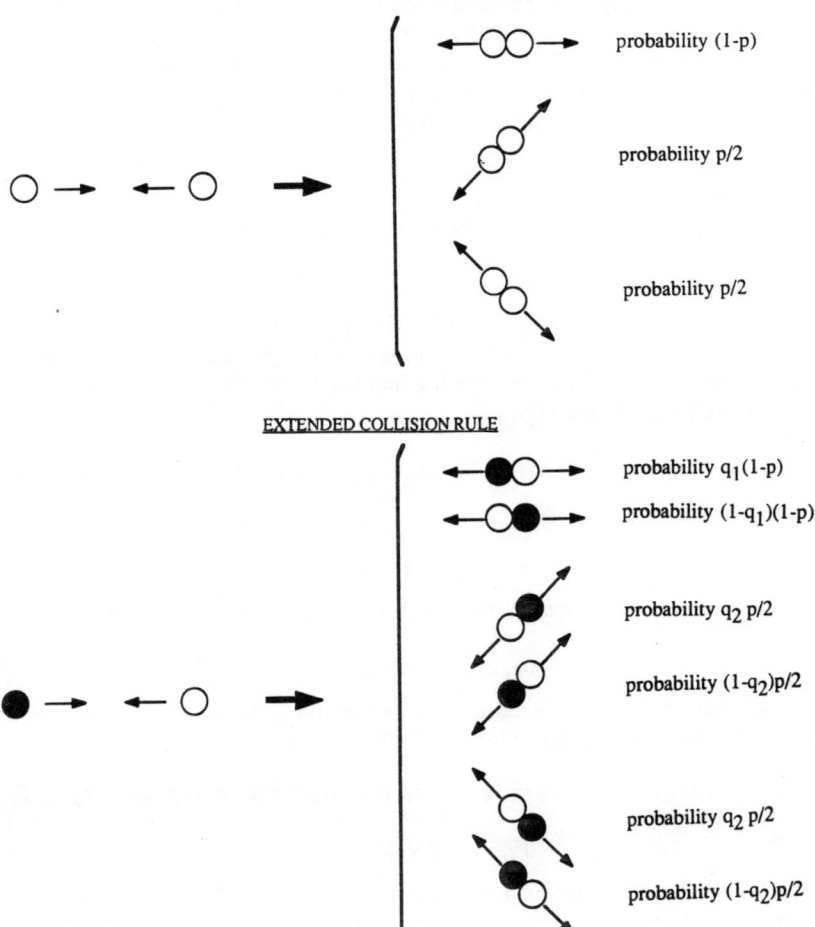

Fig. 6 Extension of the collision rules for colored particles

Using elementary kinetic theory of gases, one can study self–diffusion by analyzing the statistical properties of single particle motion in a fluid at equilibrium. With some simplifying assumptions, self–diffusion particle motion reduces to a random walk with stochastically independent steps. One then obtains the classical result that the mean–square displacement grows linearly in time, the proportionality factor being equal to the self–diffusion coefficient times a constant whose value is twice the dimension of the system.

A similar analysis can be conducted for self–diffusion in lattice gases when collision rules are extended to include passive marking of particles. For such an analysis, we invoke the Boltzmann approximation, i.e. we consider as independent stochastic events the successive collisions that a particle undergoes in a fluid at equilibrium. The sequence of displacements of a particle is noted

$$\{\mathbf{r}_1, \mathbf{r}_2, \mathbf{r}_3, \ldots, \mathbf{r}_t, \ldots\} \tag{32}$$

In the set of N permitted velocities on the lattice, these displacements have the value

$$\mathbf{r}_i \in \{\mathbf{v}_j, \quad j = 1 \cdots N\} \tag{33}$$

In all generality, in lattice gas models, this set is invariant versus spatial coordinates reversal, a property that will be used subsequently. Now the Boltzmann approximation is expressed through the independence of the successive time variations of the particles velocity, i.e.

$$\mathcal{P}[\bigcup_i (\Delta \mathbf{v}_i = \mathbf{x}_i)] = \prod_{i=1} \mathcal{P}[\Delta \mathbf{v}_i = \mathbf{x}_i] \tag{34}$$

where

$$\Delta \mathbf{v}_i = (\mathbf{r}_i - \mathbf{r}_{i-1}) \tag{35}$$

Because of time translation invariance, the velocity variation probability is time independent

$$\mathcal{P}[\Delta \mathbf{v}_i = \mathbf{x}_i] = \mathcal{P}[\Delta \mathbf{v} = \mathbf{x}] \quad \forall i \tag{36}$$

The sequence of velocity variation is a Bernouilli process, and the sequence of particle displacements is a stationary Markov chain

$$\mathcal{P}[\mathbf{r}_t = \mathbf{v}_i | \mathbf{r}_{t-1} = \mathbf{v}_j, \mathbf{r}_{t-2} = \mathbf{v}_k, \cdots, \mathbf{r}_{t-t'} = \mathbf{v}_l] = \mathcal{P}[\mathbf{r}_t = \mathbf{v}_i | \mathbf{r}_{t-1} = \mathbf{v}_j] \equiv P_{ij} \tag{37}$$

$$P_{ij} = \mathcal{P}[\Delta \mathbf{v} = (\mathbf{v}_i - \mathbf{v}_j)] \tag{38}$$

characterized by a transition probability matrix P which can be evaluated when the collision rules are set and the density is fixed. Obviously lattice symmetries induce symmetries for the matrix elements; thanks to these symmetries, the diffusion coefficient can be expressed, in many cases, in a general and rather simple form. For instance, for the FHP model without centers (no particles at rest), the symmetries impose the following structure to the transition probability matrix

$$\begin{pmatrix} p_0 & p_1 & p_2 & p_3 & p_2 & p_1 \\ p_1 & p_0 & p_1 & p_2 & p_3 & p_2 \\ p_2 & p_1 & p_0 & p_1 & p_2 & p_3 \\ p_3 & p_2 & p_1 & p_0 & p_1 & p_2 \\ p_2 & p_3 & p_2 & p_1 & p_0 & p_1 \\ p_1 & p_2 & p_3 & p_2 & p_1 & p_0 \end{pmatrix} \tag{39}$$

with
$$p_0 + 2\,p_1 + 2\,p_2 + p_3 = 1 \tag{40}$$

Except for pathological situations (e.g. close packing) the Markov chain possesses a unique invariant measure, and all probability measures converge towards this equilibrium state. We assume that such an attractor state exits, with the restriction however that this assumption be verified for each specific model application. It is also assumed that semi–detailed balance is satisfied by the collision rules. Then the matrix P is symetrical, and the equilibrium state is characterized by a distribution with equal probabilities for all velocities of the model. Note that the latter assumption is not essential; however it has the virtue to simplify the development.

For the analysis of the statistical properties of the displacement time series (32), Eq. (37), governing the Markov chain, must be supplemented with the probability distribution of the first displacement. The equilibrium state is a logical choice

$$\mathcal{P}[\mathbf{r}_1 = \mathbf{v}_i] = \frac{1}{N} \quad \forall i \tag{41}$$

We define by $\mathbf{R}(t)$ the total displacement of a single particle after some time t

$$\mathbf{R}(t) = \sum_{i=1}^{t} \mathbf{r}_i \tag{42}$$

with
$$\langle \mathbf{R}(t) \rangle = 0 \tag{43}$$

which follows from the symmetries in the dynamics, and where brackets denote an ensemble average over all possible realizations of the time series (32). The mean-square displacement (i.e. the mean–square deviation from the total displacement) reads
$$\sigma^2(t) = \langle \mathbf{R}^2(t) \rangle$$
$$= \sum_{i=1}^{t} \langle \mathbf{r}_i^2 \rangle + 2 \sum_{i<j=1}^{t} \langle \mathbf{r}_i \cdot \mathbf{r}_j \rangle \tag{44}$$

Thanks to the stationarity property of the Markov chain, (44) reduces to

$$\sigma^2(t) = t \langle \mathbf{r}^2 \rangle_{eq} + 2 \sum_{k=1}^{t-1} (t-k)\,\mathcal{C}(k) \tag{45}$$

In (45) the brackets denote an equilibrium ensemble average and $\mathcal{C}$ is the correlation function
$$\mathcal{C}(k) = \langle \mathbf{r}_1 \cdot \mathbf{r}_{1+k} \rangle_{eq}$$
$$= \frac{1}{N} \sum_{i<j} (\mathbf{v}_i \cdot \mathbf{v}_j)\,[P_{ij}^k] \tag{46}$$

Introducing the metric matrix

$$G_{ij} = \mathbf{v}_i \cdot \mathbf{v}_j \tag{47}$$

the correlation function takes the simple form

$$\mathcal{C}(k) = \frac{1}{N} Tr\,[G\,P^k] \tag{48}$$

With the assumption made above, the spectrum of the matrix P has the following structure

$$\{1, \lambda_2, \lambda_3, \ldots, \lambda_N\} \tag{49}$$

with

$$|\lambda_i| < 1 \quad \forall i = 2, 3, \ldots, N \tag{50}$$

Invariance by reversal of spatial coordinates of the velocity set (33) implies that the only eigenvector of P with eigenvalue one (corresponding to the equilibrium state) is also eigenvector of G with eigenvalue zero. Using this property, we perform the summation in (46) to obtain the general expression

$$\sigma^2(t) = t\langle \mathbf{r}^2 \rangle_{eq} + \frac{2t}{N} Tr\left(\frac{\widetilde{G}\widetilde{P} - \widetilde{G}\widetilde{P}^t}{\widetilde{1} - \widetilde{P}}\right) - \frac{2}{N} Tr\left(\widetilde{G}\left(\frac{\widetilde{P} - t\widetilde{P}^t + (t-1)\widetilde{P}^{(t+1)}}{(\widetilde{1} - \widetilde{P})^2}\right)\right) \tag{51}$$

where $\widetilde{P}$ and $\widetilde{G}$ are the $(N-1) \times (N-1)$ matrix restrictions of P and G to the sub-vectorial orthogonal to the eigenvector with eigenvalue one. This result can be generalized to models for which semi-detailed balance is violated.

As mentioned earlier, lattice symmetries induce relations between the elements of the transition probability matrix P. For certain cases, these symmetries are such that the matrix P and G can be explicitly and simultaneously diagonalized. Then the mean-square displacement (51) takes a simple form which can be expressed directly in terms of the matrix P elements. We show this for the simple FHP model without center, for wich the matrix G reads

$$\begin{pmatrix} 1 & 1/2 & -1/2 & -1 & -1/2 & 1/2 \\ 1/2 & 1 & 1/2 & -1/2 & -1 & -1/2 \\ -1/2 & 1/2 & 1 & 1/2 & -1/2 & -1 \\ -1 & -1/2 & 1/2 & 1 & 1/2 & -1/2 \\ -1/2 & -1 & -1/2 & 1/2 & 1 & 1/2 \\ 1/2 & -1/2 & -1 & -1/2 & 1/2 & 1 \end{pmatrix} \tag{52}$$

The basis which diagonalizes simultaneously (39) and (52) is given by

vector	eigenvalue for P	eigenvalue for G
(1, 1, 1, 1, 1, 1)	1	0
(-1, 1,-1, 1,-1, 1)	$1 - 4p_1 - 2p_3$	0
(-1, 0, 1,-1, 0, 1)	$1 - 3p_2 - 3p_1$	0
(-1, 2,-1,-1, 2,-1)	$1 - 3p_2 - 3p_1$	0
(1, 0,-1,-1, 0, 1)	$1 - 2p_3 - 3p_2 - p_1$	3
(-1,-2,-1, 1, 2, 1)	$1 - 2p_3 - 3p_2 - p_1$	3

In this basis, the correlation function (48) can be evaluated straightforwardly, and the mean-square displacement is obtained to yield

$$\sigma^2(t) = t + 2t\left(\frac{\lambda - \lambda^t}{1-\lambda}\right) - 2\left(\frac{\lambda - t\lambda^t + (t-1)\lambda^{t+1}}{(\lambda-1)^2}\right) \tag{53}$$

with

$$\lambda = p_0 + p_1 - p_2 - p_3 \tag{54}$$

For the FHP models including rest particles, the transition matrix has the structure

$$\begin{pmatrix} p_0 & p_1 & p_2 & p_3 & p_2 & p_1 & \alpha \\ p_1 & p_0 & p_1 & p_2 & p_3 & p_2 & \alpha \\ p_2 & p_1 & p_0 & p_1 & p_2 & p_3 & \alpha \\ p_3 & p_2 & p_1 & p_0 & p_1 & p_2 & \alpha \\ p_2 & p_3 & p_2 & p_1 & p_0 & p_1 & \alpha \\ p_1 & p_2 & p_3 & p_2 & p_1 & p_0 & \alpha \\ \alpha & \alpha & \alpha & \alpha & \alpha & \alpha & \beta \end{pmatrix} \tag{55}$$

with

$$p_0 + 2p_1 + 2p_2 + p_3 + \alpha = 1 \, , \, 6\alpha + \beta = 1 \tag{56}$$

Here α denotes a transition involving a particle at rest, and β such a transition where a center is present in the input and the output configurations. When the extended collision rules are introduced to express equiprobable choices for the various color combinations in the output configurations (see Fig. 6), the mean-square displacement is found to be 6/7 times the rhs of (53), with λ as given by (54) and

$$\lambda = 1 - \frac{7}{2}d + \frac{35}{6}d^2 - \frac{67}{12}d^3 + \frac{11}{4}d^4 - \frac{5}{12}d^5 - \frac{1}{12}d^6 \tag{57}$$

where d is the density per site ($= 1/7 \times$ density per node). When macroscopic isotropy is satisfied (as it is for the FHP models), the diffusion coefficient is given by the classical expression (in 2-D)

$$D_* = \lim_{t \to \infty} \frac{1}{4} \frac{\sigma^2(t)}{t} \tag{58}$$

Referring to the question of the lattice symmetry requirements to ensure macrodynamical isotropy, it is legitimate to ask whether diffusion takes place isotropically on the lattice. More explicitly, when the Markov chain is projected on an arbitrarily oriented direction, is the measure of the particle spread rate independent of the projection direction ? The answer is yes for the square lattice as well as for the triangular lattice in 2-D, for the FCHC in 4-D and thus for its 3-D projection also. An important consequence of the scalar nature of the diffusion equation is the validity of the square lattice regarding macroscopic isotropy requirements for diffusion; even more important is the result that the cubic lattice can be used to model and simulate diffusion processes, a property that we have exploited to study reaction–diffusion phenomena in 3-D, which we report elsewhere [17].

Acknowledgements

JPB acknowledges support from the "Fonds National de la Recherche Scientifique" (FNRS Belgium). DD and AN have benefited from a grant by the "Institut pour l'Encouragement à la Recherche Scientifique dans l'Industie et l'Agriculture" (IRSIA Belgium). This work was supported by European Community grant ST 2J-0190.

References

[1] Rapaport, D.C. and Clementi, E., Phys. Rev. Lett., **14**, 1746 (1973).
[2] Marescha, M. and Kestemont, E., Nature, **329**, 427 (1987).
[3] Hardy, J., Pomeau, Y. and de Pazzis, O., J. Math. Phys., **14**, 1746 (1973).
    Hardy, J., de Pazzis, O. and Pomeau, Y., Phys. Rev., **A13**,1949 (1976).
[4] Frisch, U., Hasslacher, B. and Pomeau, Y., Phys. Rev. Lett., **56**, 1505 (1986).
[5] Frisch, U., D'Humières, D., Hasslacher, B., Lallemand, P., Pomeau, Y. and Rivet, J.P., Complex Systems, **1**, 649 (1987).
[6] Clouqueur, A. and D'Humières, D., Complex Systems, **1**, 585 (1987).
[7] D'Humières, and Lallemand, P., Complex Systems, **1**, 599 (1987).
[8] Kadanoff, L., McNamara, G. and Zanetti, G., preprint Chicago University (1987).
[9] D'Humières, D., Lallemand, P. and Frisch, U., Europhys. Lett., **2**, 291 (1986).
[10] Hénon, M., Complex Systems, **1**, 475 (1987).
[11] Rivet, J.P., Comp. Rend. Acad. Sci. Paris, **II 305**, 751 (1987).
[12] D'Humières, D., Clavin, P., Lallemand, P. and Pomeau, Y. Comp. Rend. Acad. Sci. Paris, **II 303**, 1169 (1986).
[13] D'Humières, D., Lallemand, P. and Searby, G., Complex Systems, **1**, 633 (1987).
[14] D'Humières, D., Lallemand, P. and Searby, G., Proc. of E.P.S. Int. Conf. on Physico-chemical Hydrodynamics (La Rabida, Spain, July 1986), Velarde, M. G. ed. (Plenum Press 1987).
[15] Clavin, P., Lallemand, P., Pomeau, Y. and Searby, G., J. Fluid Mech., **188**, 437 (1988).
[16] Binder, P. and D'Humières D., to be published.
[17] Boon, J.P. and Dab, D., to be published.

CELLULAR AUTOMATA MODEL FOR THERMO-HYDRODYNAMICS

Bastien Chopard * and Michel Droz *
Department of Theoretical Physics
University of Geneva
CH-1211 Geneva 4, Switzerland

Abstract: *We study the heat conduction and the velocity of sound for a simple two-speed lattice gas model. It is found that our model give a quite sensible description of heat propagation in a fluid at rest while some nonphysical situation appears if we consider a fluid in motion.*

1.Introduction.

Lattice gas models have recently received a considerable interest due to their ability to model a fluid in 2 dimensions [1]. Moreover the existence of special purpose computer with high degree of parallelism makes them possible candidate for an alternative method of solving the Navier-Stokes equation.

These models consist in a fully discrete molecular dynamics: the particles evolve on a regular lattice with some properly chosen velocities such that at each time step they move towards a neighboring site. If several particles enter simultaneously at the same site they can undergo collisions in which they change their trajectories. The rules which govern these collisions are the usual conservation laws of mechanics, namely the conservation of mass, momentum and energy. The fact that the particles interact locally and that their dynamics is discrete make the system suitable for cellular automata simulations.

It has been shown that for particles evolving on a hexagonal lattice with all the same energy and travelling in the six different directions (FHP model), one obtains a system described by the two dimensional Navier-Stokes equation [1]. Here we shall consider the possibility of including a temperature in the model in order to obtain a system governed by the thermo-hydrodynamic equations, and we shall study some of its properties.

---

* Supported by the Swiss National Scientific Foundation

2. Simple two-speeds models.

A natural way to introduce the temperature is through the equipartition theorem, which lead us to consider a multi-speeds model where all particles do not have the same energy [2]. A simple generalization of the FHP model is a two-speeds hexagonal model in which, besides the particles travelling with velocity 1 in one of the six directions of the lattice, we add particles moving with velocity $\sqrt{3}$ from a site to one of the six second nearest neighbors. Particles of same energy undergo the usual collisions of the FHP model, while particles of different energy interact as shown in Fig. 1. in such a way that mass, momentum and energy are conserved during the collision.

For simplicity reasons, we can also consider the case of a fluid at rest or with a slow motion such that a linear expansion in the velocity is valid. It is then sufficient to consider a *square* lattice because the lack of isotropy coming from the non linear terms in velocity is not present in this case.

Our model will thus consist in two interacting HPP gases, one of slow particles with velocity v and moving in a unit time τ towards one of the four nearest neighbors, and the other of fast particles travelling along the diagonals of the lattice, with a velocity $w = \sqrt{2}v$. All particles are assumed to have a mass equal to 1 and the collision rules of the HPP model [1] hold for each kind of particles independently. Between the fast and slow particles we impose the interaction shown on Fig. 2, in agreement with the conservation laws.

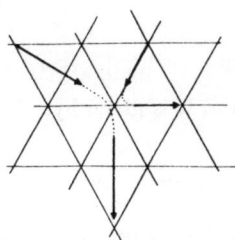

Fig. 1 : *Collision between fast and slow particles for the hexagonal model*

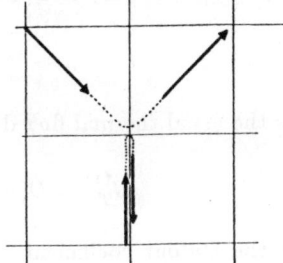

Fig. 2 : *Collision between fast and sow particles for the square model*

3. Thermo-hydrodynamical equations.

Using the standard definitions of kinetic theory [3], we define the density ρ, velocity field $\vec{u}$ and the internal energy e. The temperature T is then defined by the mean kinetic energy per particle through the relation $e = kT$, and it is convenient here to choose $k = \frac{v^2}{2}$. Following the usual procedure for deriving the hydrodynamic equations from the Boltzmann equation, and using the so-called multi-scale expansion [4], we obtain first the Euler equation which are given here to first order in $\vec{u}$ [5]:

$$\partial_t \rho + \mathrm{div}\rho\vec{u} = 0 \tag{1a}$$

$$\partial_t \rho u_\alpha + \partial_\beta \Pi^{(0)}_{\alpha\beta} = 0 \tag{1b}$$

$$\partial_t \rho T + \mathrm{div}(\rho T \vec{u} + \vec{J}^{(0)}_T) = -\rho T \mathrm{div}\vec{u} \tag{1c}$$

The local equilibrium solution gives for the momentum tensor:

$$\Pi^{(0)}_{\alpha\beta} = p\delta_{\alpha\beta}$$

with $p = k\rho T$ the pressure. $\vec{J}^{(0)}_T$ is an unusual contribution to the heat flux, given by:

$$\vec{J}^{(0)}_T = (\phi(\rho,t) - 2T)\rho\vec{u} \tag{2}$$

where ϕ is some known rational function of ρ and T.

Moreover the Chapman-Enskog solution of the Boltzmann equation governing the evolution give the dissipative part of the heat current $\vec{J}_{tot} = \vec{J}^{(1)}_T + \frac{\tau}{2}\vec{H}$, which, for a stationary situation with $\vec{u} = 0$, obeys to [5]:

$$\mathrm{div}\vec{J}_{tot} = 0.$$

$\vec{J}^{(1)}_T$ is the usual thermal flux due to collisions and it is given by:

$$\vec{J}^{(1)}_T = D(\rho,T)\mathrm{grad}\rho - \kappa(\rho,T)\mathrm{grad}T$$

where the Dufour coefficient D and the thermal conductivity κ are explicit functions of ρ and T. The term $\vec{H}$ comes from the discreteness of the lattice and is found to be:

$$\vec{H} = -v^2 \nabla\rho + 3\nabla p.$$

For a fluid at rest we have also from (1b) that $\nabla p = 0$, and the expression for the heat current reduces to:

$$\vec{J}_{tot} = -\kappa_{eff}(p,T)\mathrm{grad}T \tag{3}$$

Cellular automata simulations of this system have been performed [5] with a CAM/6 [6] and are in a quite good agreement with equation (3).

A simple physical interpretation of the term $\vec{H} = -v^2 \nabla \rho + 3\nabla p$ can be given by the following argument: we consider a horizontal section of our square lattice and then we compute the energy transferred by the motion of the particles through this section in a unit time τ (see Fig. 3). Due to their direction of motion, about one half of the fast particles crosses this section carrying an energy $\frac{w^2}{2} = v^2$, while only one quarter of the slow ones do the same with an energy $\frac{v^2}{2}$. Using that the density of fast and slow particles are respectively $\rho(T-1)$ and $\rho(2-T)$ [5], it is easy to compute the net amount of energy transferred through the section by the particles coming from above and below.

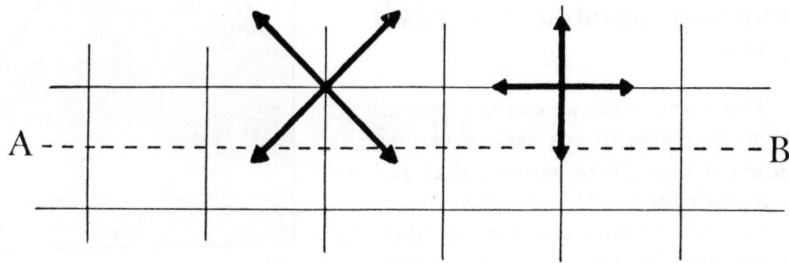

Fig. 3 : *energy transferred by the motion of the particles through a section AB of the lattice.*

These results show us that our model is suitable to describe heat propagation in a fluid at rest. We are now going to illustrate the consequences of the term $\vec{J}_T^{(0)}$ in equation (1c) by calculating the velocity of sound.

From equations (1) we obtain the following relations for the variations of density and temperature:

$$\partial_t^2 \delta \rho - k(\rho \nabla^2 \delta T + T \nabla^2 \delta \rho) = 0$$

$$\partial_t^2 \delta T - k\frac{\phi - T}{\rho}(\rho \nabla^2 \delta T + T \nabla^2 \delta \rho) = 0$$

A class of solution of these equations consists of propagating waves with velocity $c = \sqrt{\frac{v^2}{2}\phi}$. As the above equations follow from Euler's ones, we expect c to be the adiabatic velocity given by:

$$c_s^2 = \left(\frac{\partial p}{\partial \rho}\right)_s = \frac{C_p}{C_V}\left(\frac{\partial p}{\partial \rho}\right)_T$$

For our model we have $p = k\rho T$ and the specific heats are found to be:

$$C_V = k\rho \text{Volume} \quad C_p = 2k\rho \text{Volume}$$

Thus we obtain:

$$c_s^2 = \frac{v^2}{2} 2T$$

For compatibility with $c^2 = \frac{v^2}{2}\phi$ we must have $\phi(\rho,T) = 2T$. From equation (2) it means that $\vec{J}_T^{(0)} = 0$ and a vanishing heat current is precisely what we expect for an adiabatic process. Unfortunately, it turns out that this condition is never fulfilled for any permitted values of ρ and T (Fig. 4).

This means that we can not have an adiabatic process in our system as long as $\vec{u}$ is not zero. Note however that $\vec{J}_T^{(0)}$ is not associated with any entropy production. This rather nonphysical situation reflects the lack of Galilean invariance which is a general disease of lattice gas models.

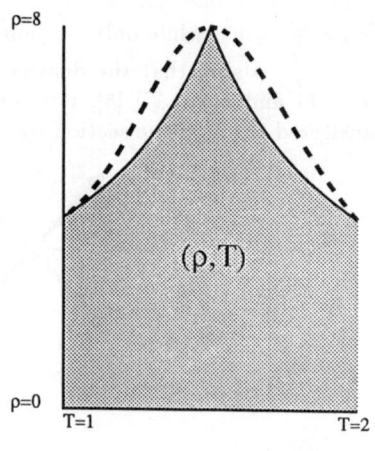

Fig. 4: *Admissible values of ρ and T. The dashed curve shows the solution of $\phi = 2T$.*

References:

[1] U. Frisch, B. Hasslacher and Y. Pomeau, Phys.Rev.Lett., 56, 1505 (1986).

[2] D. d'Humière and P. Lallemand, Europhys.Lett., 2(4), 291 (1986).

[3] H.J. Kreutzer, " Nonequilibrium Thermodynamics and its Statistical Foundations", Oxford University Press (1981).

[4] U. Frisch, D. d'Humière, B. Hasslacher, P. Lallemand, Y. Pomeau and J.-P. Rivet, in "Modern Approach to Large Nonlinear Systems" to appear in J. Stat. Phys.(1987).

[5] B.Chopard and M.Droz, to be published.

[6] T.Toffoli and N.Margolus, "Cellular Automata Machine: a new environment for modeling", MIT Press (1987).

NONEQUILIBRIUM PHASE TRANSITIONS AND CELLULAR AUTOMATA

Michel Droz * and Bastien Chopard *
Department of Theoretical Physics
University of Geneva
CH-1211 Geneva 4, Switzerland

Abstract: *A cellular automata model of surface reaction is discussed. This model describes a simple adsorbtion-dissociation-desorption process on a catalytic surface. It exhibits two second order nonequilibrium phase transitions. The stationary critical exponents β characterizing the behaviour of the order parameters near the transitions as well as the dynamical critical exponent Δ describing the critical slowing down are found to be mean field like. Qualitative physical arguments are given to explain this behaviour. Several other models showing nonequilibrium phase transitions are briefly reviewed. The problem of the existence and characterization of universality classes for nonequilibrium critical exponents is then discussed.*

1. Introduction.

It is well known that the diagram of steady states of a system far away from thermal equilibrium shows some close analogies to the phase diagram of systems in thermal equilibrium. If the steady state changes as a function of a control parameter one speaks of nonequilibrium phase transitions. Numerous examples of nonequilibrium phase transitions are discussed in the literature in different fields among which laser physics (Graham and Haken 1970), hydrodynamics instabilities (Bergé 1979), generation recombination processes in semiconductors (Scholl 1987), chemical reactions (Nitzan et al 1974; Nicolis et al. 1983), lattice gas models of fast ionic conductors (Katz et al,1984), liquid-gas transition in fluids under shear (Onuki and Kawasaki 1975).

The analogies between equilibrium and nonequilibrium phase transitions allow to define the order parameter and the control parameter. However, if

---

* Supported by the Swiss National Scientific Foundation.

there are some analogies, there are also some important differences. The equilibrium states are obtained by minimalizing the free energy but there is generally no thermodynamic potential for nonequilibrium systems which determines the steady states by an extremalization procedure. If in the framework of linear thermodynamic the entropy production plays such a role, this is no longer true in the nonlinear regime where the nonequilibrium phase transitions often occur. Another major difficulty proper to the nonequilibrium case is how to describe the fluctuations. Indeed, there is no well established dynamical Ginzburg-Landau description for the critical fluctuations. Moreover, the fluctuation-dissipation theorem does not holds anymore. It is thus difficult to justify simple phenomenological dynamical equations. However, it is clear that the fluctuations should play a crucial role in the vicinity of a second order nonequilibrium phase transition. Accordingly, one cannot expect to characterize correctly the critical behaviour of a system without treating these fluctuations properly.

In view of the above remarks and the lack of a first principle formalism, it is of interest to study particular models for which a complete solution is possible. Simulation of nonequilibrium phase transitions in terms of cellular automata offer, as we shall see, an new interesting approach to nonequilibrium phase transitions. As an example, a surface reaction model will be discussed in the next section.

2. Surface reaction model.

The surface reaction model that we consider is similar to the one proposed recently by Ziff et al. (1986). Differences between the two models will show up through the modelization in terms of CA. These reaction models are based upon some of the known steps of the reaction $A - B_2$ on a catalyst surface (for example $CO - O_2$). The basic steps in heterogeneous catalysis are the following:
i) A gas mixture with concentrations X_{B_2} of B_2 and $X_A = (1 - X_{B_2})$ of A sits above the surface and can be adsorbed. The surface can be divided in elementary cells. Each cell can absorb one atom only.
ii) The B species can only be adsorbed in the atomic form. A molecule B_2 approaching an empty cell will be dissociated into two B atoms only if another cell adjacent to the first one is empty. The two first steps correspond to the reactions

$$A \to A(\text{ads}) \quad B_2 \to 2B(\text{ads})$$

iii) If two nearest neighbour cells are occupied by different species they chemically react according to the reaction

$$A(\text{ads}) + B(\text{ads}) \rightarrow AB$$

and the product of the reaction is desorbed. This final desorption step is necessary for the product to be recovered and for the catalyst to be regenerated. However, the gas above the surface is assumed to be continually replenished by fresh material. Thus its composition is constant during the whole evolution.

Note that this model neglects several features which may be present in a real situation (see Chopard and Droz (1987) for a more detailed discussion).

Cellular automaton realization

In view of the above description, it seems natural to represent the problem in term of a two-dimensional cellular automaton.

The cells of the automaton correspond to the elementary cells of the catalyst. Each cell j can be in four different states $|\psi_j\rangle = |0\rangle, |A\rangle, |B\rangle$ or $|C\rangle$.

$|0\rangle$ correspond to an empty cell, $|A\rangle$ to a cell occupied by an atom A, $|B\rangle$ to a cell occupied by an atom B. The state $|C\rangle$ describes the conditional occupation of the cell by an atom B. Conditional means that during the next evolution step of the automaton, $|C\rangle$ will become $|B\rangle$ or $|0\rangle$ depending upon the fact that a nearest neighbour cell is empty and ready to receive the second B atom of the molecule B_2. This conditional states is necessary to describe the dissociation of the B_2 molecules on the surface.

The time evolution of the CA is given by the following set of rules, fixing the state of the cell j at time $t+1$, $|\psi_j\rangle(t+1)$, as a function of the state of the cell j and its nearest neighbors (Von Neumann neighborhood) at time t:

<u>R1</u> : If $|\psi_j\rangle(t) = |0\rangle$ then

$$|\psi_j\rangle(t+1) = \begin{cases} |A\rangle & \text{with probability } X_A \\ |C\rangle & \text{with probability } (1 - X_A) \end{cases}$$

<u>R2</u> : If $|\psi_j\rangle(t) = |A\rangle$ then

$$|\psi_j\rangle(t+1) = \begin{cases} |0\rangle & \text{if at least one of the nearest neighbour cell} \\ & \text{of } j \text{ was in the state } |B\rangle \text{ at time } t \\ |A\rangle & \text{otherwise} \end{cases}$$

R3 : If $|\psi_j\rangle(t) = |B\rangle$ then

$$|\psi_j\rangle(t+1) = \begin{cases} |0\rangle & \text{if at least one of the nearest neighbour cell} \\ & \text{of } j \text{ was in the state } |A\rangle \text{ at time } t \\ |B\rangle & \text{otherwise} \end{cases}$$

R4 : If $|\psi_j\rangle(t) = |C\rangle$ then

$$|\psi_j\rangle(t+1) = \begin{cases} |0\rangle & \text{if none of the nearest neighbors was} \\ & \text{in the state } |C\rangle \text{ at time } t \\ |B\rangle & \text{otherwise} \end{cases}$$

This rule expresses the fact that the atoms of B_2 can be adsorbed only if they had been dissociated on two adjacent cells, that is if at least two adjacent cells were empty at time $t-1$.

Rules R1, R4 describe the adsorbtion-dissociation mechanism while rules R2, R3 describe the reaction desorption process. As discussed by Chopard and Droz (1987) the above rules does not reproduced exactly the physics described above but retain the main features.

As a function of the concentration X_A of the gas, different stationary states of the catalysis surface can be foreseen. If X_A is large the surface will completely be covered by the A atoms after some time. The small fraction of B atoms originally adsorbed will rapidly be eliminated through the desorption of AB. The stationary state will consist of a "poisoned" surface with pure A. If X_A is small we have the opposite situation and the stationary state will be "poisoned" with pure B. Obviously, once the surface is poisoned by a species, the reaction rate for the creation of AB is zero.

For an intermediate value of X_A, a more interesting situation can occur in which the stationary state is a mixed state composed of a fraction X_A^a of atoms A, a fraction X_B^a of atoms B and a fraction X_o^a of empty cells. The adsorption-dissociation mechanism compensate for the desorption process. The reaction rate is obviously finite in this intermediate region.

The diagram of steady states is then obtained by numerical simulation.

3. Results of the numerical simulations

The CA model defined in Section 2 has been simulated on a special purpose machine having 256×256 sites with periodic boundary conditions. At time $t = 0$ a randomly prepared mixture of gas with fixed concentration of one species

(X_A) sits on top of the surface. All the cells are initially empty (state $|\phi\rangle$). The evolution starts following the rules defined in Section 2. The rules are iterated many times until a stationary state is reached. The stationary state is a state for which the mean coverage fractions X_A^a and X_B^a of atoms of type A or B does not change in time, although microscopically the configurations of the CA can still change.

Typical configurations are shown on Fig. 1 a and b. The black dots represent the adsorbed B atoms.

The resulting phase diagram is shown in Figure 2. For $X_A \geq 0.6515 \pm 0,0004 = X_{A1}$ the stationary state is "poisoned" with A. For $X_A \leq 0,5761 \pm 0,0004 = X_{A2}$ the stationary state is poisoned with B. At X_{A1} and X_{A2} one has two non equilibrium phase transitions in the sense that, in the mixed phase, the coverage concentrations varie continuously as a function of X_A at the transitions.

We have also determined how the coverage fraction X_A^a was going to zero (or one) near the transition points. The distances from the critical points are determined respectively by the parameters $\varepsilon_1 = (X_{A1} - X_A)/X_{A1}$, $\varepsilon_2 = (X_A - X_{A2})/X_{A2}$.

Fitting the data in the range $8 \cdot 10^{-4} < \varepsilon_1 < 5 \cdot 10^{-3}$ one finds that

$$\psi_1 \equiv X_A^a(X_{A1}) - X_A^a(X_A) \sim \varepsilon_1^{\beta_1} \tag{1}$$

with $\beta_1 = 0.55 \pm 0.05$.

ψ_1 can be identified with the order parameter. β_1 is then the stationary critical exponent for the order parameter.

For the second transition, associated to the order parameter ψ_2, one has :

$$\psi_2 \equiv X_A^a(X_A) - X_A^a(X_{A2}) \sim \varepsilon_2^{\beta_2}$$

with
$$\beta_2 = 0,45 \pm 0,05 \quad \text{for} \quad 2 \cdot 10^{-3} < \varepsilon_2 < 10^{-2}. \tag{2}$$

Interesting information can also be obtained concerning the dynamics. In equilibrium statistical mechanics, it is known that a system initially in a nonequilibrium state will decay towards its equilibrium one usually exponentially. The decay is characterized by a relaxation time τ. When approaching a second order phase transition, this relaxation time diverges (critical slowing down) like :

$$\tau \sim \varepsilon^{-\Delta} \tag{3}$$

where ε is the reduced temperature, i.e. the distance from the critical point and Δ the dynamical critical exponent.

A similar situation is present in nonequilibrium phase transition. The critical slowing down shows up explicitly in Fig. 1a and 1b. In the case a, one has $\varepsilon_1 = 0.0018$ while in the case b, $\varepsilon_1 = 0.0116$. The pictures are sensibly the same but in the case a they correspond to respectively 0, 100 and 400 evolution steps while in the case b they correspond to 0, 500 and 2000 steps. The evolution is thus slower when one is closer to the critical point.

The numerical procedure adopted to compute the critical dynamical exponents is the following (Chopard and Droz (1987)). For a fixed value of X_A, one computes $X_A^a(t)$ or $X_B^a(t)$ depending of the critical point considered. Then the data are fitted with the following decay law :

$$X_\alpha^a(t) = [X_\alpha^a(0) - X_\alpha^a(\infty)] \exp[-t/\tau(X_A)] + X_\alpha^a(\infty) \tag{4}$$

with $\alpha = A$ or B.

Near the transition points, the relaxation times are fitted with the following law :

$$\tau_j(X_A) \sim \varepsilon_j^{-\Delta_j} \tag{5}$$

where $j = 1, 2$ labels the two transitions.

The values obtained for Δ_j are the following. When approaching the transitions from the "disordered" phase, i.e. the poisoned ones one finds, for $9 \cdot 10^{-4} < \varepsilon_{1,2} < 3 \cdot 10^{-2}$:

$$\Delta_1 = 0,98 \pm 0,08 \quad \Delta_2 = 1,03 \pm 0.10. \tag{6}$$

When approaching the transitions from the "ordered" phase, i.e. nonpoisoned, one finds :

$$\Delta_1 = 1,02 \pm 0,15 \quad \Delta_2 = 0,84 \pm 0,15 \tag{7}$$

4. Mean field theory.

It is natural when studying the critical behaviour of a system to start with a mean field like approximation. The basic idea behind such an approximation is to average out the spatial inhomogeneities. Different levels of approximation are possible. The simplest one is the site approximation in which all spatial correlations between sites are ignored. One writes equation of motion for the concentrations of the different chemical species. A better approximation consists in taking into account the nearest neighbour pair correlations. One then writes evolution equations for the bond concentrations. Such approximations have been worked out in details for the Ziff model by Dickman (1987). The extension to our model is immediate. The mean field values of the critical exponents are $\beta = \frac{1}{2}$ and $\Delta = 1$.

Coming back to the numerical values obtained in our simulation, we see that both the "static" and dynamic exponents are, within the precision of the simulation, compatible with the mean field predictions. This surprising result demands some explanation. What is the reason for which the critical behaviour is mean field like? One important ingredient of our model is the gas mixture sitting above the catalytic surface. This gas has its own independent dynamics. Accordingly a cell on the surface sees above it, during the time evolution of the CA, a random succession of A and B species (with however a given probability). This kind of stirring seems efficient enough to suppress the short range correlations between the cells of the surface, hence, explaining a mean field behaviour.

5. Comparison with other models for nonequilibrium phase transitions

The validity of above argument concerning the mean field nature of the transition is strengthened by the results of van Beijeren and Schulman (1984). They considered a slightly different version of the model proposed by Katz, Lebowitz and Spohn (1984). Those are stochastic lattice gas models with particles conserving hoping dynamics in a nonuniform external field. They modelized fast ionic conductors. The interest of the van Beijeren et al. model is the fact that one can find exactly the stationary states. It turns out that the "static" nonequilibrium critical exponents are mean field like. In this case, the field stirs the degrees of freedom of the system so effectively that the nearest neighbor correlations are almost completely suppressed.

This mean field character is encountered in other nonequilibrium phase transitions as well. Onuki and Kawasaki (1979) have studied the nonequilibrium steady state of critical fluids under shear flow using a renormalization group approach. They showed that the introduction of the shearing, however small, changes the character of the liquid-gas transition to mean field type. In this case, distortion and suppression of the long wavelength fluctuations by the flow becomes essential.

If several nonequilibrium phase transitions have a mean field character, there are some other cases which are not mean field like. A typical example is given by the Schlögl model (1972) of chemical reaction. Elderfield and Vvendensky (1985) have showed that this model can be mapped on a Reggeon field theory (Abrabanel et al. 1976). The critical exponents of these models are not mean field in dimensions $d < d_c = 4$.

6. Universality classes in nonequilibrium phase transitions

One of the most important concept in the theory of equilibrium phase transitions is the one of universality classes. It is well established that the static critical exponents characterizing an equilibrium phase transition as well as the dynamic exponent z characterizing the critical slowing down near an equilibrium critical point, can be grouped in universality classes. The dimensionality of the system d, the number of components of the order parameter n, the range of the interactions σ and the conservation laws entering into the dynamical equations suffice to characterize the universality classes (Hohenberg and Halperin 1977).

It is thus legitimate to ask if such universality classes exist in the framework of nonequilibrium phase transitions. A concept which is certainly still valid is the one of critical upper dimensionality d_c. If the dimensionality d of the system is larger than d_c, the mean field predictions are correct for the exponents. Below d_c, the fluctuations play an important role in the critical domain. Two questions then arise. What is the value of d_c and how large is the critical region?

It is difficult to give a general answer to those questions partly in view of the fact that a first principle approach to the description of nonequilibrium systems is lacking. However, several remarks can be made. As far as the width of the critical region is concerned, it has been argued in some particular examples that the critical region could be very small and not accessible to the experiments (Switf and Hohenberg 1977).

As far as the value of d_c is concerned, two cases could be distinguished. In the first one, the system is submitted to external fields and forces which strongly uncorrelate the degrees of freedom of the system. In this case, d_c could be very low ($d_c = 1$ or 2) and thus in dimensions 2 or 3, the critical behaviour is mean field like. In the second case, the external fields do not uncorrelate the degrees of freedom and d_c could be large (4 or more). Thus corrections to mean field are expected in dimensions 2 or 3. In this case, the characterization of the universality classes is certainly not as simple as for the equilibrium situation. Complications proper to the nonequilibrium aspect should be expected. For example, Onuki and Kawasaki (1979) have showed, in their study of critical fluid under shear flow, that the renormalization group recursion relations were not separable in a static and a dynamic part as it is usually the case in critical equilibrium dynamics.

In conclusion, the definition of universality classes for nonequilibrium phase transitions is a difficult and open problem. By analogy with the equilibrium case, the formulation of a Ginzburg-Landau criterion for nonequilibrium phase transitions would be very useful. Having that, one could hope to be able to go one step further in the characterization of the universality classes.

References:

Abarbanel H.D, Bronzan J.B, Schwimmer A. and Sugar R.L., (1976), Phys.Rev. D14, 632.
Bergé P., (1979), in "Dynamical Critical Phenomena and Related Topics", 288, Springer Verlag Lecture Notes in Physics, eds. C.P. Enz.
Chopard B. and Droz M., (1987), J. Phys. A, to appear.
Dickman R., (1987), Phys. Rev. A. to appear.
Elderfield D. and Vvedensky D.D. (1985), J.Phys. A18, 2581.
Graham R. and Haken H., (1970), Z. Physik 237, 31.
Hohenberg P.C., and Halperin B.I., (1977), Rev. Mod. Phys. 49, 435.
Katz S., Lebowitz J.L., and Spohn, H., (1984), J. Stat. Phys. 34, 497.
Nicolis G., Baras F. and Malek Mansour M., (1983), in "Nonequilibrium cooperative phenomena in physics and related fields", M.G. Velarde eds., Plenum Press, New-York.
Nitzan A., Orloleva P., Deutch J. and Ross.J., (1974), Jour. Chem. Phys. 61, 1056.
Onuki A. and Kawasaki K., (1979), Ann. Phys. 121, 456.
Schlögl F., (1972), Z. Physik 253, 147.
Schöll E., (1987), "Nonequilibrium phase transitions in semiconductors", Springer-Verlag, Berlin.
Swift J. and Hohenberg P.C., (1977), Phys.Rev. 15A, 319.
van Beijeren H., and Shulman L.S., (1984), Phys. Rev. Lett. 53, 806.
Ziff R.M., Gulari E., and Barshad Y., (1986), Phys. Rev. Lett. 56, 2553.

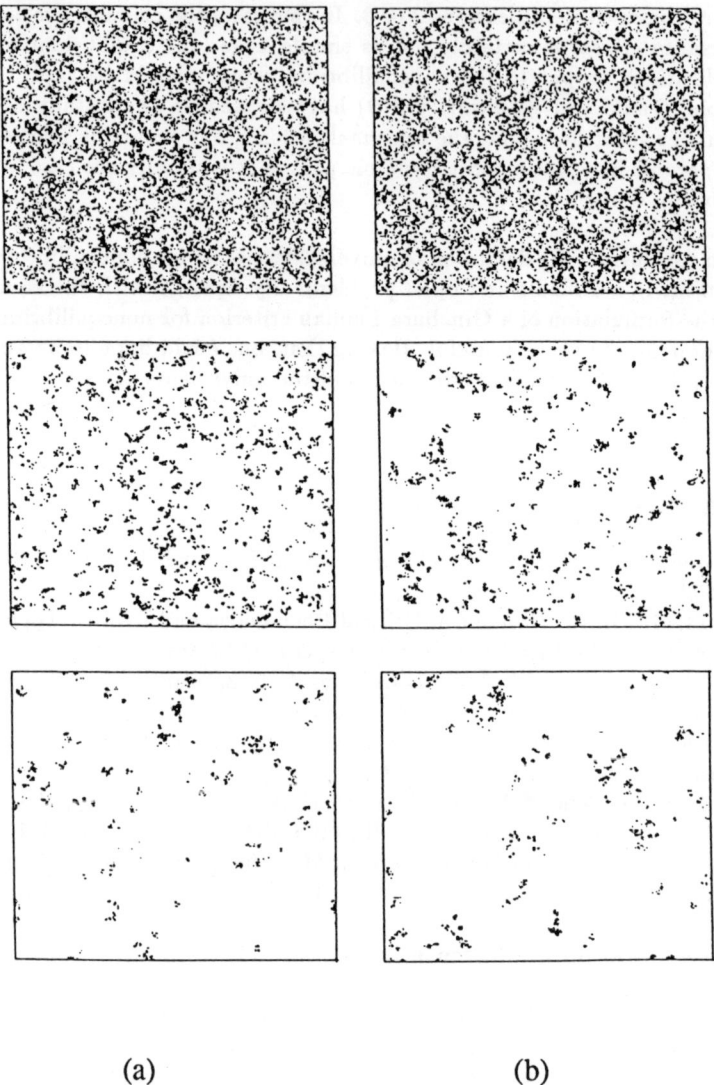

(a)   (b)

Fig. 1: *Number of particles on the catalytic reaction plane. Each dot represents a particle of species B.*
(a) For $\varepsilon_1=0.0018$ and at times $t=0, 500, 2000$ (from above to below).
(b) For $\varepsilon_1=0.0116$ and at times $t=0, 100, 400$ (from above to below).

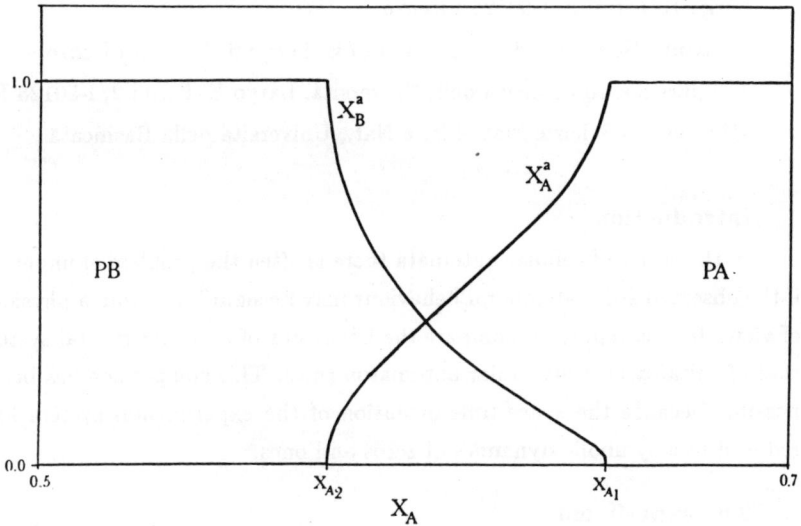

Fig. 2: *Stationary state diagram. X_A is the concentration of the A species of the gas sitting above the catalyst. X_A^a and X_B^a are respectively the fractions of A and B adsorbed. X_{A1} and X_{A2} are the two critical concentrations. PA and PB mean phases poisoned in respectively A and B.*

Cellular Automaton Model for a Fluid Experiment

F. Bagnoli[1], S. Ciliberto[1,2], A. Francescato[1],
R. Livi[1,3], S. Ruffo[1,4]

1) Istituto Nazionale di Fisica Nucleare Sez. di Firenze,
Largo E. Fermi 2, I-50125 Firenze

2) Istituto Nazionale di Ottica, Largo E. Fermi 6, I-50125 Firenze

3) Dipartimento di Fisica dell' Universitá, Largo E. Fermi 2, I-50125 Firenze

4) Facoltá di Scienze Mat. Fis. e Nat., Universitá della Basilicata

Introduction

In the study of cellular automata there is often the problem of understanding if the observed spatio-temporal behaviour may be significant from a physical point of view. In this report we compare the behaviour of an experimental system with that of suitably chosen cellular automaton rules. This comparison has been made possible because the space-time evolution of the experimental system has been reduced to a symbolic dynamics of zeros and ones.

The experiment

The system of interest is an annular fluid layer confined between two horizontal plates and heated from below (Rayleigh-Benard convection). When the temperature difference ΔT exceeds the threshold value ΔT_c a steady convective flow consisting of radial rolls (roll axes along radial directions) arises. In the annular geometry the spatial pattern has periodic boundary conditions. Furthermore the sizes of the cell constrain the convective structure to an almost one dimensional chain of rolls. The inner and outer diameters of the annulus are 6 cm and 8 cm respectively and the depth of the layer is 1 cm. The working fluid is silicon oil with a Prandtl number of about 30. The critical value of ΔT at the onset of convection is $\Delta T_c = 0.06°C$. The space-time evolution of the system has been characterized by using $u(\theta, t) = \frac{1}{r_0} \frac{\partial T(r_0, \theta, t)}{\partial \theta}$ that is the component of the temperature gradient perpendicular to the roll axis. The function $u(x,t)$, where $x = \theta/(2\pi)$, is sampled at 128 points in space on the circle of radius $r_0 = 3.6$ cm. When the regimes are time dependent $u(x,t)$ is recorded for at least 5000 times at intervals of 1 sec, that is about 1/10 of the main oscillation period of our system. More details on the experimental apparatus are given in Ref. [1].

Analyzing the fluid behaviour as a function of $\eta = \Delta T/\Delta T_c$ we observe that, for η around 1, the spatial structure has 24 rolls. This number increases with η and reaches 38 at η around 200. The spatial structure remains stationary for $\eta < 183$ where a subcritical bifurcation to the time dependent regime takes place.

The regime to be compared with the cellular automaton begins at $\eta = 200$. It presents, at the same time several domains where the spatial periodicity is completely lost (we will refer to them as turbulent) and other regions (that we call laminar) where the spatial coherence is still maintained.

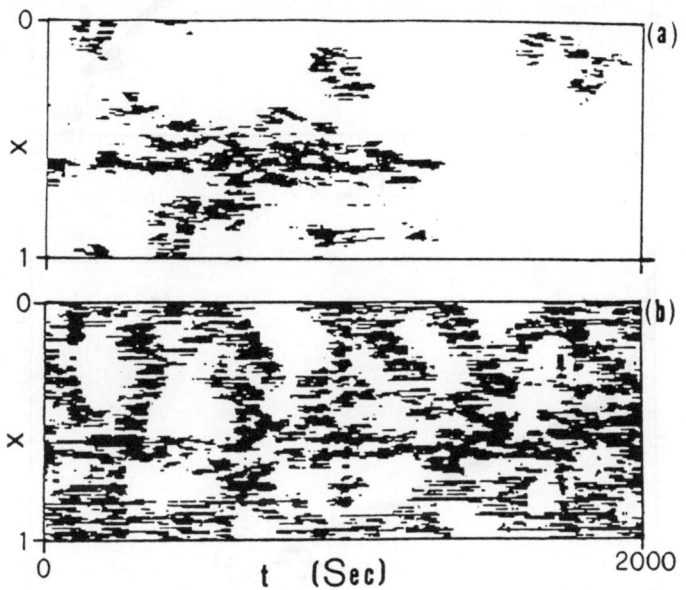

Figure 1: Binary representation at $\alpha = 1.5°C/cm$ of the space-time evolution of $u(x,t)$ at $\eta = 216$ (a) and $\eta = 248$ (b). The dark and white areas correspond to turbulent and laminar domains, respectively.

In the turbulent domains the time evolution is characterized by the appearance of large oscillatory bursts, that locally destroy the spatial order. Instead, in laminar regions the oscillations remain very weak. Thus the two regions can be identified by measuring the local peak to peak amplitude, for a time interval comparable with the mean period of the oscillation. Choosing a cutoff α and making black all the points where the oscillation amplitude is above α, we can easily represent the dynamics of turbulent and laminar regions.

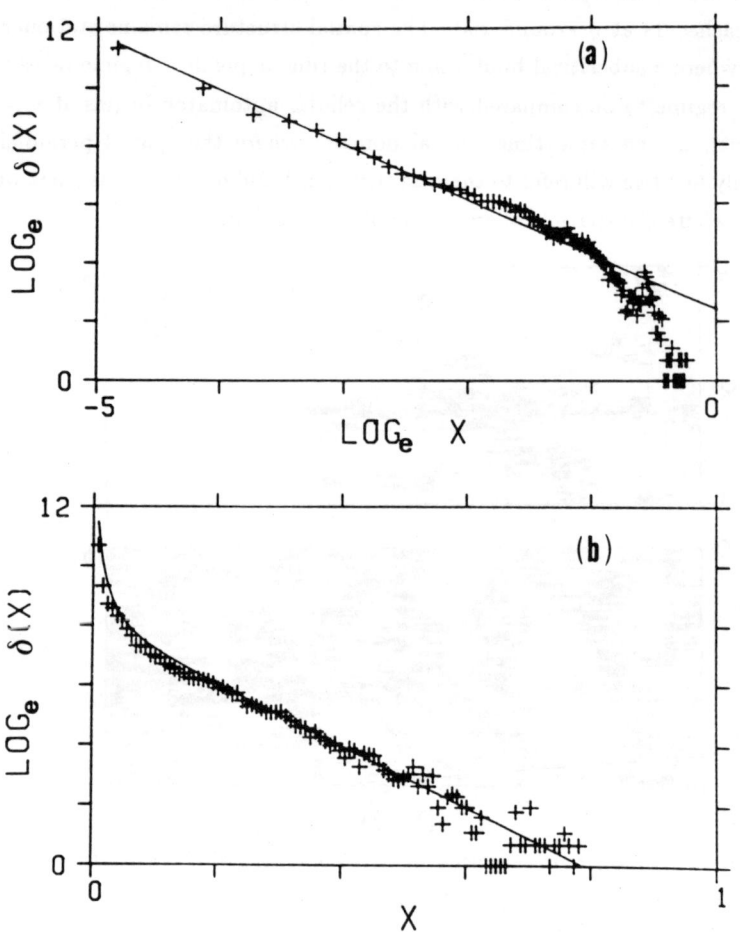

Figure 2: Distribution $\delta(x)$ of the laminar domains of length x; (a) $\eta = 241$, the algebraic decay with exponent 1.9; (b) $\eta = 310$, and $\alpha = 1.6°C/cm$ exponential decay with a characteristic length $(1/m) = 0.10$. The solid lines are numerical fits.

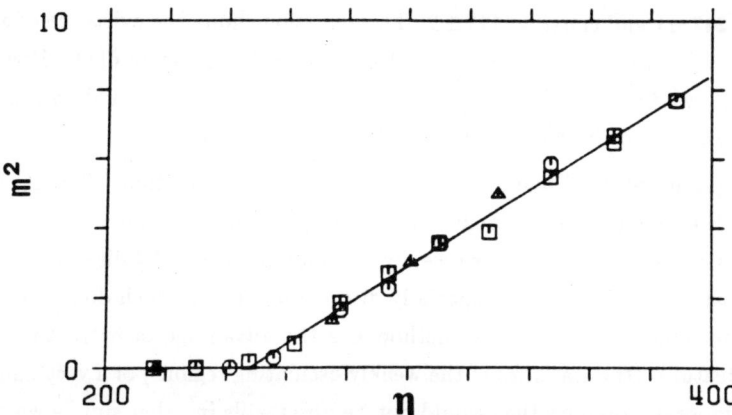

Figure 3: Dependence of m on η, the different symbols pertain different sets of measurements done either increasing or decreasing η.

As an example of such a code we show the spacetime evolution of $u(x,t)$ at $\eta = 216$, in Fig.1a, and $\eta = 248$ in Fig.1b. We remark that the qualitative features of these pictures are rather independent of the precise value of the cutoff. At $\eta = 216$ (see Fig.1a) a wide laminar region surrounds completely the turbulent patches that remain localized in space, after their appearance. Furthermore, the nucleation of a turbulent domain has no relationship with the relaxation of another one. In contrast, at $\eta = 248$ (see Fig.1b), the turbulent regions migrate and slowly invade the laminar ones. Such a behaviour has been quantitatively characterized by computing, over a time interval of $10^4 sec$, the distribution $\delta(x)$ of the the laminar domains of length x. The existence of two different regimes is clearly shown in Fig.2a and Fig.2b which display $\delta(x)$ versus x at $\eta = 241$ and $\eta = 310$, respectively .At $\eta = 241$ (Fig. 2a) $\delta(x)$ can be fitted with a power law. The exponent ,equal to 1.9 in Fig. 2a, does not depend, within our accuracy, either on α or on η. Its average value is $\rho = 1.9 \pm 0.1$. On the other hand, at $\eta = 310$, Fig. 2b, the decay for $x > 0.1$ is exponential with a characteristic length $1/m = 0.10$. The dependence of m as a function of η is reported in Fig.3 . The linear best fit for $\eta > 246$ of the points of Fig.3 gives the following result

$$m(\eta) = const.(\eta - 1)^{\frac{1}{2}} \qquad (1)$$

with $\eta = 247 \pm 1$ and $const. = 117 \pm 2$. This equation shows the existence of a well defined threshold η for the appearance of an exponential decay in $\delta(x)$. However, for $\eta \leq 247$, we find that the minimum of m is of order .1 as it should be expected because $1/m$ cannot be bigger than the size of the system.

This point will be discussed in detail in the following sections. The choice of a simple threshold criterion to reduce the spatio-temporal dynamics to a symbolic one merits a special comment because, by this method, many details of the system may be lost. Specifically all the spatially disordered regions which do not oscillate are not detected. However the method has the advantage of extracting some relevant features (the statistics of the weakly oscillating regions) of a very complex spatio-temporal dynamics that would not be observable in other simple ways.

The cellular automaton model

The reduction of the experimental signal to a binary sequence leads in a natural way to a modelization in terms of a boolean cellular automaton (C.A.). From now on our convention is 0=laminar, 1=turbulent. The characteristic geometry of the experiment allows us to restrict to a one-dimensional C.A. with N cells and periodic boundary conditions (p.b.c.). The C.A. model we are going to discuss can be thought as a simplified microscopic model of the distribution in space of fluid turbulence. It shows in the thermodynamic limit ($N \to \infty$), a phase transition for the density of laminar (turbulent) regions. This transition produces drastic effects on the other statistical properties of the spatial distribution of laminar and turbulent regions. Let us briefly discuss the steps which lead to the choice of the model, implying both theoretical and experimental arguments. A qualitative inspection of the experimental patterns shows that the system does not evolve according to some deterministic rule: each neighborhood evolves in a probabilistic fashion. For the sake of simplicity, we want to deal with only one control parameter. One easy choice is to mix two deterministic rules with normalized probabilities.

Some heuristic considerations may help in the choice of the two rules. First of all we observe experimentally a region of the control parameter where turbulence is spatially localized. If a turbulent region appears it does not spread and eventually dies. The simplest localization mechanism is produced by a majority rule. If we restrict to nearest-neighbor interaction this is rule 232 in the notation by Wolfram

[2]. A random initial configuration evolves, after a short transient, to a stable pattern of stripes of 0's and 1's. In the average the initial densities are conserved. In another region of the control parameter turbulence is no more localized and spreads over space. This behavior was also found in coupled map lattices [3], introduced for the first time in ref.[4]. An interpretation in terms of directed percolation [5] has been suggested, but the transition does not seem to belong to the same universality class [3]. Rule 90 is a good choice to represent a percolation process [6,7]. In fact, a strict correspondence among probabilistic growth processes in 1-D and 2-D statistical models on disorder varieties has been derived [7].

Our proposal is, therefore, to consider a probabilistic C.A., which is a mixture of rule 90 with probability P and rule 232 with probability Q. At each step the updating of each site is performed according to this probabilistic recipe (we are considering the "annealed" case [8]).

The experimental data show that a turbulent region may originate from a laminar one. This fluctuation effects due to thermal noise can be introduced in our C.A. model by using "illegal rules" [2], which generate a site with value 1 from a triplet [000] (odd number rules in the notation of Wolfram). Instead of choosing an illegal rule randomly (say the largest illegal rule 255) we have considered the rules which are complementary to 90 and 232, i.e. rule 165 and 23 respectively. We have assigned to them a much smaller probability, according to the relations:

$$p_{23} = \nu P \quad ; \quad p_{165} = \nu Q \qquad (2)$$

with $\nu = O(10^{-2})$. The normalization condition involves the following relation between P, Q and ν : $(P+Q)(1+\nu) = 1$. With this choice the number of turbulent (laminar) outputs is constant for any input configuration. Moreover, the probability of a mostly turbulent region to return laminar is non-zero but $O(\nu)$. These two requests seem physically reasonable.

Mean field theory and numerical simulations

Let us consider the evolution equation for the probabilities of local configurations

$$p_1(x;t+1) = \sum_{z,v,w} \tau(z,v,w|x) p_3(z,v,w;t) \qquad (3)$$

where $p_n(x_1, \cdots x_n; t)$ is the probability of the configuration $x_1, \cdots x_n$ at time t with $x_i = 0, 1$ and $\tau(x_1, \cdots x_n | x)$ is the transition probability from time t to time $t+1$ given by the evolution rule. A simple mean field hypothesis consists in approximating multiple probabilities $p_n(x_1, \cdots x_n; t)$ with independent factorized probabilities of lower order. As a first approximation we impose

$$p_n(x_1, \cdots x_n; t) \sim \prod_i p_1(x_i; t) \qquad (4)$$

Taking into account the normalization condition $p(0; t) + p(1; t) = 1$ one arrives, for the probabilistic C.A. defined in the previous section, at the following recursion relation for $p(0; t)$ when $\nu = 0$:

$$p(0; t+1) = 2p^3(0; t)(P-1) + p^2(0; t)(3-P) - 2p(0; t)P + P \qquad (5)$$

The dependence of the fixed point $p(0)$ of this map as a function of the control parameter P is shown in Fig.4 .

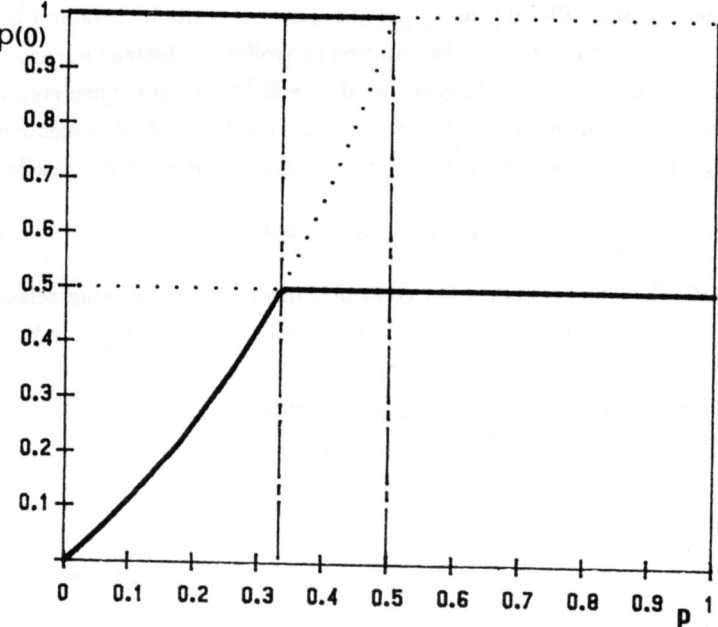

Figure 4: $p(0)$ as a function of P.

For any value of P there are two stable (full line) and one unstable (dotted line) fixed points. For $P > .5$ the upper stable fixed point is not shown in Fig.4, since it goes in the non-physical region $p(0) > 1$; its analytical expression is $p(0) = P/(1-P)$. The numerical simulations have been performed for the noisy case ($\nu = 5 \cdot 10^{-3}$) and show that $p(0)$ follows the upper fixed point $p(0) = 1$ until $P = .5$; above this value $p(0)$ relaxes to the other stable fixed point value $p(0) = .5$; $p(0)$ is shown by the full thick line in Fig.5 .

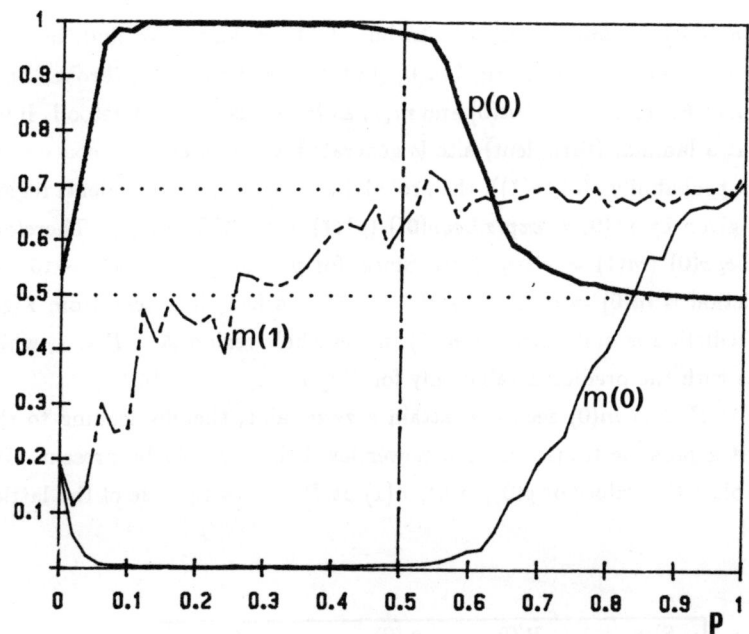

Figure 5: $p(0), m(0), m(1)$ as a function of P with $\nu = 5 \cdot 10^{-3}$

Even the behavior of $p(0)$ near $P = 0$ ($P \leq .1$) can be understood: our simulations were started with an initial value for the density $p(0;0) = .5$ and since rule 232, corresponding to the $P = 0$ value in Fig.5, preserves the initial density it is clear that the curve must approach the value $p(0) = .5$ as $P \to 0$. Near $P = .5$ the effect of noise can be estimated from mean field theory, which can partially explain the disagreement with the noise-less prediction: around $P = .5$ the theory predicts that fluctuations due to noise get enhanced. Refinements of the mean field theory, including probabilities of higher order, are under investigation.

Another quantity of interest is the probability distribution $\delta_0(x)$ of laminar regions of size x. In ref.[3], where this quantity was introduced, a transition was found for coupled-map lattices, where this quantity changes from an exponential to a power law behavior. This transition is not present in our C.A. model. Let us introduce the parametrization

$$\delta_0(x) \sim exp(-m(0)x) \quad ; \quad \delta_1(x) \sim exp(-m(1)x) \tag{6}$$

for the probability of laminar (0) and turbulent (1) regions. We find, in fact, exponential laws and we show in Fig.5 $m(0)$ (full thin line) and $m(1)$ (broken line) as a function of P. The value of $m(0)$ and $m(1)$ as $P \to 1$ can be understood. If we suppose that a laminar (turbulent) site is generated at any step with space-time independent probability $p(0)$ ($p(1)$), the probability of a laminar (turbulent) region of size x is given by $p^x(0) = exp(xLogp(0))$ ($p^x(1) = exp(xLogp(1))$). Therefore $m(0) = -Logp(0)$ ($m(1) = -Logp(1)$). Since, for $p > .5$, $p(0) \sim p(1) \sim .5$ our naive prediction is $m(0) = m(1) = Log(.5) = .69$. In fact, it is seen from Fig.2 that this prediction is quite nice for $m(1)$ in the whole region $.5 < P < 1$, while $m(0)$ agrees with the predicted value only for $P \lesssim 1$.

For $.1 < P < .5$ $m(0)$ seems to attain a zero value, thereby leading to the suspect that a possible transition to a power law behavior may be present. We show in Table 1 the values of $p(0), m(0), m(1)$ at $P = .4$ as the size of the lattice is increased.

Size	P(0)	m(0)	m(1)
500	.9959	2.43E-3	3.81
1000	.9891	4.30E-3	5.24E-1
2000	.9944	4.30E-3	5.73E-1
5000	.9922	3.58E-3	5.99E-1
10000	.9934	3.57E-3	5.74E-1
20000	.9916	3.70E-3	5.61E-1

Table 1: $p(0), m(0) and m(1)$ at $P = .4$ as a function of the lattice size.

We cannot draw a definite answer, but the most likely interpretation of these data is that both $m(0)$ and $m(1)$ reach a non-zero value in the thermodynamic limit. It could be noted that $m(0)$ reaches a value which is of the order of the noise ν. One could suppose that $m(0)$, and even $m(1)$, reaches zero as $\nu \to 0$. We have no numerical result which could support this hypothesis. However, we want our model be "realistic" from the point of view of the fluid experiment, and in the fluid we clearly observe $\nu \neq 0$.

Therefore, although a transition is present in the density of laminar (turbulent) regions, we observe no transition in the functional form of the probability distribution of laminar regions of length x, as it happens in directed percolation (for "wet" sites) [3].

The experiment on the fluid confirms our result for the transition in the density of laminar (turbulent) regions. However, it cannot discriminate a case where $m(0)$ is very small from one where $m(0) = 0$. Therefore an "apparent" transition to a power law behavior of the probability distribution of laminar regions may be due both to experimental uncertainty and to finiteness of the lattice of physically relevant points (in the experiment $N = 128$). Indeed the curvature for $Log x > -1$ in Fig. 2a may be related to these effects.

References

[1] S. Ciliberto and P. Bigazzi, Phys. Rev. Lett. **60**, 286 (1988).
[2] S. Wolfram, Rev. Mod. Phys. **55**, 601 (1983).
[3] H. Chaté and P. Manneville, Phys. Rev. Lett. **58**, 112 (1987) and Saclay - Preprint 1987.
[4] K. Kaneko, Prog. Theor. Phys.,**74**, 1033 (1985) and G.L. Oppo and R. Kapral, Phys. Rev. **A33**, 4219 (1986).
[5] Y. Pomeau, Physica **D23**, 3 (1986).
[6] W. Kinzel, Z. Phys. **58**, 229 (1985).
[7] A. Georges and P. Le Doussal, Ecole Normale preprint LPTENS 87/21 Oct. 1987 and P. Rujan, J. Stat. Phys. **49**, 139 (1987).
[8] G. Vichniac, P. Tamayo and H. Hartmann, J. Stat. Phys. **45**, 875 (1986).

Biology, Evolution Theory and Biological Networks

INFORMATION FLUX AND CONSTRAINTS IN DEVELOPMENT AND EVOLUTION:
A CRITICAL REVIEW

Marcello Buiatti,
Department of Animal Biology and Genetics
Via Romana, 17
50125 Florence, Italy

ABSTRACT

A critical review of the present knowledge of biological dynamical processes is given with the aim of discussing the adequacy of "classical" information theory for the description of such processes and the construction of models.

An analysis is then carried out of information structure and change at three hierarchical organisation levels (molecular, organismal, population and species) and of the interplay between information flux and form and function constraints.

The conclusion is then drawn that, due to the equivocal expression of genetic information, the unpredictable and irreversible nature of biological processes, the heterogeneous meaning of "information" in biology, only "local" modelling can be foreseen at least with the present level of mathematical tools and concepts.

1. INTRODUCTION

Biology is, as other scineces, going through a period of conceptual turmoil, at the very moment when its technical tools are becoming more and more powerful and allow a faster than ever progress in the empyrical knowledge of life.

It is indeed the tremendous amount of data wich have been collected in the last few years which has shown the inadequacy of previous models and theories for a coherent explanation of the living processes.

Historically, biologists have either refused modelling, or borrowed concepts, models and mathematical tools from other sciences. Probability theory and statistics helped giving the foundation to Mendelian and later, population genetics, information theory was consciously or not at the basis of modern explanations of code functions and thereby of development and evolution, catastrophe theory and field theory are being currently used along with thermodynamics to attempt an explanation of complexity etc.

Unfortunately, two range of problems seem to remain unresolved. Firstly, very seldom models are constructed an the basis of up to date knowledge of the empyrical data and the analysis of their conceptual meaning. Secondly, available mathematical tools appear increasingly inadequate for biology, a science with is often based on the study of qualitative rater than quantitative events.

The aim of this paper, then is not to build a new model but to give an overview of biological processes mainly from the point of view of information structure, expression and change analysing with a critical eye the possibilities of use in this field of concepts taken from other disciplines.

Particularly, the models still prevailing in biology schools and

text books, based on the acritical extension to biology of "classical" information theory will be discussed. These models, are generally based on the following assumptions:
1) Organisms are assembled according to a program written on DNA in a readable language whose alphabet is known.
2) Information is unequivocally transferred from DNA to RNA, to protein in a colinear way (Central Dogma).
3) Forms and functions essentially derive from the sequential, additive expression of read information.
4) Evolution is the result of cumulative effects of (adaptive) changes in relative frequencies of different versions (alleles) of information units (genes).

To help the discussion I will formulate the following relevant questions:
1) Can the term "information" be used in a classical sense (information theory) in the case of hereditary and developmental processes?
2) Can we speak of a heritable "assembly" program in the case of organisms?
3) Which role(s) does DNA play in determining form and function?
4) How does it express itself and change during history (development and evolution)?
5) How do organisms change during development and evolution?

The discussion will then use the following working hypoteses:
a) Life is organised on different hierarchical interaction levels[1], the rules of which will be analyzed.
Major levels are: molecules (genetic information), cells, multicellular organisms, populations, ecosystems.
b) Units of each level are continuously fluctuating due to quasi random processes of change.
c) Each level imposes its constraints on lower levels and obeys to

higher level constraints.

2. THE MOLECULAR (INFORMATIONAL) AND CELLULAR LEVELS

2.1. Definition of information units

The crucial questions to be asked at the molecular level are, in my opinion, if and how information units can be unequivocally defined and if their meaning for this and higher levels resides solely on the sequence of letters(A,T,G,C) which compose the read message in such a way that information content can be measured according to Shannon's equation ($- \Sigma \, p_i \ln_2 p_i$). Historically the definitions of genetic information (message) units has been changed several times.

In mendelian times the unit was called a "gene" and could contain the information for one character; later it was shown that one gene would code for one enzyme, many enzymes being involved in the definition of one character. With Benzer's [5] discovery that more than one sequence could be involved in the synthesis of one enzyme the genetic unit was called cistron and shown later[6] to be a DNA sequence colinearly coding for one polypeptide.

Now, this definition only seems to held true for one of the different kinds of information units which may be defined as follows.

a) Structural genes: These are the "classical" genes. Their meaning derives, as previously stated, directly from their code which is transcribed into RNA and translated into proteins essentially according to central dogma. However, not all structural genes have the same influence on the expressed information at higher hierarchical levels (form and function); on the contrary some genes are more "important" than others as they are at the top of a cascade of regulatory events on which we will discuss more at length later on.

Moreover, within the genes there are sequences which are essential for the functioning of proteins translated from them and which have been suggested as evolutionary units[7]. These sequences (domains) are higly conserved and present in proteins with different functions. If this is true, then evolutionary units would be different from functional units. Finally, "classical" genes are continuous in prokaryotes, discontinuous in eukaryotes, in the sense that a part of them is not translated, and in some cases may be read in more than one way. This is the case of "overlapping genes" where the same sequence may be read, with different meanings (different proteins synthesized) according to the "open reading frame" (the point of the sequence from which reading is started).

b) Regulatory genes: These sequences code for proteins whose function is the switching on and of and/or the quantitative modulation of structural genes or which,without being transcribed or translated are a part of one of the different, complex systems of regulation of the action of structural genes. These sequences and their functions are rather heterogeneous[8,12] and cannot be described at lenght in this paper. In all cases however they have a quantitative meaning based on yes/no binary rules. The continuosly varying algebraic sum of yes and nos allows a continuous modulation of the expression of structural genes thus regulated.

The amount of the expressions deriving from regulatory switches is however controlled by factors external to regulatory sequences which may be acting at the level of their translation into regulatory proteins, alter their structure, etc. Interacting hierarchical chains of controls may then be formed which are in turn influenced by conditions external to the system (exchanges of matter and energy with the environment). In some cases specific, highly conserved domains may be recognized also in these sequences. One example is the "homeo box" domain in enkaryotes whose coded aminoacid sequence is probably critical for the binding of the regulatory protein to the regulated "structural" DNA.

c) Intervening sequences: They are the "non translated part" of eukaryotic structural discontinuous genes. Their meaning is again indirect in the sense that their length and their sequence influence the way and the amount of translated sequences. Of great importance are some short sequences within them ("consensus sequences") which "say" where the transcribed full lenght messenger RNA should be cut and, thereby which part of it should be translated. In a way intervening regions may be considered "within gene" regulatory sequences.

d) Amplifiable sequences and (selfish) mobile elements. These are sequences of varying lenght (from a few nucleotides to thousands) which may or not be transcribed but are seldom translated.

Their meaning depends in order of importance from their number of copies, their position on the chromosome, their sequence. Particularly in eukaryotes, the number of copies may vary from the mendelian two, to thousands. The process of increase in number of copies is called "amplification" and may lead to the spread of these sequences throughout the genome with both direct regulatory effects on the expression of genes [14,15] and indirect influences on organismal behaviour of great evolutionary importance due to the consequent increase in total DNA amount independently from the qualitative nature of the sequences amplified. Quality is almost irrelevant also in "mobile elements" first discovered by Barbara Mc Clintock[16] in maize but now known to be present in most organisms. These sequences as many of the fore mentioned units "jump" from one position to another throughout the genome and may alter gene expression when getting close to structural regions or inactivate them by jumping within them.

On the other hand, DNA sequence is important in the sense that only few variations are allowed, for some long repetitive sequences which are transcribed but not translated and may be part of the protein synthesizing machine. One good example of such sequences is that coding for ribosomal RNA^{17}.

The amount of repeated, amplifiable sequences is very low in prokaryotes, but in eukaryotes may range from 20% up to 95% of the total DNA.

2.2. The causes of variation in genetic information

Genetic information and its meaning may be modified by a serie of different events. Firstly, classical "gene mutations", i.e. changes in the quality of the coide of any sequence may occur with consequences which vary according to where the change has occurred within the gene and to the importance of quality for the expression of the particular meaning (in the sense discussed before) of the affected gene. In addition to that rearrangements of sequence subunits (domains), changes in copy number (and/or in the number of entire sets of informations, the chromosomes) modification of genes or groups of genes on the chromosomes, insertion of "alien" genes may all occur with varying frequencies. As already discussed not only qualitative but also quantitative changes in the number of copies of the same gene and positional changes may alter in a heritable fashion gene expression and, hence, what is classically called genetic information.

It should be stressed that, although they are seldom discussed in attempts to construct theoretical models for the genetic structure of organisms examples of both positional and quantitative "mutations" have been known for a long time as shown by the cases of variegations due to position effects in Drosophila[18] and the human syndromes due to changes in chromosome numbers the first of which (Down's syndrome) was discovered back in 1959[19].

All kinds of mutations can occur in higher organisms both in somatic cells and in the germ line and thereby affect the development and/or be transmitted to the sexual or asexual progeny. In the case of plants, where asexual propagation is rather frequent semipermanent physiological states may also be transmitted to the vegetative progeny like the tendency to early flowering in plants regenerated from petals[20].

Mutations are a quasi-random process, in the sense that all sequences can be changed but the frequencies with which mutations occur may be partially modulated by environment. Not only physical and chemical mutagens can influence mutation frequencies and spectra but also general environmental conditions like stress[21] or nutrients availability. The last is the case for the preferential induction of the heritable amplification of sequences leading to lower or higher growth by specific chemical soil compositions shown and studied at length in flax by Durrant, Cullis and their group [22]. Due to the wide range and high frequency of all kinds of genetic changes we may well conclude then that quasi-random fluctuation of the genome is a permanent character of life and that, as consequence, not only individuals but also cells within individuals are highly genetically heterogeneous.

2.3. Constraints on genetic information and relational structure

Quasi-random fluctuations of genetic information and their interaction with environment produce in an open system like a cell a very high number of possible combinations which are then subject as a whole to environmental challenges and intercellular (inter-system) competition.

It should be stressed that the intricate network of metabolic path ways within a cell is connected in such a way that no quantitative or qualitative change can be induced by changes in genetic information and/or its expression without starting a chain of events which will alter in an impredictable way the cell's behaviour. In other words, small, quasi random single changes can induce large whole-system modifications. However, not all cell functional configurations (phenotypes) are possible but only those compatible with internal and external constraints. Internal constraints are overall cell economy, and the need of an inter-sequence hierarchical organization (some genes and their products have to be more active than others).

External constraints range from cell integrity of form and function, to cell to cell competitive factors, to constraints deriving from antogenetic (developmental) and evolutionary rules. The interaction between internal constraints, the fluctuating information and cell environment results in a finite number of living "relational structures"[23]. A change in one of such structures may be recovered (the structure "folds back" to itself), lead to a saltational change to another structure or, alternatively to the "schizoid" disruption of relations and hence to what we call "death" (absence of relational structure).

3. THE ORGANISMAL LEVEL

As we have seen a single cell may be considered a relational structure of lower order. Cells, however, assemble together to form more or less organized colonies, and may be part of an organism. At the cell level units of information, heterogeneous in structure and meaning as they are, nevertheless are all located on strings of nucleic acids in a linear way.

At the organismal level cells may be themselves thought as information units as they behave, within higher order constraints the rules of wich are partially determined by genetic informations, as heterogeneous, fluctuating individuals actively involved in the building of the organism. Their genetic heterogeneity has been now definitely proven both in animals and in plants.

In animals, genetic somatic variability due to quasi-random rearrangements of DNA sequences is the basis of immune responses (see Parisi, this volume). In plants, progenies of individuals regenerated from single cells set free from organismal constraints and grown <u>in vitro</u> show a very high frequency or mutations (of all the types described earlier)[24]. These mutations, clearly partially already present in cultured cells before regeneration, which may affect over one third of analyzed individuals, very frequently involve gross changes in developmental patterns. Such "hopeful monsters"[25] seem then to be selected against during the development of "normal" plants while being favoured when in vitro cell proliferation is free. During normal ontogenesis selection operates on somatic genetic variability be it artificially induced[26] or spontaneous in stress conditions.

Cell selection operates also in animals but probably to a lesser extent due to the earlier differentiation of specific cell lineages.

When trying then to define some of the rules which control

development, cell selection should be considered as one of them. A second role which has to be taken into account is the role of canalization i.e. the loss of degrees of freedom in cells and cell lineages with their age.

Canalization has been shown to occur in regeneration experiments both in animals and in plants. In animals, canalization ("determination") as been assayed by transplanting nuclei from cells of different ages into fertilized eggs and thus analyzing the loss of totipotency (capacity to regenerate a whole animal)[27] or by studying the capacity of specific groups of cells known to differentiate into specific organs in normal development (the "imaginal discs" in Drosophila) to swhitch to another developmental pattern when transplanted in different regions of the larval body[28]. In plants, plant development from single cells deriving from different tissues and individuals of different ages was studied with the same aims[29]. In all cases age was shown to bring about canalization and loss of totipotency. It can be said then that cells, during development, "chhose" specific developmental fates (or are chosen for them by somatic selection) and that the choice becomes more and more obligate with age. The possible choices, i.e. the "epigenetic landscape"[30] are determined by the interaction with external stimuli of a hierarchy of genes some of wich, for their ubiquitary regulatory role are clearly more important than others. "Homeotic" genes in Drosophila and other animals and genes for phytohormone sintheses in plants are the most studied from this point of view. Homeotic genes can be divided into two gross categories. "Maternal genes" active in females of Drosophila, influence the formation of a concentration gradient of specific proteins in the egg and early embryos. Specific threshold concentrations of these proteins determine the activation of a serie of other morphogenetic genes ("gap", "pair-rule", "segment polarity" genes) which specify for the formation of body segments and their functions[31].

Formation of gradients along the plants axis and the presence of threshold phytohormone concentrations have been shown in plants and found to be controlled by the interaction between growth conditions and a few genes involved in the biosynthetic pathways of major regulators (auxins and cytokinins)[32]. The presence of additional copies of these genes integrated into the plant genome through modern recombinant DNA techniques has moreover been shown to result in "catastrophic" changes in development (tubers in the place of leaves, loss of ability to form roots or shoots) comparable to those found in "homeotic mutants" in Drosophila.

Quantitative regulation of gene actions in time and space more than the quality (sequence) of structural genes, threshold processes leading to "catastrophic" changes in developmental patterns along with cell selection under these rules seem then the main processes which determine the "shaping" of organisms. Again, as for cells, due to the complex net of interactions between cells and tissues and between their functions, not all developmental patterns are allowed, disturbances leading again to back folding or the appearance of discontinously new forms and functions.

4. THE POPULATION AND SPECIES LEVELS

As discussed in 2. changes in genome structure (qualitative and quantitative) if present in gametes lead to heritable changes in organismal structure which are then subject to external selection pressure. At variance with what was thought before such changes in somatic cells can pass through meiosis and thus contribute to evolution both in animals[34] and plants[15,21,33], being the barrier between soma and germ line not absolute[35].

As stressed by several authors like Waddington, organisms, not genes nor genomes, are the units of evolution with their discontinuous shapes and functions which derive from the complex process just described. Adaptation, moreover is clearly time-bound in the sense that changing environments favour changing organisms. Evolution, as cell and organism development, is unpredictable as at every level small changes may fold back but also create new structures, and irreversible in the sense that an apparent inversion of strategy never results in "previous" forms and functions. However, again, not all structures and behaviours are allowed but only those which are compatible with "organismal coherence". Constraints include also "average" parameters like total DNA amount (only finite multiples of certain amounts are allowed)[36] or mean (6+C) contents[37,38].

Evolution is multistrategic and no single strategy is ever-winning. Microorganisms adapt through mutations (changes in the genetic code) which can rapidly spread through population due to their short generation time. Animals and plants with longer lives, utilize somatic plasticity as a strategy and rely more on quantitative regulatory changes and somatic mutations in code and number.

It is then rather difficult to establish winning trends in evolution or even temporal trends. It may be only said that evolution is a continuous/discontinuous process which, with time, has been leading to the appearance of new organisms with a general trend to increase DNA amounts (particularly "non structural information"), developmental plasticity, divergence in forms, functions and, hence, adaptational strategies.

5. SOME CONCLUSIONS

It seems clear from what we have been discussing so far that the general rules of life which result from recent research suggest the need of a profound revision of generally accepted models. This comes from a serie of considerations.

Firstly, biological information cannot be treated according to "classical" information theory.

More than that, it seems extremely difficult to evaluate such parameters as information content or even entropy in a living organism at least if the level which the analysis is being carried out is not specified.

At the molecular level, although quantitative data on the amount of DNA and the nature of sequence variability are easily obtained an extrapolation from these data to information amounts seems to be entirely fallacious because of the different values of different sequences and the interacting constraints on quality and amounts of genetic information.

Moreover passages from (potential) information written in DNA and "expressed" information in terms of forms and functions is not only not linear but largely unpredictable due to the rules described and to the continuous fluctuations of all the elements of living systems. The same being true for all levels described in this paper any unifying theory not taking into account fluctuations, unpredictability, irreversible behaviours and aiming at the unequivocal quantitative prevision of biological processes should be discarderd. Modelling of specific processes, starting from a clear definition of the hierarchical level at which the proposed model is assumed to be operative, the meaning of parameters studied at this level and external conditions (constraints from other levels), may be more fruitful and offer at the same time opportunities for the construction of the new mathematical tools needed for a modern description of parts of life mechanisms.

1) N. Eldredge and S.N. Salthe: Hierarchy and evolution Oxford Surveys in evolutionary Biology, Oxford University Press, Oxford 1: 184-209 (1984).
2) Allen T.H.I. and Starr T.B.: Hierarchy: Perspectives for ecological complexity, University of Chicago Press, Chicago (1982).
3) Pattee H.H.: The physical basis and origin of hierarchical control In: Hierarchy Theory: The challenge of complex systems, Braziller, New York (1973);
4) Urba E.S., and Eldredge N.: Individuals, hierarchies and processes: towards a more complex evolutionary model. Paleobiology 10:30-44 (1984).
5) Benzer S.: On the topography of the genetic fine structure Proc. Nat. Acad. Sci. 47: 403-417 (1961).
6) Yanofsky C., Carlton B.C., Guest J.R., Helsinki D.R., Henning U.: On the colinearity of gene structure and protein structure. Proc. Natl. Acad. Sci. U.S. 51: 266-272 (1964).
7) Coggins J.R., G.E. Hardie: The domain as the fundamental unit of proteins structure and evolution in: Multidomain proteins, structure and evolution Elsevier Science Publishers, Amsterdam: 2-11 (1986).
8) Jacob F. Monod J.: Genetic regulatory mechanisms in the synthesis of protein J. Mol. Biol. 3: 318-356 (1961).
9) O'Neill M.C., Amass K., de Crombrugghe B.: Molecular model of the DNA interaction site for cyclic AMP receptor protein. Proc. Natl. Acad. Sci. USA 78: 2213-2217 (1981).
10) Yanofsky C.: Attenuation in the control of expression of bacterial operons Nature 289: 751-758 (1981).
11) Darnell J.E. Variety in the level of gene control in eukaryotic cells Nature 297: 365-371 (1982).
12) Mirault M.E., Delwart E., Southgate R.: A DNA sequence upstream of Drosophila HSp70 genes is essential for their heat induction in monkey cells in Heat shock. Tissieres Ed. Cold Spring Harbour Symposia (1982).

13) Sharp P.A.: Speculations on RNA splicing Cell 23: 643-646 (1981).
14) Davidson E.H., Jacobs H.T., Britten R.J.: Very short repeats and coordinate induction of genes, Nature 301: 468-470 (1983).
15) Buiatti M.: DNA amplification and tissue culture. In: Reinert and Bajaj S. (eds.) Plant cell tissue and organ culture Berlin, Springer: 358-374 (1977).
16) Mc Clintock B.: Chromosome organization and genic expression Cold Spring Harbor Symp Quant. Bicl. 16: 13-25 (1952).
17) Ritossa F.: The bobbed locus. In "The genetics and biology of Drosophila melanogaster", Ashburner and Novitzki Eds.: Academic Press New York Vol. 1B: 801-946 (1976).
18) Lewis E.B.: Genes and gene complexes in: "Heritage from Mendel",R.A. Brink and E.D. Styles Eds., University of Wisconsin Press, Madison: 17-47 (1967).
19) Lejeune J., Turpin R., Gautier M.: le mongolisme, premier example d'aberration autosomique humaine Ann. Genet. Hum. 1: 41-49 (1959).
20) M. Polsinelli, M. Buaitti, E. Ottaviano, F. Ritossa: Genetica Sansoni Ed. Florence: 688-689 (1985).
21) Mc Clintock B.: The significance of responses of the genome to challenge Science 226: 792-801 (1984).
22) Walbot V., Cullis Ch. A.: Rapid genomic change in higher plants. Ann. rev. Plant Physiol. 36: 367-396 (1985).
23) Goodwin B.C., A relational or field theory of reproduction and its evolutionary implications In: Beyond neo-Darwinism Eds Maewan-Ho and P.T. Saunders Academic Press Inc., London: 219-241 (1984).
24) Semal J.Somaclonal variation and crop improvement M. Nijhoff Publishers, Amsterdam (1986).
25) Goldschmidt R.B. Physiological Genetics. Mc. Graw Hill, New York (1938).

26) Buiatti M., Tesi R., Molino M. A developmental study of induced mutations in Gladiolus Rad. Bot. 9: 34-45 (1969).
27) Gurdon J.B.: The control of gene expression in animal development. Clarendon Press, Oxford (1974).
28) Postlethwait J.H., Scheidermann H.A.: Developmental genetics of Drosophila imaginal discs Ann. Rev. Genetics 7: 381-433 (1973).
29) Halperin W.: Alternative morphogenetic events in cell suspensions Amer. J. Bot. 53: 443-453 (1966).
30) Waddington C.H.: A catastrophe theory of evolution in: Mathematical analysis of fundamental biological phenomena Ann. N.Y. Acad. Sci. 231: 32-42 (1974).
31) Macdonald M.P., Sthuhl G.: A molecular gradient in early Drosophila embryos and its role in specifying the body Nature 324: 537-545 (1986).
32) Nacunias B., S. Ugolini, M.D. Ricci, M.G. Pellegrini, P. Bogani, P. Bettini, D. Inzé, M. Buiatti: Tumor formation and morphogenesis in different Nicotiana spp. and hydrids induced by Agrobacterium tumefaciens T-DNA mutants Dev. Genetics 8: 61-71 (1987).
33) Cullis C.A.: Phenotypic consequences of environmentally induced changes in plant DNA: Trends in Genetics 2: 307-309 (1986).
34) Holiday R.: The inhevitance of epigenetic defects Science 238: 163-170 (1987).
35) Pollard J.W.: Is Weismann's barrier absolute? in: "Beyond Neo-Darwinism" M. Wan Ho and P.T. Jsaüders Eds., Academic Press, N.Y.: 291-314 (1984).
36) Narayan R.K.J.: Discontinuous DNA variation in the evolution of plant species J. Genet. 64: 101-109 (1985).
37) Zuckernhandl E. Polite DNA: functional density and functional compatibility in genomes J. Mol. Evol. 24: 12-27 (1986).
38) Bernardi G., Bernardi G.: Compositional constraints and genome evolution J. Mol. Evol. 24: 1-11 (1986).

ORIGINS OF ORDER IN EVOLUTION: SELF ORGANIZATION AND SELECTION

Stuart A. Kauffman

University of Pennsylvania, School of Medicine, Department of Biochemistry and Biophysics, Philadelphia, PA 19104 - 6059 U.S.A.

ABSTRACT

I have tried to suggest that evolution is a complex combinatorial optimization process on some form of rugged fitness landscape, (20). This leads us to realize that adaptation on such landscapes tends to become trapped on local optima, or diffuse away from such optima if mutation rates are large enough with respect to selective forces. We were led to wonder if there might be quite general statistical features of rugged landscapes holding over many combinatorial optimization problems.

Adaptive evolution confronts the profound opportunities afforded by the self organized properties of complex dynamical systems. I have here broached only genetic regulatory networks, but similar issues appear to arise in models of the origin of life, (15-18), of connected metabolisms, (16), and morphogenesis, (19). This surely implies that we must build a larger theory which marries Darwin's idea of selection to the self organized properties of the entities selection was privileged to operate upon. Here we confront the possible limitations in the power of selection to avoid the typical properties of such systems, hence potential universals in biology, and an escape from the epistemological necessity of full reductionism. But beyond this, selection appears able to fashion the entities it operates on to be ones with useful landscapes. We thus need a still broader theory, encompassing this possibility, and then the probability that within such a "selected" ensemble, selection will still not be able to avoid the typical properties of that selected ensemble, and again an escape from necessary reductionism in our explanations.

Only the surface has been touched. We want a true theory of biological evolution embracing self organization, selection, and historical accident.

1. INTRODUCTION

This article is written as a prolegamena, both to a research program, and a forthcoming book discussing the same issues in greater detail[1]. The suspicion that evolutionary theory needs broadening is widespread. To accomplish this, however, shall not be easy. The new framework I shall discuss here grows out of the realization that complex systems of many kinds exhibit high spontaneous order. This implies that such order is available to evolution and selective forces for further molding. But is also implies, quite profoundly, that the spontaneous order in such systems may enable, guide and LIMIT selection. Therefore, the spontaneous order in complex systems implies that selection may not be the sole source of order in organisms, and that we must invent a new theory of evolution which encompasses the marriage of selection and self organization.

The overall themes I wish to explore are these:
1) Complex systems, such as the genomic regulatory networks underlying ontogeny, exhibit powerful "self organized" structural and dynamical properties.
2) The kinds of order which arises spontaneously in such systems is strikingly similar to the order found in organisms.
3) This raises the plausible possibility that the spontaneous order found in such complex systems accounts for some or much of the order found in organisms.
4) The existence of strongly self ordered properties in complex systems implies that selection must be acting on systems with their own "inherent" properties; hence at a minimum, what is ultimately selected may often reflect a compromise between selection and the spontaneous properties of the class of systems upon which selection is acting.

5) Such compromises reflect the fact that selection is in general a combinatorial optimization process in a rugged fitness landscape with many peaks, ridges and valleys. The typical structure of such landscapes and population flow upon them under the drives of mutation and selection, assure that attaining and maintaining high optima, with properties which are very rare in the class of systems under selection, usually cannot occur.

6) The typical failure of selection to be able to AVOID the typical properties of the class of systems under selection is therefore expected to sustain points 3)· and 4) above. The spontaneous properties of complex systems under selection will often be similar to those generic in the class of systems under selection, not BECAUSE of selection, but DESPITE it.

7) In turn, if many features of organisms reflect the generic properties of an entire class of systems under selection, not the particular successes of selection, then non-reductionistic theories based on analysis of those generic properties can be expected to be predictive of features of organisms. We should, in short, be able to predict many features found in organisms without needing to know the details.

8) The capacity of selection to achieve rare highly functioning forms is governed by the statistical features of the adaptive landscape over which selection tries to pull an adapting population. As we shall see, many landscape are "bad" in the sense that adaptation is likely to become trapped on mediocre local optima which become ever more mediocre as the complexity of the entities under selection increases. We must envision yet a further broadening of evolutionary theory to encompass the possibility that selection has achieved entities with the internal properties which allow them to adapt on "good" adaptive landscape with high optima. Beginning conditions for this to occur will be discussed as well.

In preliminary summary: complex systems are self ordered, promising relief to selection as the sole source of order in biology. The relief must be purchased, however, by a broadened theory which considers the capacity of such systems to adapt on rugged fitness landscapes, and the capacity of selection to alter the kinds of entities which exist, hence the kinds of landscapes they evolve upon. Evidently, we must take Darwin's central profound idea profoundly seriously, but move beyond it.

This article is organized as follows. In the second section I introduce the concept of an adaptive landscape, and characterize briefly the properties of adaptive walks to local optima in rugged landscapes. In the third section, I introduce a class of models which generate a family of tunably rugged landscapes and seek generic statistical features of adaptive walks on such landscapes. In the fourth section I discuss briefly the emergence of spontaneous order in models of genomic regulatory systems and the parallels between the generic properties of this class of systems and features of ontogeny. In the fifth section, I report briefly on the capacity of selection to act on genomic regulatory systems, and the limitations due to the statistical features of their rugged landscapes. In the sixth section I return to the issue of whether selection may achieve entities wich have the internal properties allowing them to adapt on "good" fitness landscapes, and discuss the extent to which this may circumvent or modify the general theme that the selection cannot avoid the typical properties of the class of systems upon which it operates.

2. RUGGED ADAPTIVE LANDSCAPE

The adaptive landscape metaphor in evolutionary biology is at least as old as Wright[2]. The metaphor is limited in certain respects. Thus, it is well known in population genetics that a population undergoing selection on two or more loci, with two or more alleles per locus, may not flow "up" the fitness landscape due to recombinatorial constraints between the two loci. Density and frequency dependent effects may also limit the simplest vision of a fitness landscape, where fitness is a function of genotype and attendant phenotype[3]. More generally, the fitness of an organism is a function of its own phenotype and the number of others of the same or other phenotypes, hence not a pure function of one phenotype alone.

Despite these and other limitation, the fitness landscape image is powerful, basic, and a proper starting point to think about selection. We conceive, next, of a very simple space of objects, namely peptides, and analyse the character of adaptive walks via 1-mutant fitter variants to local or global optima for a defined functional property of the peptides[4,5].

2.1. Sequence Space

Consider the set of all peptides length 10. With 20 amino acid types, there are 20 sup 10 possible peptides with 10 amino acids. Define the 1-mutant neighbors of a peptide to be all those sequences which can be obtained by changing one amino acid to one of the 19 other possibilities. Thus, each peptide has 19N = 190 1-mutant neighbors. We are then concieving of a peptide sequence space with all 20 sup 10 possible peptides length 10, ordered in a high dimensional sequence space in which each peptide is a point, and is connected by a line to its 190 1-mutant neighbors. This space, in short, is a sequence space. Distance between points along connected lines passing through other points correctly represents the minimum number of changes in amino acids to convert one sequence to another.

We next need to define a fitness landscape. We do so based on the capacity of each peptide to perform some function. For example, we might measure the affinity with which each peptide binds a specific hormonal receptor on a cell surface. This measured affinity then can be thought of as a measure of the "fitness" of each peptide with respect to that function. Because each peptide has a measured affinity, and the peptides are arranged as points in an ordered sequence space, the measured affinities constitute a fitness landscape over the sequence space.

Consider next the simplest possible caricture of an adaptive walk in peptide space. We imagine that the adaptive process begins with some arbitrary peptide. Next we imagine that at each generation one or more mutant variants of that peptide are produced. In the simplest case, a single mutant variant is produced, and it is mutant in a single amino acid. This 1-mutant neighbor may be fitter than the initial peptide. If so, let the adaptive process step to this improved variant. If not, let the process remain "at" the initial peptide and try another 1-mutant variant. (I show below that this simplest case corresponds to a well defined reasonable population genetic situation).

In this simplest image, an adaptive WALK begins at an arbitrary peptide, and on each trial samples a 1-mutant variant of the current peptide, and "moves" to that variant only if fitter. Thus, the process is constrained to pass via 1-mutant variants until it reaches a peptide which is fitter than all its 1-mutant neighbors.

Obviously, the character of an adaptive walk depends upon the disposition of the fitness values in the fitness landscape. That distribution might range from very smooth, such that neighboring peptides have highly similar fitness values, to rugged landscapes with many peaks ridges and valleys but still substantial correlation between the fitness values of 1-mutant neighbors, to fully uncorrelated landscapes in which the fitness values of 1-mutant neighbors were completely unrelated to each other.

Given any fitness landscape, the immediate natural questions which arise are: 1) How many improvement steps are taken on an adaptive walk before arresting on a local optimum? 2) How many local optima exist? 3) How many alternative local optima are accessible via branching adaptive walks from an initial arbitrary peptide? 4) How does that alter with the fitness of the initial peptide? 5) How "fit" are the local optima with respect to the mean fitness in the space of peptides? 6) How do these properties depend upon the ruggedness of the landscape and the complexity of the entities (here peptides) under selection?

Given any fitness landscape, the actual flow of a population across it depends not only upon its structure, but the population sized, mutation rate, initial dispersion across the landscape, and other factors[3]. Thus, even if we understand the structure of such landscapes, we still will not understand population flow upon them.

Yet deeper questions arise: What kinds of entities, peptides or otherwise, have given types of fitness landscapes and why? Are some landscapes easier to evolve upon than others? Can selection achieve entities which "live on" good landscapes? I return to moot these below.

2.2. The Fully Uncorrelated Landscapes

Simon Levin and I, have recently analysed the case of adaptive walks on fully uncorrelated landscapes[6]. These are constructed by assigning each peptide a fitness value drawn at random from a fixed underlying fitness distribution. To be concrete, we may take that distribution to be uniform between 0.0 and 1.0. Since the adaptive process we are considering passes via 1-mutant fitter variants, the actual fitness values are not important, and can be replaced by the RANK ORDER of the peptides, from worst to best.

We have established the following features of adaptive walks on uncorrelated landscapes:

1) The number of local optima is very large. For peptides with 20 amino acids it is $20^{N/(19N + 1)}$. That is, the expected number of local optima is just the number of entities in the space divided by 1 plus the number of 1-mutant neighbors to each entity.
2) Walks to local optima via fitter 1-mutant variants are very short: $\log_2 (19N)$. That is, the number of fitter variants encountered on a walk is just the logarithm base two of the number of 1-mutant neighbors to any peptide in the space. This implies that, for peptides length 10, walk lengths are on the order of 7 or 8, while for proteins of length 100, walks are on the order of length 11.
3) Only a small fraction of all local optima are accessible from a single initial peptide via branching adaptive walks.
4) As fitness increases at each step along an dapative walk, the number of fitter 1-mutant neighbors decreases by a half. Thus branching walks to alternative optima initially may have alternative routes upward, and these dwindle to single routes which wind upward to local optima. Such branching trees, in short, are bushy at the base and dwindle to single lineages.

2.3. Brief Justification

I have idealized a walk as sampling 1-mutant neighbors one at a time, and constrained to pass via fitter neighbors. This idealization corresponds quite closely to a population with a low mutation rate, in which the rate of finding a fitter variant is very low, while the fitness differentials between improved variants and "wild type" is moderate. In this case, on a slow time scale, the population uncovers a fitter mutant, and on a fast time scale fitter mutant sweeps through the population. Thus, roughly, the population "hops" to a fitter 1-mutant neighbor. Gillespie[7] has shown that this limiting case corresponds to a continuous time, discrete state Markov process, with the population passing from one state to a neighboring fitter state.

Obviously, in reality, population may harbor 1-mutant, 2-mutant,... J-mutant variants at the same time. Actual population flow upon a landscape is more complex than this simplest case, which I have adopted as a means to analyse the structure of the landscape itself, rather than flow upon it.

2.4. Universal Features of Long Jump Adaptation

Real population may be able to "jump" long distances in peptide spaces in single mutational events. For example, recombination is a mutational process which may substitute a large number of amino acids simultaneously into a peptide or protein. Such a mutation can be thought of as jumping long distances across peptide space.

Consider next any correlated but highly rugged landscape, such as the Alps. If a mutational process jumps BEYOND the correlation length of that landscape, then the fitness of the point reached is fully uncorrelated with the fitness of the point left. This corresponds to the fact that altitudes of points 50 kilometers from any point in the Alps are essentially uncorrelated with the altitude of the point left. Thus, long jump processes in correlated but rugged landscapes encounter an uncorrelated landscape.

Levin and I, (ibid), established several universal features of such adaptive processes:

1) After each improved variant is found, the waiting time to find the next improved variant DOUBLES. Thus, the rate of improvement slows rapidly. The consequence is that the expected number of improvement steps, S, after G long jump trials, is:

$$S = \log 2 \, G$$

2) As is the case described above, the number of alternative fitter long jump variants is halved on average at each improvement step, thus branching lineages are bushy at the base and the rate of branching dwindles as fitness increases.

2.5. A Complexity Catastrophy Limits Selection as Complexity Increases

We have uncovered a fundamental and previously unexpected limitation of adaptation on rugged landscapes via long jumps. The same limitation applies to adaptation via 1 or few mutant variants on fully uncorrelated landscapes: As the COMPLEXITY of the entities under selection increases, there is a marked tendency for the ATTAINABLE LOCAL OPTIMA or the states attained after long jump walks of any fixed length to FALL toward the mean fitness of the space of entities! In the current case, the complexity of the entities is just the length of the peptides. As N increases, the number of local optima increases, and the number of 1-mutant or J-mutant variants of each peptide increases, but the actual fitness of local optima, or fitness attained after a fixed walk lenght, DECREASES. This constitutes a kind of COMPLEXITY CATASTROPHY.

This limitation is terribly important. We have already assumed in our idealization that selection is ALWAYS STRONG enough to pull a population to any optimum and hold it there, an assumption which is generally false. Even with this assumption, this new limitation says that as entities under selection become more complex, then in the limits of uncorrelated landscapes and walks via neighboring fitter variants, or for rugged landscapes and long jump adaptational walks, the fitness achieved FALLS. More complex means more mediocre. This is a powerful limitation. We must ask whther there may be conditions which circumvent this limitation. Those conditions would be upon the character of the ADAPTIVE LANDSCAPE ITSELF. We must ask whether there are conditions such that entities under selection can become more complex, yet the adaptive peaks do not dwindle. The answer is "yes", as we discuss in the next section.

3. GENERAL EPISTATIC INTERACTIONS AND RUGGED LANDSCAPES

What properties of complex systems control the ruggedness of the fitness landscapes upon which they may evolve? Do peptides adapt on higly correlated, or very rugged landscapes? What of large proteins? What of the coupled genomic regulatory system where the activities of structural genes are regulated by a variety of control genes in complex circuitry? What of the morphology of an organism? In general, the answers are unknown. In fact, to my knowledge, the question of the relation between the character of entities under selection and the ruggedness of their adaptive landscapes has never been addressed at all.

To approach this issue, I here introduce a general model of "epistatic" interactions among traits in an organism. This model is similar in spirit to spin glasses[8], and as we shall see, appears to have wide applicability.

Consider an organism with N "traits". For the moment let us restrict attention to the case where each trait is simply present or absent, denoted by a 1 or 0. With N traits, there are therefore 2^N possible combinations of traits which might be present or absent. I wish to consider the fitness contribution of each of the N traits. In general, the fitness contribution of each trait may well depend upon the presence or absence of others of the N traits. Such dependencies are called epistatic interactions in population genetic models where one thinks of the fitness contribution of the gene at one locus, which may depend not only on the allele of that gene, but depend epistatically on the alleles present at some other loci. Because I wish to study the effect of RICHNESS of epistatic interactions on the ruggedness of fitness landscapes, I shall assume that the fitness contribution of each trait depends upon iself and K other traits among the N. Increasing or decreasing K then alters how many traits affect the fitness contribution of each trait. In addition to the constraint that each trait's fitness contribution depends upon the presence or absence of K other traits, I shall add one further constraint for the moment. I assume that if trait "i" matters to trait "j", then trait "j" matters to trait "i". Whithin these constraints, I shall assign the K traits mattering to each trait, i, randomly.

The fitness contribution of each trait, i, thus depends upon the presence or absence of itself plus K other traits. Therefore the fitness contribution of the i-th trait must be specified for each of the $2^{(K+1)}$ combinations of presence or absence of the K+1 traits which affect trait i. To do so, I shall draw a different random number from a uniform distribution between 0.0 and 1.0 and assign it to each of the $2^{(K+1)}$ combinations of traits. The same is done for all N different traits. Therefore, for any particular combination of N traits being present or absent, the fitness contribution of all N traits is specified. The fitness of the entire combination of N traits is defined as the sum of contributions of all traits, divided by N. This yields an overall fitness for each combination of traits which lies between 0.0 and 1.0.

The model I have defined is a random epistatic fitness model, tuned by N and K. It is open to many interpretations. For example, the "traits" might instead be considered amino acids in a peptide. In the present case we would conceive of only two types of amino acids. Given a spatial interpretation of the N bits as amino acids, we might choose the K which matter to each site from neighboring or from random sites in the peptide. Obviously the model can be generalized to 20 rather than 2 kinds of amino acids. Under the interpretation that the "traits" are genes, each open to 2 or many alleles, the model is interpretable as a model of epistatic interactions of K loci upon each site, with the K distributed in some defined order on neighboring or random sites on the chromosome. This genetic interpretation corresponds to a haploid model with a single copy of each chromosome.

Given N, K, and the distribution of K among the N, what do the resulting fitness landscapes look like? return to the simple case where each "trait" is present or absent. Then the space under consideration is the ordered N dimensional Boolean hypercube, where each vertex or point is a particular combination of the presence or absence of each of the N traits. Further, each point is a 1-mutant neighbor of N other points, attained by changing one trait from present to absent, or vice versa. Each point has a well defined fitness. Therefore a well defined fitness landscape exists over the Boolean hypercube.

K = 0 Corresponds To A Fully Correlated Landscape With One Optimum.

Consider the case where K = 0. Here the fitness contribution of each "trait" is independent of all other traits. Since the fitness of the entire system is just the sum of the fitness of the N traits, this model is identical to a simple additive fitness population genetic model with N genes, each with two alleles and no epistatic interactions among the genes. In the present case, each trait, by chance, makes a higher fitness contribution if present, or if absent. Thus a single global optimum exists in which each trait is in its more valuable state, present or absent. Note that any other combination of traits is less fit than the global optimum. Further, any such suboptimal combination of traits lies on a connected pathway via fitter variants leading to the single (global) optimum; Each trait in its less valuable state need merely be "flipped" to its more valuable state. Note next that the fitness of 1-mutant neighbors will be nearly the same. This follows since flipping a single "trait" will, on average change the total fitness less than 1/N. That is, the K = 0 landscape is a highly CORRELATED landscape.

Deeper insight into the present K = 0 case is readily attained. Because each fitness contribution has, in the present case, been drawn at random from a uniform distribution between 0.0 and 1.0, and the fitter value is the fitter of two random draws, on average the less fit value is 0.333 and the more fit value is 0.666. In a system with N "traits", the expected fitness of the global optimum is 0.666. Furthermore, the expected fitness of the global optimum is obviously INDEPENDENT of N. As N increases, the fitness of the global optimum remains 0.66666. If an adaptive walk begins with an arbitrary combination of traits, its fitness will be about 0.5. Half the traits will be in their less valuable state, thus on the order of N/2 improvement steps occur on and adaptative walk to the single and global otpimum. Along that walk, the number of fitter neighbors decreases by one on each step upward.

K = N Corresponds To a Fully UNCORRELATED Fitness Landscape

At the opposite extreme, suppose the fitness contribution of each trait depends upon all traits. Then $K = N - 1$, or for large N, K effectively equals N. Consider a particular combination of N traits. Change one trait from present to absent. Then the context of all N traits has changed. For each we shall have specified a fitness value at random from the uniform distribution. Therefore flipping a single trait yields a 1-mutant neighbor whose fitness is a entirely new sum of N random samples from the uniform distribution. That is, the fitness values of 1-mutant neighbors in the space are entirely random with respect to one another. The $K = N$ case therefore corresponds to a fully UNCORRELATED fitness landscape.

We already understand the character of adaptive walks on such a landscape. There are a very large number of local optima. Walks to local optima take on the order of $\log_2 N$, rather than $N/2$ for the $K = 0$ limit. The number of fitter neighbors is halved after each improvement step. From any initial combination of traits only a small fraction of the local optima are accessible.

Among the most critical features of $K = N$ landscapes is that they exhibit the COMPLEXITY CATASTROPHY described above. As the number of traits, N, increases, the fitness of attainable local optima DECREASES toward 0.5, the mean of the space.

$0 < K < N$. Tunably Rugged Fitness Landscapes.

Clearly, increasing K from 0 to N, corresponding to increasing the epistatic interactions among the traits in the system, increases the ruggedness of landscapes from fully correlated to fully uncorrelated. The intent of the model now becomes clear, for the parameters N, K, and the distribution of K among the N, generate a FAMILY of expected landscapes with statistically characterizable adaptive walks. It may well be the case that this family covers a vast spectrum of "real" adaptive landscapes which arise in a variety of combinatorial optimization problems from the Traveling Saleman problem[9], to walks in protein spaces during maturation of the immune response[6]. At a minimum, this family of landscapes should offer beginning intuition about the requirements on system construction and the ruggedness of the resulting landscapes.

To apparoach this issue, my collegue E. Weisberger and I have written a general computer program. In figures 1 and 2 I show the results of numerical Monte Carlo trials for different values of N and K, and for two different distribution rules on the K. In the first, we conceived of circular peptides (to avoid end effects) and required the K sites bearing on each site to be its K/2 neighbors on either side. In the second set (figure 2) we removed the restriction that K be reflexive, and allowed each site to be affected by K randomly chosen other sites. The results do not appear to depend upon this difference strongly. Thus, N and K determine the major features of the class of landscapes.

The numerical results confirm the general expectations derived above:

1) Both figures confirm that for K = 0, the single global optimum has a fitness of 0.66666, and that the optimum is independent of N. Both figures confirm the expectation for the K = N case that as N increases, the fitness of local optima DECREASE towards 0.5. The complexity catastrophy on uncorrelated landscapes is real.

2) For K = 0, walk lengths are on the order of N/2, and the number of fitter neighbors declines by 1 per step. For K = N, walk lengths are on the order of log 2 N, and the number of fitter neighbors declines roughly exponentially.

But for values of K between 0 and N, new features emerge.

1) Note for K small and fixed, for example, K = 2,4 or 8, that as N increases, the fitness of optima actually increases towards an apparant asymptote which becomes independent of N, and which is LARGER than the global optimum for K = 0.

That is, a small amount of epistatic interaction actually HELP CREATE A LANDSCAPE WITH OPTIMA OF HIGHER FITNESS! And the fitness appears to become independent of N for large N. Thus, for K fixed (and perhaps small) complexity the sense of increasing N does not lead to a complexity catastrophy. The space retains optima of high fitness.

2) What happens if K increases proportionally to N? Both tables show that for K = N/2, K = N/3, and K = N/4, as N increases the fitness of optima first increase above the K = 0 case, then decline below it, and presumably fall toward 0.5 as N grows sufficiently large. Thus, if K is proportional to N, it appears that a complexity catastrophy occurs. Increasing complexity leads to landscapes with lower optima. K proportional to N would mean, intuitively, that as the number of traits in an organism increases, or number of amino acids in a peptide, that a constant fraction of those traits affect the fitness contribution (ie. function) of each trait or amino acid.

These last two observations carry the following implication. Consider K as a function of N. For K constant as N increases, the complexity catastrophy is averted. For K proportional to N, it is not averted. Thus for K equal to some monotonically increasing function of N a transition must occur between landscapes which do not and do exhibit the complexity catastrophy. What that boundary is remains unknown.

How generalizable are these results? In the present simplest model we have sampled fitness values from a uniform distribution between 0.0 and 1.0. We do not yet know, but strongly conjecture that the qualitative results are independent of the underlying distribution, which might be uniform, Gaussian, exponential, etc. so long as the fitness contribution for each combination of "traits" is sampled from the SAME distribution. Indeed the results may generalize to a process which samples from a set of different distributions.

To summarize tentative conclusions: If as N grows large the number of "traits" which impinge epistatically on any trait is bounded below some constant or slowly increasing value, then such systems adapt on rugged landscapes which avert the complexity catastrophy. Beyond some critical rate of increase in K, the complexity catastrophy sets in. I suspect that the implication of this general argument holds quite generally. For complex entities to evolve, the number of parts which directly impinge upon any part probably must grow very much more slowly than the number of parts in the whole. I turn in the next section to describe the self organized dynamical behavior of model genomic regulatory systems. There it turns out that genetic networks in which each gene is directly affected by only a few other genes exhibit marked order reminiscent of real ontogeny. In the subsequent section I present evidence that genomic systems of low connectivity also adapt on correlated landscapes which avert the complexity catastrophy. Thus, the same property, low K, appear to abet orderly behavior, and capacity to evolve well!

4. SELF ORGANIZED BEHAVIOR IN GENETIC NETS OF LOW CONNECTIVITY

Metazoans such as mammals have on the order of 100.000 different structural and regulatory genes. Since Jacob and Monod discovered that products of one gene could activate or repress the activities of other genes, biologists have come to think of the genomic system as a kind of biochemical computer. The structural and regulatory genes are linked into some kind of circuitry, regulating and coordinating one another's behavior. It is a cannon of current developmental biology that all the diverse cell types of an organism contain the same set of genes coordinated by the same regulatory circuitry. While there are minor exceptions to this cannon, it is close enough to truth for current purposes. In terms of this cannon, the fundamental problem of cell differentiation is: Why are cells different from one another, despite harboring identical genomic systems? The answer, of course, is that cells differ because different subsets of genes are expressed in different cell types. It follows that it is the genomic circuitry and regulatory system which coordinates and engenders these different patterns of gene expression.

The presumptive complexity of a genomic system with about 100.000 genes regulating one another's activities bodes serious epistemological problems. Molecular genetics is fast uncovering the molecular machinery by which one gene acts on a second gene. Structural genes, which code for proteins, typically are flanked by specific DNA sequences which are bound by the protein products of other genes. This binding can turn on or turn off transcription of the strucrtural gene to messenger RNA, thence to protein. Such adjacent DNA sequences, because they regulate adjacent genes on the SAME chromosome, are called CIS acting regulatory genes. The upstream genes whose products bind to cis acting sequences are themselves called TRANS acting, because they can exert regulatory influences on genes located on different chromosomes. Regulation is more complex that mere control of transcription. Different points in maturation of messenger RNA, and translation to protein are also subject to regulation. For our purposes I shall lump all these into the term "genomic regulatory network".

The genomic regulatory network is subject to evolution in two quite different ways. First, of course, point and recombinational mutations may alter DNA sequences leading to new useful structural genes and proteins. Some of these altered proteins may themselves be trans-acting proteins, hence their alteration may alter the effect of the regulating gene upon its target genes. But beyond such mutations, chromosomal mutations literally shuffle genes from one to another position in the set of chromosomes. Such mutations include duplications and various processes which disperse duplicated genes to new locations such as translocations, transpositions, inversions, conversions. Consider the obvious consequences. If gene A is adjacent to a cis acting sequence X, and a transposition moves a new gene B into position between X and A, then X may now control B as well as A. If B happens to be a trans acting gene regulating a downstream cascade, then X now regulates the same cascade via B. In short, chromosomal mutants literally scramble the circuitry in the genomic regulatory system.

From the foregoing, it follows that EVOLUTION OF NEW CELL TYPES involved both the evolution of new genes, and new circuitry among old and perhaps new genes. Thus our intellectual task is foreboding. We want to understand both how the genomic system in an organism underwrites its ordered ontogeny, and we want to understand how such regulatory systems evolve.

4.1. An Ensemble Approach

It may be the case that to understand the genomic system in Man, or any other organism, its detailed structure, component by component, will require elucidation. Indeed, our intuitions based on complex systems such as current computer programs suggests this might be the case. Current computer programs are fragile. It two instructions are exchanged, typically the program is drastically altered. Nevertheless, I strongly believe that this intuition misleads us with respect to understanding genomic regulatory systems and their capacity to evolve. Instead, as I shall soon point out, complex systems of interacting genes can exhibit spontaneous order.

Two of the most readily accessible features of genomic regulatory systems are the numbers of genes which typically directly act on and regulate a single gene, and the kind of regulated responses which occur. With regard to the first issue, genes in bacteria and viruses typically are directly regulated by 0 to perhaps 5 or 6 other genes and products[10-12]. That is, the CONNECTIVITY of real genomic regulatory systems in these organisms is LOW. It now appears, but remains to be shown in more detail, that this is true in higher eukaryotes as well. What of the regulated behavior of a gene? For the moment, let us idealize the behavior of a gene and think of it as either active, 1, or inactive, 0. Then the behavior of any gene with respect to its regulatory inputs is given by a Boolean switching rule. For example, a gene might be active at the next moment if either its first, or its second or both inputs were active the moment before. This is the OR function. Alternatively, a gene might be active only if its first input were active and the second inactive the moment before. This is the NOT IF function. Within the idealization to 1 and 0 values, examination of bacterial, viral and eukaryotic genes shows that different genes are governed by different Boolean functions.

This leads to the following question. Suppose we consider models of genomic regulatory systems in which each gene is either active or inactive, hence a binary switching variable. If ALL we know about such networks is that, on average, any gene is directly regulated by only a few other genes, can we say anything about the expected structure and behavior of large genomic networks? Otherwise stated, does the LOCAL feature of LOW CONNECTIVITY imply anything about the GLOBAL behavior of such genomic regulatory systems? If so, that local feature would predict those global features, hence might explain them without need for detailed reductionistic analysis.

The form of this question is an ENSEMBLE QUESTION. That is, there is an enormous number of genomic regulatory systems with N binary genes, each regulated by K other genes. This set of systems constitutes an ensemble of all systems which might be built given N and K. Thus, what we want to know is the typical or GENERIC properties of members of this esenmble. To answer this question by numerical simulation, we should then construct model genomic systems entirely at random within the constraints that there be N genes, each regulated by K other genes. Specifically, this means deciding at random which among the N genes are the K inputs to each gene, then deciding at random for each gene which among the 2^{2^K} possible Boolean functions of K inputs shall govern its behavior. That is, both the wiring diagram and the logic are chosen at random. Once chosen, the structure is fixed.

The results of many numerical simulations for different values of N and K have shown that for small K, for example K = 2, such networks exhibit powerful order which is strongly reminescent of real cells. Briefly, such a network is a finite automation. Each combination of activities of the N genes is a state. Thus there are 2^N states. At each moment, the net is in a state. The genes each examine the activities of their inputs, consult their switching rule, then all genes synchronously assume their designated new activity value. Thus the net passes from a state to a state. Over time the system passes through a succession of states. Since there is a finite number of states, eventually the system reenters a state previously encountered. Thereafter, since the net is deterministic, it cycles repeatedly through this reentrant cycle of states. It is critical that many states may lie on sequences of states which flow to the same state cycle. But the network may harbor more than one state cycle. If so, some states flow to one of these asymptotic attractors, while others flow to the other state cycles. Wherever the system is released, it will flow to one state cycle attractor. Thus ultimately, the system comes to cycle about one or another of its state cycles.

The order which emerges spontaneously when K = 2 includes the fact that the lengths of state cycles are small, on the order of square root N, the number of states cycles in a net's behavioral repertoire is also about square root N, each cycle is stable to transient reversal of the activity of most genes one at a time, and if unstable will flow to only a few of the remaining state cycles.

I have for a number of years wanted to see in this spontaneous order a deep similarity to real ontogeny. I make a single interpretation: I identify a CELL TYPE with a STATE CYCLE ATTRACTOR of a genomic network. That is, a cell type is a stable recurrent pattern of gene expression engendered by the genomic logic. Given this interpretation, the theory makes a number of predictions which are surprisingly accurate. Thus, the number of cell types in an organism should increase as a square root function of the number of genes! This is surprisingly close to what is observed across many phyla[10,12]. If a cell type is a state cycle attractor, then differentiation is the passage from one attractor to another. These models imply that any cell type can only flow to a FEW NEIGHBORING CELL TYPES, and from thence to a few other cell types. But this implies that ontogeny should occur along BRANCHING DEVELOPMENTAL PATHWAYS. Indeed, all contemporary multicellular organisms develop along just such branching cell lineage pathways. Presumably this has remained true for the past 600 million years. I return to this below, for I shall want to ask whether this feature of organisms reflects selection, or such a deep property of genomic systems that it is IMPERVIOUS to selection.

This class of models predicts many other features of current organisms, such as the similarity in gene expression patterns in different cell types (about 90% TO 95% overlap), the existence of a large core of genes active in all cell types, the typical cascading consequences of deleting a single gene, and so forth[10,12].

Without belaboring the detailed results, the main poin to cull from all these studies is that even randomly constructed model genomic systems with few inputs per gene exhibit order which is strongly reminiscent of that seen in ontogeny. This implies that our intuition about complex systems is wrong. Order emerges spontaneously. From this it follows that this order may account for the origin and persistence of such order in organisms. The spontaneous order is at least a handmaiden to selection. But of course, selection operates continuously. This brings us to the central problem I believe faces evolutionary theory. We need to broaden evolutionary theory to understand how selection can act upon, with and thorugh systems which have their own strongly self organized properties. How is selection enabled, guided and limited by those properties? Note that no area of science has dealt with this problem. I turn to beginning steps in the next section.

5. SELECTION AND ITS LIMITS FOR DESIRED CELL TYPES

Can selection, operating both on structural genes and the "circuitry and logic" of a genomic regulatory system, mold such a regulatory system to achieve arbitrary "good" cell types? Obviously, to broach this question requires first of all some model of a cell type, and the ways in which mutation and selection can modify cell types. The ensemble of genomic network models described above provides just such a framework.

To be concrete, I shall ignore evolution of new structural genes coding for new useful proteins with novel enzymatic, structural or other features. Rather, I shall focus on evolution of the circuitry and logic to alter the coordinated patterns of expression among a constant set of structural genes. Within that limitation, and in the frawework of the ensemble theory developed above, evolution can proceed via mutations in the regulatory CONNECTIONS between genes, thus altering the "siring diagram" of the genomic circuitry; or evolution can proceed by altering the regulated behavior of a gene as its inputs alter activities. That is, mutations can affect the Boolean function characterizing the response of any regulated gene. To be concrete, chromosomal mutations may alter the wiring diagram, or such mutations and point mutations may alter the local rule. For example, a mutation which prevents a repressor protein from binding to its cis acting site may render an adjacent regulated gene constitutively (constantly) active. The gene now realizes the "Tautology" logical function, always active.

5.1. Evolution Explores an Ensemble of Regulatory Networks

In the second and third sections of this article I discussed the structure of fitness landscapes, where the points in the space were conceived to be proteins, each a 1-mutant neighbor of those other proteins obtainable by altering one amino acid in the protein's primary sequence. The concept of rugged fitness landscapes is far broader however. Consider a genetic network with N genes, and K inputs per gene. Each such network is a member of a vast ensemble of all networks constructable with N and K as constraints. Each network is a 1-mutant neighbor of all networks which can be attained by altering one regulatory connectioon, or one "bit" in the Boolean function regulating one gene. Therefore, we can, again, define the concept of a high dimensional space, where each point is an entire genetic regulatory network, and its one mutant neighbors are all those genetic networks accessible by minimal changes of wiring diagram or logic. Thus, we can consider evolution as occuring across a space of genetic regulatory networks.

Any property of such networks might serve as a property upon which selection acts. To be concrete, let us define the fitness of a genetic regulatory system by how closely the patterns of gene activity occurring on one or another of its state cycles match to an arbitrary pattern of gene expression. That is, we shall define the "fitness" of a genetic network of N genes by how closely one of its cyclic attractors comes to having a pattern matching an arbitrary "desired" or "good" pattern of gene expression across the N genes. For example, if the arbitrary pattern among the N genes is (101010101010...), then the fitness of a given network is given by the fraction of genes whose activities match this pattern on the best matching state of one of the state cycle attractors. This therefore assigns a fitness between 0.0 and 1.0 each network, hence generates a fitness landscape over the space of genomic systems.

As before, this fitness landscape may be more or less rugged. Also as before, in the simple case of adaptive walks constrained to pass via fitter 1-mutant variants, we can ask how many steps occur on the way to local optima, the number of alternative local optima accessible, the similarities of those optima. Of course we are interested in how closely such optima match to the arbitrary "good" pattern for which we are selecting. That is, can selection achieve ARBITRARY GOOD cell types? And we are interested, as the complexity of the genomic system,

N, increases, whether selection leads to less fit optima. Does the complexity catastrophy creep in?

I note in passing that these selection studies are a root form of learning in massively parallel processing systems. Thus, in the neurobiological context, models of distributed content addressable memories use attractors as internal memories of external events[13]. Asking whether mutation and selection can achieve desired attractors is a form of asking whether mutation and selection can be used to evolve parallel processing networks to have desired "memories", (14). It is a further point of interest that any such system, if placed in a varying environment, will spontaneously classifiy its environments into equivalence classes given by those environments which leave a given attractor unaltered. That is these kinds of model genomic systems, if placed in an environment, naturally classify the SAME environment in different ways according to which attractor the system attains, and classify different environments as the SAME if those environments leave a given attractor unaltered. Nonlinear systems with multiple attractors exhibit the basis of classification and a kind of cognition of their environments. Selective evolution for useful behaviors is thus a start towards evolving systems which classify their environments, recognize and act upon them.

Numerical Studies of K = 2 and K = 10 Boolean Networks.

The general results of numerical simulations of such adaptive walks are these[14]:

1) For Boolean networks with N = 50 and N = 100, and K = 2 or K = 10 the first major conclusion is that selection never achieves networks with fitness 1.0. Rather, fitness increases from an initial 0.5 and typically arrests at a local optimum with values of about .65 or .7. This means that, in general, adaptation by 1-mutant neighbors becomes

trapped on local optima far below the logically possible perfect network. This is not abetted significantly by allowing 2-mutant or 5-mutant variants to be sampled.

2) The second general result is that the rate of finding fitter variants slows rapidly. This implies that, as fitness increases, the fraction of fitter neighbors dwindles rapidly.

3) The third general results is that LONG JUMP adaptation, in which up to half the connections or one fourth the bits in Boolean functions are altered at once, reveals the exptected general features of long jump adaptation on rugged landscapes. Thus, the waiting time to find fitter variants doubles on average after each improvement step. And as N increases, the fitness attained after any fixed number of tries FALLS towards the mean fitness in the space of systems, 0.5.

4) The fitness landscape for $K = 2$ is higly correlated. One mutant alterations to connections of logic make little difference to fitness. By contrast, the fitness landscape for $K = 10$ networks is far more rugged. Altering one connection, and even 1 bit in a Boolean function, typically alters a net's fitness drastically. Thus $K = 2$ nets adapt on a more correlated landscape than do $K = 10$ networks.

5) Given 4), it is interesting to ask if $K = 2$ networks exhibit the complexity catastrophy for adaptation via 1 or 2 or 5 mutant neighbors. (They do for long jump adaptation, of course). The answer is NO. As N increases, the fitness of local optima remains essentially constant. The results for $K = 10$ networks are not available.

Point 5 is particularly interesting. Recall from section III that for that NK spin-glass like model, we found that for K small, optima were insensitive to N, while for $K = .5N$ or $K = N$, the complexity catastrophy sets in as N increases. Thus there and here, low connectivity yelds a correlated landscape and the capacity to avoid the complexity catastrophy as complexity, N, increases. I stress, of course, that the two models are not at all identical. In section III, the NK models are taken as general epistatic models. In section VI, the NK models are DYNAMICAL MODELS of Boolean switching networks with N genes, and K inputs per gene. Landscapes in the switching networks are defined by attractors. Nevertheless there may be deep homologies between them with respect to ruggedness of landscape.

A major conclusion of both NK models, then, is that the complexity catastrophy can be averted by systems having the proper properties to adapt on correlated landscapes where optima do not whither toward the mean as complexity increases. In both cases this seems to require low connectivity among a rich system of many interacting parts. I turn next to wonder if selection can select for systems which adapt on "good" landscapes.

Two even more fundamental observations about these simulations must be stressed:
- First, the fact that strong selection, that is selection which is always able to pull an adapting population "to" a fitter variant, becomes trapped on local optima far from the logically possible Boolean network which matches a given pattern of gene expression precisely, very strongly implies that real selection acting on real organisms CANNOT sculpt arbitrary patterns of gene expression as the cell types of an organism. Whatever our intuitions have been, cell types almost certainly are not "precise" in that sense. Evolution must make do with local optima, the best attainable on adaptive walks from whatever the starting point may be. The "developmental program" beloved of developmental biologists is almost certainly not able to evolve to arbitrary logics from any given starting point.
- Second, and perhaps most fundamentally, selection in rugged landscapes typically cannot AVOID the typical features of members of that

landscape. Adaptive walks in a space of genomic regulatory systems: 1) become arrested on local optima; 2) in the limit of long jump adaptation climb higher at an ever slower rate remaining far from global optima; or 3) in the face of a high mutation rate which disperses a population away from attained optima, wander ergodically among suboptimal genomic systems. In all cases, it appears, selection is limited in its capacity to attain genomic systems with features which are extremely untypical of the space of genomic systems in which adaptive evolution is occuring. But if true this bears the deepest consequences. For if selection can typically not AVOID the typical or generic features of the genetic regulatory systems in the ensemble explored, then those typical features should "shine through", and be found in organisms, not because of selection, but despite it. Here then is a major theme. If selection cannot avoid the generic features of complex genomic regulatory systems which are members of some definable ensemble, then those generic properties may prove to be biological universals. Their explanation would not lie in details of structure, nor details of selective forces in branching phylogenies, but in mere ensemble membership. As phase transitions in water are understood as a form of critical phenomena in a general class of systems without precise accounting for locations of water molecules at the transition, so some or many ordered properties of organisms may find their explanation in the generic properties of the ensemble of which the organisms are members.

I return, in this light, to the fact that model genomic systems with few inputs per gene have the property that each state cycle attractor, which models a cell type, is a "neighbor" of only a FEW other model cell types. Thus, this broad class of genomic models predicts that any cell type can differentiate into only a few other cell types, and they in turn to a few others. As noted earlier, this implies that ontogeny must be based about BRANCHING cell differentiation pathways, as is indeed the case. Does this organization of ontogeny, presumably present since the late Precambrian, reflect an achievement of selection? Or is it-a property which is so deeply embedded in the entire ensemble of genomic regulatory systems accessible to selection, so deep a property of parallel processing nonlinear dynamical systems, that selection cannot avoid this property? I suspect the latter. And add that it is no trivial property, for it is the same property which assures the homeostatic stability of cell types in the face of perturbations, which assures that a variety of specific and non specific "inductive" agents induce the SAME differentiation transformation, hence which underlies the subsequent elaboration of the logic of development. The possibility that such branching pathways reflect self organization, not selection, I suspect, is the harbinger of many possible new universals reflecting the balance between self organization and selection.

Bu all of this brings us to a yet harder issue. Is selection doomed to wander among systems in a given ensemble? Or can selection actually change the way genomic systems are constructed, hence change the ensemble being explored, its generic properties, and the carachter of fitness landscapes upon the ensemble? Can selection find ensembles with "good landscapes"?

6. SELECTION FOR "GOOD" LANDSCAPES

What kinds of systems have "good" landscapes? It is clear that fully uncorrelated landscapes are not "good" in at least two senses:
1) First, in a correlated landscape, advantage may be taken of the

correlation structure to "look" where the looking is good. For example, many landscapes may prove to be self similar. Spin-glass energy landscapes are an example (8), and my first NK models in section III may also be self similar. That means that the landscape is fractal, with small hills and local optima located in the "sides" of similarly shaped, but wider and higher hills and optima, which in turn are on the sides of still larger wider hills with higher optima. In such a correlated landscape, not only does it pay to look in the vicinity of modest local optima by jumping just far enough away to avoid trapping, but it may be the case that local optima carry global information about the general region of space with yet higher otima. Thus, one optimum is a good starting place to find yet higher optima. In contrast, in a fully uncorrelated landscape, locations of optima can carry no information about good regions of space. The search may as well be entirely random.

2) In fully uncorrelated landscapes, the complexity catastrophy creeps in. As systems become more complex, the optima attained relapse towards the mean fitness in the ensemble being explored.

In contrast, we have so far encountered only one class of entities which adapt on landscapes which avoid these problems. Such entities appear to have the property than any component directly interacts with a small number of other components in the system, and that small number remains bounded small in some as yet unknown way as the number of components in the system increases. Can selection "tune" K and keep it small?

Let us consider two different levels of systems upon which selection may act. Consider proteins first. Small peptides, that is sequences of amino acids with about 20 or fewer members, probably adapt on very uncorrelated landscapes, for substitution of any single amino acid seems likely to have marked effects on the behavior of the peptide.

By contrast, in large proteins, many amino acid substitutions have little effect on the overall architecture and function of the protein. Thus, large proteins almost certainly evolve on more correlated landscapes than small peptides. Evolution has opted for large proteins, by and large. This may reflect the commonly held view that large proteins can fold better, hence form better enzymes and structural components of cells. But it may also reflect the fact that in selecting for large proteins, selection has also achieved the kinds of entities which adapt on more correlated lanscapes hence can actually ADAPT BETTER. Large proteins live on landscapes with connected walks to higher optima than peptides do. And probably the linear structure of proteins assures that the dominant interactions of any amino acid is with its few neighbors in the primary sequence. To a lesser extent, folding brings distant regions into contact, but still any amino acid interacts directly with only a few others. Thus, as the length of a protein becomes large, the number of strong interactions bearing on the contribution of any amino acido remains bounded and small. Proteins are thus probably well set up to adapt on correlated landscapes by their inherent nature.

What of genetic regulatory systems? What is low connectivity? Nothing other than the property that the one gene is directly regulated by only a few other genes or their products. But this is a restatement of MOLECULAR SPECIFICITY. Enzymes and other liganding agents with high specificity will, in general, discriminate among a vast number of molecular variables and bind only a few close cognates. Selection for high specificity is also selection for low connectivity in a genomic regulatory system. And selection for low connectivity is simultaneously selection for systems which exhibit the globally ordered dynamics of homeostasis so reminiscent of actual cell types. In turn, it happens that selection for low connectivity and consequent homeostatic dynamical behavior also yields genomic regulatory networks which adapt on higly correlated landscapes which avoid the complexity catastrophy!

Two points warrent attention. First, by selecting for large proteins, or high specificity, selection may be changing the adaptive landscapes of the kinds of entities which are evolving. Thus we do seem to confront the fact that selection can act on entities to achieve those with "good" landscapes. This would not seem to require any group selection argument. This implies that we need to broaden evolution theory not only to include the balancing forces of selection operating on systems with self organized properties, but further, to begin to understand how selection may operate to cull entities better able to adapt because they adapt on good landscapes.

Second, in both the case of selection of large proteins and genomic systems of low connectivity, the drive towards good primary function - proteins which are good enzymes and genomic systems with stable homeostatic dynamics - appears to bring with it almost gratuitiously selection for entities which happen to adapt on "good" landscapes. Is this fortituous? Or is there some deeper reason hinting that complex systems that "work well" also "adapt well"?

Quite obviously, if we take the Kantian stance and ask "What must organisms be such they can evolve adaptively?", we become aware of how very much we have to learn. And if we look beyond evolution in biology to evolution in technological economies and society at large, we may well wonder whether there be found general principles of adaptation in complex systems.

Bibliography and Notes

1. Kauffman, S. Origins of Order: Self-Organization and Selection in Evolution. Book in preparation.
2. Wright, S. The Roles of Mutation, Inbreeding, Crossbreeding, and Selection in Evolution. In: Proceedings 6th International Congress on Genetics, I, 356 (1932).
3. Ewens, W.J. Mathematical Population Genetics (1979). Springer Verlag, New York, Heidelberg.
4. Eigen, M. Macromolecular Evolution: Dynamical Ordering In Sequence Space. In. Emerging Syntheses in Science, Proceedings of the Founding Workshops of the Santa Fe Institute (1985) Santa Fe.
5. Smith, John Maynard, Natural Selection and the Concept of a Protein Space. Nature 225 563 (1970).
6. Kauffman, S. and Levin S. Towards a General Theory of Adaptation on Rugged Fitness Landscapes. In Press. J. Theoret. Biol, (1987).
7. Gillespie, J.H. Theoret. Population Biology. 44 167 (1974).
8. Anderson, P.W. Spin Glass Hamiltonians: A Bridge Between Biology, Statistical Mechanics and Computer Science. In Emerging Synthesis in Science, Proceedings of the Founding Workshops of the Santa Fe Institute. Santa Fe N.M. (1985).
9. Lin, S and Kernighan, B.W. An effective heuristic algorithm for the traveling salesman problem. Oper. Res. 21, 498 (1973).
10. Kauffman, S. Metabolic Stability and Epigenesis in Randomly Constructed Genetic Nets. J. Theoret. Biol. 437 (1969).
11. Kauffman, S. The Large Scale Structure and Dynamics of Gene Control Circuits: An Ensemble Approach. J. Theoret. Biol. 44 167 (1974).
12. Kauffman, S. Emergent Properties in Random Complex Automata. Phisyca 10D, 145 (1984).

13. Kauffman, S. and Smith R.G. Adaptive Automata Based On Darwinian Selection. In Evolution, Games and Learning. Models for Adaptation in Machines and Nature. eds. Farmer, Lapedes, Packard and Wendroff. North Holland, Amsterdam (1986).
14. Hopfied, J.J. Proc. Natl. Acad Sci. $\underline{79}$ 254 (1982).
15. Eigen, M. and Schuster, P. The Hypercycle. Springer Verlag, New York, Heidelberg (1979).
16. Kauffman, S. Autocatalytic Sets of Proteins. J. Theoret. Biol. 119 1 (1986).
17. Dyson. F. Origins of Life. Cambridge University Press, London (1985).
18. Rossler, O. Chemical Automata In Homogeneous and Reaction Diffusion Kinetics. Notes in Biomath. $\underline{B4}$ 399 (1974).
19. A very large number of articles an books based on A. Turing's reaction-diffusion model of morphogenesis, Proc Roy Soc 1954, exists by authors including: Goodwin, Brian; Murray, James; Kauffman, Stuart; Meinhart, Hans; Harrison, Lionel, over the past decade in Science, Nature, J. Theorét. Biol., Developmental Biology. Other articles by Wolpert, Lewis; Oster, George; Bryant, Peter, and Bryant, Susan, among others are scattered through the same journals.
20. A large literature on combinatorial optimization exists, but not often from the evolutionary perspective. In part this is based on simulated annealing: Kirkpatrick, S, Gelatt, C.D. Jr. and Vechi, M P Optimization by Simulated Annealing. Science, $\underline{220}$ 671 (1983).

N\K	8	16	24	48	96	
0	.646	.6502	0.656	.0659	.665	Fitness of optima
	5.5	9.59	13.59	25.31	49.77	Steps to optima
2	.697	.699	.699	.703	.705	Fitness
	5.14	9.13	12.22	23.49	46.20	Steps
4	.6983	.7052	.696	.7011	.699	Fitness
	4.58	7.61	10.35	20.27	38.31	Steps
8	.659	.679	.680	.688	.682	Fitness
	3.67	5.73	8.66	16.29	28.70	Steps
16		.646	.655	.660	.658	Fitness
		4.33	5.63	11.02	28.86	Steps
24			.627	.641	.643	Fitness
			4.48	8.65	15.75	Steps
48				.598	.613	Fitness
				5.02	9.90	Steps
96					5.66	Fitness
					6.10	Steps

Fig. 1: Results of Monte Carlo trial for different values of N and K with K sites bearing on each site being its k/2 neighbors on either siede.

K \ N	0	8	16	24	48	96	
0	.649	.662	.657	.659	.664		Fitness of optima
	5.6	7.54	12.67	25.31	48.61		Steps to optima
2	.676	.679	.687	.690	.691		Fitness
	5.48	9.15	13.68	25.60	49.11		Steps
4	.690	.700	.712	.714	.717		Fitness
	4.79	8.62	13.17	24.23	47.94		Steps
8	.676	.693	.702	.712	.715		Fitness
	3.62	6.64	9.91	19.55	37.57		Steps
16		.644	.653	.669	.686		Fitness
		4.2	5.9	11.67	24.83		Steps
24			6.28	.649	.663		Fitness
			4.5	9.14	17.61		Steps

Fig. 2: Results of Monte Carlo trial for different valves of N andk with each site affected by Krandomly chisen other sites.

Statistical Mechanical Models of the Emergence of Biological Order

Luca Peliti

Dipartimento di Scienze Fisiche and Unità GNSM-CISM
Università di Napoli, Mostra d'Oltremare, Pad.19, 80125 NAPOLI (Italy)
Associato INFN, Sezione di Roma

Living beings are very complex. Even the simplest bacteria exceed in complexity most, if not all, manufactured objects. This complexity appears to be irreducible, in the sense that it is hard to conceive a substantially simpler being which would perform the activities we associate with life. Yet we have the preconception that life evolved out of simpler, non-living things.

We need therefore to identify a series of systems which could interpolate between inanimate matter and living beings. One should also be able to argue convincingly that the transition from each level to the following one could have been accomplished in a comparatively short time[1].

But of course we would not expect that each step had necessarily to occur in the way it has occurred. The path actually followed in the development of life was probably determined by a series of random events, which could have driven the following evolution in one way instead of another. We are thus forced to consider the problem in a more general setting, i.e. to consider the set of all possible paths that could have been followed in the emergence of life.

We have therefore to consider an *ensemble* of possibilities, described by some probability distribution in a suitable space, and to show that a generic element of this space does exhibit the features we are looking for.

In this line of thought one might find it likely that methods developed in statistical mechanics could find useful application in the theory of biogenesis, in the way they have found other biologically relevant applications[2].

As pointed out by Anderson[3], living structures are characterized by *stability* and *diversity*. Stability in the sense that not any structure is allowed, and slightly perturbed structures recede to a nearby stable condition. Diversity, since there are many stable structures, and it is the historical development which leads to one or another of the many possible ones. "Stability and diversity together are the key, and this suggested the one system in equilibrium statistical mechanics that has both of these properties: the spin glass. This possibility was suggested by Hopfield's ideas on associative memory..." (Anderson, ref.3). In the spin glass model, the system can lay in any state corresponding to a minimum of its free energy (hence its stability). But the interaction among the different units which form it is frustrated, in the sense that any unit can receive contradictory influences from the units with which it is connected. Hence the number of free energy minima is high: diversity. Frustration is a generic feature of disordered systems, if inhibitory beside excitatory interactions are allowed. It probably plays a fundamental role in allowing for a multiplicity of stable states in systems like neural networks, ecological systems, or the immune system.

I shall discuss a few models of the emergence of biological order which lay their roots in statistical mechanics. The model proposed by Dyson[4] is the simplest. It focuses on the emergence of metabolism, i.e. of the organization of mutually catalyzing chemical reactions which sustain themselves out of equilibrium. It considers metabolism as independent of replication, and indeed as having arisen earlier. Its aim is to understand the conditions which may have allowed the formation of "active" networks of chemical reactions.

It suggests that a population (an "island") made of a few thousand monomers, rather loosely organized into chemically active proto-enzymes of comparatively small efficacity, could have represented the first step towards living systems. Such systems may appear in a large variety of composition, and can mutate into one another quite freely. Although one may devise some mechanisms for natural selection to start acting on a population of Dyson's islands, they are not within the scope of the model. Natural selection acts on systems which are sufficiently stable that their characteristics do not vary during the time in which selection may operate (a few generations). It cannot operate on Dyson's islands unless some mechanism prevents them from drifting around at each iteration step.

Anderson[3] has the following view of the key problem of prebiotic evolution: "From essentially uniquely characterized simple molecules, which even when autocatalytical generated do not have any resemblance to a self-replicating information string such as is necessary if evolutionary choice is to begin to operate, how can such information strings develop?" This very formulation focuses on the activity of proto-polynucleotides, precursors of the nucleic acids. To simplify matters, let us consider a soup of such polymers, made up two types of monomers, A and B, complementary to each other, and let us assume that each such string can catalyze its reproduction by means of template replication, like in present-day DNA. We may describe such a proto-polynucleotide chain as a system made up of N units, such that the nature of each unit is identified by a variable S_i, taking the value +1 if the corresponding monomer is A, and -1 if it is B. The structure of the polymer is thus defined by a sequence $S = (S_1, S_2, ..., S_N)$ of $\pm$ 1's. We associate with S a function $H(S)$ with a large number of minima, and we postulate that the polymer is hydrolized ("dies") with a probability which is a sigmoid function of $H(S)$; e.g.

$$(1 + e^{-\beta H(S)})^{-1}, \qquad (1)$$

where β is a parameter, reminescent of the inverse temperature, which controls the sharpness of the selection. In this way only the chains with comparatively small values of $H(S)$ are likely to survive (stability), but the large number of essentially equivalent minima will guarantee diversity.

One still has to define $H(S)$. Since the interactions governing the survival of such structures are very complex, we may as well take $H(S)$ to be a random function. Any function of binary sequences S may be expanded in the following way:

$$H(S) = - \sum_i h_i S_i - \sum_{(i,j)} J_{ij} S_i S_j - ... \qquad (2)$$

It is crucial to stop at least at the second term, and indeed to let it dominate the first. The first term has in fact only one global minimum, whereas the second one may have a large number of minima, due to the occurrence of *frustration*[5]. This phenomenon may arise when the couplings J_{ij} can take on both positive and negative values. If for a cycle of indices (ijk...m) the product $(J_{ij} J_{jk}...J_{mi})$ is negative, all interactions cannot be simultaneously satisfied. If the number of frustrated cycles is large, there will be a large number of essentially equivalent minima of H(S). To achieve stability it is also important that J_{ij} does not decrease too rapidly with the distance $|i-j|$ of the interacting units along the string: the simplest hypothesis is to take all J_{ij}'s as independent, identically distributed random variables with a symmetric distribution. In this way H(S) becomes a spin glass hamiltonian in the mean field limit, of which a great deal is known[6], at least in the large N limit. The system settles quite rapidly to a quasi-equilibrium in which a few different kinds of chains, near to the minima of H(S), are present.

I have the impression that this situation is not stable in the long run. Once this quasi-equilibrium is reached, we may simplify the description of the system by labelling each quasi-species by an index $\alpha(=1,2,...,n)$ and considering the evolution of the population x_α of chains composing each one of them. Different chains interact only because they share a common source of "food", i.e. of monomers. In this situation it is likely that the system can be well described by the equations of population dynamics under constant flux conditions[7]. To simplify the analysis we may as well consider constant population conditions: this does not change the results. The equations read therefore:

$$\frac{dx_\alpha}{dt} = A_\alpha x_\alpha - b \left(\sum_\beta A_\beta x_\beta \right) x_\alpha \ . \tag{3}$$

The coefficient A_α arises from the difference between a reproduction rate and a death rate, and depends therefore on the quasi-species one considers. The first term represents therefore the natural population increase of a quasi-species when no competitors are present. The second term imposes the fixed population constraint. It is easy to see that the total population,

$$X = \sum_\alpha x_\alpha \ , \tag{4}$$

reaches an equilibrium value equal to 1/b. We have neglected the (small) probability that a member of one quasi-species mutates into that of another, because we have seen that quasi-species are well separated. It is easy to see now that in the long run only the quasi-species with the largest value of A_α survives. If we introduce in fact

$$A(t) = \frac{1}{X} \sum_\alpha A_\alpha x_\alpha(t) \ , \tag{5}$$

we may rewrite eq.(20) as follows:

$$\frac{dx_\alpha}{dt} = [A_\alpha - A(t)] x_\alpha \ , \tag{6}$$

and we see that only those quasi-species in which A_α exceed $A(t)$ increase their population. But this has the effect of pushing A towards higher values, until it reaches its maximum possible value A_{max}, corresponding to the "optimal" quasi-species with the largest value of A_α. But at this moment only one quasi-species has survived.

A possible mechanism to keep diversity within this approach had already been suggested by Stein and Anderson[8] building on Eigen's ideas: to consider the death function of a single chain to be determined by the whole composition of the soup. Stein and Anderson focus on the emergence of Eigen's hypercycle, and distinguish therefore between proto-polynucleotides and proto-polypeptides, in which the first may give rise to the second by some mechanism of template synthesis, while the second assist in the self-replication of the first. This possibility is worth exploring, but no simulation results are yet available to me. But I wonder if, once this step is taken, it is at all necessary to postulate the existence of a chemical species, like the proto-polynucleotides, capable of template replication.

Abbott[9] proposes a model of autocatalytic replication built again on a spin glass framework. As in Stein and Anderson's model, the unit of selection is a whole soup made up, say, of n chains, each of length N. The nature of these chains is yet to be specified. For simplicity, we assume as usual that each unit, e.g. the i-th unit of the chain α, may be of just two types, and its nature is identified by a binary variable S_i^α:

$$S_i^\alpha = \pm 1; \quad i = 1,2,...,N; \quad \alpha = 1,2,...,n. \tag{7}$$

The soup evolves by discrete steps. The unit S_i^α at step t+1 is determined by the structure of the whole chain at step t by a rule analogous to a spin glass rule at zero temperature:

$$S_i^\alpha(t+1) = \text{sgn}(\sum_j J_{ij} S_j^\alpha(t)). \tag{8}$$

The coupling matrix J_{ij} is the same for all chains. We assume it to be symmetric ($J_{ij}=J_{ji}$). Then, if the couplings J_{ij}'s were fixed, the chains would evolve towards the minima of the hamiltonian

$$H(S) = -\frac{1}{2} \sum_{ij} J_{ij} S_i S_j \tag{9}$$

Hence, after a few iteration steps, we expect eq.(8) to reach a fixed point, depending on the starting configuration of the chain one is looking at.

But, as we have mentioned, we have to assume that the J_{ij}'s are determined by the composition of the soup. The simplest mechanism is Hebb's rule, as applied e.g. in Hopfield's model of associative memory[10]:

$$J_{ij} = \sum_\alpha S_i^\alpha S_j^\alpha . \tag{10}$$

The trouble with this prescription is that it makes each chain present at any instant in the soup to generate catalytically itself - and this leads to homogeneity in the long run. Just as frustration causes diversity within one chain, we need frustration among chains to keep diversity in the soup. This leads to the following modification of the rule:

$$J_{ij} = \sum_{\alpha} B_{ij} S_i^{\alpha} S_j^{\alpha}, \qquad (11)$$

where B_{ij} is a random symmetric matrix with elements of either sign. This does not rule out, of course, the possibility that J_{ij} also contains a constant, random contribution A_{ij}, supposed to reflect the stability of the various free chains, when isolated. The most interesting results appear when this constant contribution is quite small.

Replication is the outcome, in this model, of the growth of soup size and of random splitting, as in Oparin's coacervate philosophy[11]. It arises therefore as a byproduct of stability, since the structure last in the splitting must be regenerates by the remaining members of the two daughter soups. Stability arises since each of the chains must correspond to a minimum of H(S), with the J_{ij}'s determined by the whole soup composition. In view of eq.(11), this corresponds to a four-body interaction. The interaction is not completely general, however, since chains are interchangeable, and no explicit dependence on α should appear.

To get the actual Abbott model one should add a few details. On the one hand, to avoid the instability which arises if a chain can catalyze itself, one considers in eq.(8) and effective coupling J'_{ij} given by:

$$J'_{ij} = J_{ij} - B_{ij} S_i^{\alpha} S_j^{\alpha}, \qquad (12)$$

On the other hand, one then considers a growth-and-splitting protocol defined as follows. One starts with $n=n_f/2$ chains in a random configurations, and lets the soup evolve by repeated application of eqs.(8) and (11) till it reaches a fixed configuration. Then a "food chain" (chosen in a random, but fixed, configuration) is added to the soup and the process is repeated. In this way the soup population is increased by one chain at a time till its size has reached $n=n_f$. The soup is then split at random in two daughter systems of $n_f/2$ chains each, and the whole protocol is repeated. This constitutes one "generation".

Computer simulations show that the soup composition does remain more or less stable from generation to generation, i.e. that replication is achieved without the need of templating. We may expect indeed quite a strong stability - at the expenses of diversity - because not *any* soup composition will be able to replicate, but only the quite stable ones. The need of a sophisticated replication machinery, able to replicate any input chain, arises only when selection operates on further consequences of the chain structure than its immediate catalytic activity. We thus expect, with the "same degree", of catalytic activity, that this mechanism of limited scope is *more exact* than "general purpose" templating.

Kauffman[12] has shown that the emergence of autocatalytic sets of polypeptides is a generic property for any fixed probability of catalysis, if the chain length is sufficiently large. The mechanism considered by Abbott shows a way in which these sets may stabilize and evolve, and shows the relevance of repressive interactions to provide stability and diversity. As such, this mechanism appears one of the most promising pathways to explore the land between the emergence of the first organic monomers and that of more organized structures like the hypercycle.

I think that an understanding of the behavior of Abbott's model - or some similar, more general, ones - from the point of view of population dynamics and selection would be most welcome.

Acknowledgments

This is an abstract of the lectures given by the author at the first Taller Internacional sobre Teoría de los Sistemas Desordenados en Modelos Biológicos, held at the Centro Internacional de Física, Bogotá (Colombia), September 1987. He is grateful to the Centro for having hosted the Taller. Enlightening conversations with G.Arango, J.Bower, S.A.Solla and C.Tsallis are gratefully acknowledged. I thank M.Mézard and G.Weisbuch of the Ecole Normale Superieure (Paris) for useful suggestions, and F.Di Liberto, E.Dogliotti, G.Monroy, M.A.Virasoro and especially M.F.Macchiato for encouragement and criticism.

References and Footnotes

1. "The oldest 'possible' evidence for life is that from Isua [Schidlowski et al., *Geochim. Cosmochim. Acta* **43** 189 (1979)] (3.7×10^9 yr old); the oldest 'probable' evidence is from the Pilbara stromatolites reported in this issue of *Nature* [3.4-3.5×10^9 yr old: Lowe, *Nature* **284** 441 (1980); Walter et al., *Nature* **284** 443 (1980)]; and the oldest 'compelling' evidence from the superb late Archaean stromatolites in Canada [2.7×10^9 yr old: Henderson, *Can. J.Earth Sci.* **12** 1619 (1975)] and near Belingwe [Bickle et al., *Earth Planet. Sci. Lett.* **27** 155 (1975)].[...] It is difficult to imagine that the Earth was hospitable to life before, say, 4.2-4.3×10^9 yr ago. [...] A few hundred million years is not long on a geological time scale". [E.G.Nisbet, *Nature* **284** 395 (1980)].

2. See e.g. E.Bienenstock, F.Fogelman-Soulié and G.Weisbuch (eds.), *Disordered Systems and Biological Organization*, NATO ASI Series in Computer and System Sciences (Berlin: Springer, 1986).

3. P.W.Anderson, *Proc.Natl.Acad.Sci. USA* **80** 3386 (1983).

4. F.J.Dyson, *J.Mol.Evol.* **18** 344 (1982); *Origins of Life* (Cambridge: Cambridge University Press, 1985).

5. A recent reference is: M.Mézard, G.Parisi, M.Virasoro (eds.), *Spin Glass Theory and Beyond* (Singapore: World Scientifics, 1987).

6. G.Toulouse, *Commun.Phys.* **2** 115 (1977).

7. See the first of ref.8, and B.L.Jones, R.H.Enns and S.S.Rangnekar, *Bull.Math.Biol.* **38** 15 (1976). A detailed discussion is contained in: B.-O. Küppers, *Molecular Theory of Evolution* (Berlin: Springer) p.46 ff.

8. D.L.Stein and P.W.Anderson, *Proc.Natl.Acad.Sci. USA*, **81** 1751 (1984).

9. L.F. Abbott, A Model of Autocatalytic Replication, Boston University preprint, Feb. 1987.

10. J.J.Hopfield, *Proc.Natl.Acad.Sci. USA* **79** 2554 (1982).

11. A.E.Oparin, *The Chemical Origin of Life* (Springfield: Thomas, 1964).

Immunological memory in a network perspective

Giorgio Parisi

Dipartimento di Fisica

Universita' di Roma II, "Tor Vergata"

Via O. Raimondo, Roma 173

and

INFN, sezione di Roma.

Abstract

In this note we present a model of immunological memory which is based on the idiotypic network. Some of the similarities with spin glass model and neural networks are discussed.

It is well known (also to the layman) that the introduction of a given amount of antigen (e. g. a protein) inside a high level animal (e. g. a mouse) stimulates the production of antibodies direct against that given antigen; more precisely the immune system produces gammaglobulines (IgM and IgG) which have a high affinity (a typical value of K is 10^5) for the antigen (some basic notions on the immune system can be found in Alberts et al. 1983 or Hood et al. 1984). If the animal is stimulated with the same antigen at a later time (e. g. three weeks later), the amount of IgG produced is much higher.

This phenomenon is at the basis of vaccination. The immune system of the animal has learned to produce high quantities of the antibody directed against that given antigen. Moreover the antibody is produced by B cells which have (in average) a life span of two days; those B cells which have an high affinity with the antigen are stimulated and expand exponentially (clonal expansion). Only one B cell is enough to start the clonal expansion (a typical number in a mouse is 20 cells with given specificity per spleen). After the first exposure to the antigen, memory B cells are produced; they have a much higher life span, may be one month, (a study of the lymphocyte population kinetics can be found in Jerne (1984) and Freitas et al. (1986)).

The precise number of antibodies, that an organism (e. g. a mouse) is able to produce at a given moment (i. e. the available repertoire), is of order 10^6-10^7 and the number of antibodies, which are actually produced (the actual repertoire), is likely a factor 10 smaller[F1]. As far as the number of different antibodies the immune system may produces is very high, it is usually said that the repertoire is complete (if we neglect holes), i. e. the immune system may react against any possible protein (Coutinho 1980).

Although this picture is essentially correct, it strongly simplifies the real process. Indeed it is crucial that the organism must not react (too much) against its own proteins (if this happens, serious diseases, like *diabetus mellitus*, may arise). It is established that the distinction among self and non self is not encoded in the germ line; in other words the immune system is able to produce antibodies directed against the self and during the ontogenesis the immune system must learn which antibodies it should not produce.

The following phenomenon, which is the opposite of vaccination, certainly plays a crucial role. If vaccination is typically induced by the injection of 100µg of antigen, the injection of a few micrograms of antigen produces tolerance; i. e. the amount of IgG produced three weeks later after a second injection of the same antibody is much smaller than in absence of the first injection. In other words the response of the immune system to an antigen may be or immunity (vaccination) or tolerance. Low doses of antigen normally induce tolerance and medium doses induce immunity (high doses too induce tolerance). Both the low dose tolerance and the immunity are related to the proliferation of T cells, suppressors in the first case and helpers in the second case, which have a negative or a positive effect on the antigen B cells.

When the immune system is stimulated two pathways are open: tolerance or immunity; the choice of the pathway is crucial and depends on many factors, the amount of antigen, the way it enters in the organism (if endovenously, tolerance is preferred, if intramuscularily, immunity is preferred, for a discussion of this point see Monroe et al. (1984)), the age (tolerance is much easier in the new born). It is quite likely the the time dependence of the antigen concentration plays a crucial role: common sense tells us that a slow increase should push the balance toward tolerance and an abrupt increase should favor immunity; unfortunately practically no data are available: the time dependence of the antigen concentration is normally neglected.

Summarizing we have immunological memory for both suppression and for immunity and also T cells plays a crucial role in both phenomena.

A third effect, idiotypy (Oudin & Michel 1963), is at the basis of network theories (Jerne 1974[F2])of the immune system. The antibody elicited directly by the antigen (let it call Ab1) is for all practical purposes a new protein; it reasonable that it will elicit the production of a new antigen (Ab2), which induces Ab3, which induces Ab4, which induces Ab5 and so on...

This phenomenon, the idiotypic cascade, can be studied experimentally in the following way: it is reasonable to assume that the different waves (Ab1, Ab2, Ab3...) are separated in time by a delay of one week, so that the first antibodies, the mouse produces after stimulation, are Ab1. These antibodies may be injected in an genetically identical mouse and in this way we obtain Ab2, similarly we obtain Ab3 and so on... Sometimes Ab3 looks like Ab1 and it binds to the same antigen of Ab1 (Cazenave 1977, Urbain & al. 1977).

It is naturally to assume that this idiotypic cascade happens inside the same organism and it plays a crucial role in the regulatory phenomena. However we must be very careful in these identifications for many reasons: (i) the antibody concentration in the same organism increases relatively slowly while it jumps istantaneously when the antibody is injected into an other organism: the differences in both the protocol and the time dependence may be responsable for changing from tolerance to immunity; (ii) it is possible to transfer the antibodies which have been produced (Ab2), but it is not possible to study in this way those antibodies which are no more produced as effect of the increase in the Ab1 concentration; (iii) genetical identical mice may have a different idiotopyc enviroment and this may account for different behaviours of the same antigen in genetical identical, but different mice.

Up to this moment we have shortly reviewed some more or less extablished facts (for a dissentient opinion see Cohen 1986). We now discuss how the network may be functionally useful as far memory is concerned (obvously the network may be relevant in other contexts like self-non-self recognition, this point will not be addressed in this paper). The idea is very simple: after that the production of Ab1 starts the environment of B and T cells is modified by the presence of Ab2 in such way that the life span of Ab1 producing B cells is increased and also the population of helper T cells specific for Ab1 is increased: Ab3 must have a strong component which coincide with Ab1 or it is functionally equivalent to Ab1. In other words we suppose that the internal image of the antigen (Ab2) remains after that the antigen has disappeared and its presence induces the survival of memory B cells directed against the antigen.

We now try to formulate a model which we want to be as simple as possible. Although it is clear from the previous discussions that the number of B and T lymphocytes of given specificity play a crucial role, the actors of our model will be the antibody concentrations and it is understood that their interaction is mediated by lymphocytes.

For a given antibody (i) we assume that in absence of external antigen its concentration (c_i) may have only two values which conventionally we take 0 or 1. We will also assume that in presence of antigen the concentration c_i may become much greater than one. The status of the immune system is determined by all the values of the c_i for all possible antibodies (i. e. i=1,...N , where N is of order 10^7). We assume that at equilibrium (i. e. in absence of external antigen) the following equations are satisfied:

$$h_i = \sum_{k=1,N} J_{i,k} c_k \qquad (J_{i,i}=0)$$

$$c_i = \theta(h_i) \qquad (1)$$

where the function $\theta(x)$ is zero for negative x and 1 for positive x; $J_{i,k}$ represent the influence of the antibody k on the antibody i; if $J_{i,k}$ is positive the antibody k elicits the production of the antibody i, on the contrary, if $J_{i,k}$ is negative, the antibody k suppresses the production of the antibody i; the absolute value of $J_{i,k}$ represent the efficacity of the control of the antibody k on the antibody i. If the excitatory effects of the antibody is greater that the suppressive effect, h_i, which represent the total stimulatory (or inibitory) effect of the network on the ith antibody, will be positive and c_i will be one; c_i will be zero otherwise.

The simplification of only two levels of concentrations (0 or 1) bypasses the problem of the choice of the pathway (immunity or tolerance): we assume the protocol in which antibodies are produced is such to induce only a given pathway independently of the initial antibody concentrations. If this hypothesis fails, the precise value of the antibody concentration is a relevant variable and more complex nonlinear differential equations for the time evolutions should be written and the study of the system becomes more difficult.

In our model the antibodies for which c_i is positive are those which are actually produced by the system and the others are suppressed. Here we completely neglect the suppression due to clonal abortion and we consider only the active suppression which selects which cells of the available repertoire are transferred into the actual repertoire. We are not interested here to study the physiological level at which the interaction between the different antibody happens; our aim is to obtain a global, as simple as possible, functional description of the immune network; this is the main reasons for which we are neglecting the role of T cells in inducing the interaction among different antibodies.

The whole memory of the system (immune state and suppressed states) is encoded in the network in the sense that the knowledge of the concentrations of all antibodies different from a given one completely determines the concentration of that given antibody.

Now we must do some hypothesis of the J's; we assume that:
(a) there are J's of both sign.
(b) the J's are symmetric, i. e.

$$J_{i,k} = J_{k,i} \qquad (2)$$

(c) the J's are random and they are equidistributed in the interval -1,1.

The last assumption is clearly an oversimplification: is more reasonable that $\ln|J_{i,k}|$ (which should be related to the chemical affinities) is equidistributed; moreover the antibodies are not random protein's. The probability distribution of the J's can be modified in later refinement of the model without changing the qualitative predictions of the models. Here we stick to assumption (c) just for its simplicity.

Let us discuss if assumptions (a) and (b) are physiological plausible.

If (a) is correct, we must have antibody which suppress the production of other antibodies; this effect is well known (Cosenza & Kohler 1972, Hart et al. 1972, Rajewsky & Takemori 1983). Moreover, it has been recently reported that in some cases, where Ab1 is producing using as an antigen a virus, the injection of Ab2 before the infection with the same virus, strongly increases mortality at medium doses [F3] (Finberg & Ertl 1986).

On the contrary, if all J's are positive, the only solution of equation (1) is that all concentration of antibodies are equal to 1. In this model inhibition plays a crucial role in regulating the immune network.

The most crucial and controversial point is (b); if both J's are positive, there is a clear experimental evidence of the correctness of the picture (Hoffman 1980, Cooper Willis & Hofmann 1983, Rajewsky 1983), if one of the J's is negative, the situation is less clear; however we shall assume for simplicity that (b) is strictly satisfied, indeed as we shall see later the correctness of (b) when both J's are postive, is enough for most of our qualitative conclusions. If (b) fails when some of the two J's is negative, the main new feature is the possible existence of time oscillations in the antibodies concentrations also after the exposure to the antigen [F4].

If we assume for simplicity that Ab1 is monoclonal (let us label it with an i), we must have that the concentrations in Ab2 ($C^{(2)}_k$) and Ab3 ($C^{(3)}_k$) are given by:

$$C^{(2)}_k = J_{k,i} \tag{3}$$

$$C^{(3)}_k = \Sigma_{m=1,N} J_{k,m} C^{(2)}_m = \Sigma_{m=1,N} J_{k,m} J_{m,i}$$

If we consider $C^{(3)}_k$, we see that for k≠i the sum should be incoherent and terms with different sign should cancel one with the other, while all the terms are positive and add coherently if i=k. Ab2 is polyclonal, Ab3 is mainly monoclonal and coincides with Ab1, on the contrary if we use monoclonal Ab2, only one term in the sum is present; a polyclonal response is produced. As a consequence monoclonal Ab2 is not efficient as polyclonal Ab2 in inducing the production Ab1 or of other antibodies with the same specificity. The different effects of monoclonal and polyclonal Ab2 been observed: for example it has been shown that sometimes monoclonal Ab2 does not elicit Ab1-like molecules, but rather a heterogeneous responce similar to the heterogeneous responce to monoclonal Ab1 (Saks et al. 1982, Bottomly 1984).

In any model based on eq (1) or its generalizations, the presence of both positive and negative J's is crucial: if all J's are positive the modification of a single antibody would enforce (by the idiotypic cascade) a modification of all the idotypic environment: the cancellation in $C^{(3)}$ for $k \neq i$ (and the symmetry of the J's) implies that Ab3 looks like Ab2, Ab4 looks like Ab2. A more careful analysis shows that the conclusions are similar, if only the positive J's are symmetric and the symmetry is lost when the J's are negative; however for simplicity and for avoiding technical problems we will stick to the case where the J's are symmetric.

Having defined the model, we can now analyze it[F5]. It is well known that the number of equilibrium configuration is very high and it increases exponentially with N. It is natural to assume that only a tiny fraction of all the antibodies have a physiological relevance; in other terms there are $M = \alpha N$ antibodies which should have a preassigned concentration, some of them should have a zero concentration, others a non zero concentration; the value of M is likely much less than N, i. e. $\alpha << 1$. The natural question is the following: there exist one (or more) equilibrium states with a preassigned value of the M concentrations? The answer depends on M; handwaving arguments strongly suggest that such a state exist for $\alpha < \alpha_c$ (i. e. $M < \alpha_c N$), where α_c should not too far from .3 (Mezard & Parisi 1988). In this context, the storage capacity ($M_c = \alpha_c N$), is the total number of antibodies whose concentration may be assigned in a way compatible with eq. (1). It is not clear how many antibodies have been actually learned by the immune system, however this number is certainly high: it is clear that a N independent storage capacity would be a serious disaster for this model.

If we compare this idiotypic model with the neural models (Hopfield 1982, Mezard et al. 1987), we see that the J's plays the same role of the synaptic strengths of the neural network; however in this case they cannot be modified[F6]. There is no associative memory in this idiotypic net. We only ask the net to have a high enough number of equilibrium states in which some of the component are preassigned. Here learning means to jump from an equilibrium state to an other one.

The way in which learning happens should be investegated more carefully; as we already said, the values of the concentrations are 0 or 1 only in the absence of the antigen; it is reasonable that in presence of the antigen the concentrations becomes much higher; the idiotypic cascade starts and the concentrations of many antibodys are changed; when the external pressure is removed, we find ourselves in a new equilibrium state. A more detailed model with continously changing concentration is needed to investigate propertly the way in which learning may happens; however there are two main features that should be model independent:

(a) Each time we learn something, we modify the concentration of some of the antibodies and therefore we forget something else (Bernabe et al. 1981, Slaui et al 1986): the antibodies, which are the target for being easyly forgotten, are those having a small h_i in eq. (1). The total memory capacity of antibodies, which will not be easily forgotten, depends on the details of the learning process but it will be certainly much smaller than the maximal one.

(b) Repeated exposure to the antigen is quite likely to increase the value of h_i and therefore to strenghen the memory. It is satisfactory that one apparent characteristics of auto-anti-idiotypic antibodies is that they are produced mostly effectively by means of repeated immunization (Bottomly 1984)

At the present moment it is a wild speculation to ask why this kind of network theory is used by the immune system; a possible answer is that the immunological memory is much more robust, if it is distributed in many clones, and the decrease in the total storage capacity is a reasonable price to pay for this increased robustness.

It is a pleasure for me to thank C. Brezin, A. Coutinho, M. Mezard and M. Virasoro for useful discussions and suggestions on the immune network and related topics.

Footnotes

[F1] We recall the precise definitions (Holmberg et al. 1986): the available repertoire is composed of all Ig molecules expressed as membrane bound receptors by resting B lymphocites at a given moment in the life time of the individual; the actual repertoire is composed by all Ig molecules which, at any given time, circulate in the blood, lymph, tissues and mucosal surfaces, avter being secreted by plasma cells or other activated B lymphocytes.

[F2] There is a very large literature on the idiotypic cascade and the network theory. A good starting point may be the two issues of Immunological Reviews (79 and 90) dedicated to these subjects.

[F4] If we push this idea (of reversing the sign of the interaction), we should have some antibodies which at low dose are immunogenic and at medium dose are tolerogenic; there are some indications that this effect may exist (Hiernaux et al. 1981, Rubistein et al. 1983, Bona et al. 1984), however a more careful discussion of this point is needed. It may interesting to study if this approach may be used to understand why preimmunization of a a graft recipient with blood from the donor leads to prolonged allograft survival (Gorczynski & Robillard 1986).

[F4] The discussion presented here is very short and does not take into account many of the features of the immune system; for example we have compleately overlooked the fact that Ab2 antibodies may be functionally classified into 4 major categories (Bona & Kohler 1984) or the detailed studies of the different class of antibodies in idiotypic cascade (see for example Cazenave & Roland 1984 or Slaoui et al. 1986); in principle one should also take into account the results on antibodies reperoire before exposure to antigen (Hoemberg et al. 1986) and on the network regulation among T cells (Cooper et al. 1984). However the aim of this note is to discuss some general features of the immune network without enter so much into details.

[F5] If the J's are symmetric, this model coincided with a very well known and widely investigated model for spin glasses (for a review see Mezard et al. 1987).

[F6] We neglect the role that somatic hypermutation has in producing new antibodies and consequently modifying the J's. It is usually believed that somatic hypermutation is crucial in producing antibodies with higher affinities with the antigen and it is not relevant from the point of view of the network. It is possible that this point of wiew is simply wrong: the esperiments of Lanzavecchia on the B cells - T cells interaction in vitro (Lanzavecchia 1987) seem to suggest that (at least in framework of his vacum tube model) the value of the affinity is not a crucial quantity in controlling the clonal expansion. Somatic hypermutation may be a crucial (unfortunately neglected) element of the immune network.

References

Alberts B., Bray D., Lewis J., Raff M., Roberts K. & Watson J. D. (1984) Molecular Biology of the Cell. Garland, New York.

Bernabe R., Coutinho A., Cazenave P.-A. & Forni, L. (1981) Suppression of a "recurrent" idiotype results in profound alterations of the whole B cell compartment. Proc. Natl. Acad. Sci. (USA) 78, 6416.

Bona C. A. & Koher H. (1984) Anti-idiotype antibodies and the internal image in monoclonal and anti-idiotype antibodies: probes for receptor structure and function. Allan R. Liss, Inc. New York.

Bona C. A., Victor-Kobrin C., Manheimer A. J., Bellon B. & Rubistein L. J. (1984) Regulatory arms of the immune network Immunol. Rev. 79, 25.

Bottomly k. (1984) 1984: all idiotypes are equal, but some are more equal than others. Immunol. Rev. 79, 45.

Cazenave P. A. (1977) Idiotypic antiidiotypic regulation of antibody synthesis in rabbits. Proc. Nat. Acad. Sci. 74, 5122.

Cazenave P. A. & Roland J. (1984) Internal images of rabbit immunoglobulin allotopes. Immunol. Rev. 79, 139.

Cohen M. (1986) The concept of functional idiotype network for the immune regulation mocks all and comforts none. Ann. Immunol. (Ins. Pasteur) 137C, 64.

Cooper J., Eichmann K., Fey K., Melchers I., Simon M. M. & Weltzien H. U. (1984) Network regulation among T Cells: qualitative and quantitative studies on suppression in the non-immune state. Immunol. Rev 79, 63.

Cooper-Willis A. & Hopffmann G. W. (1983) Symmetry of effector funtion in the immune system network. Mol. Immunol. 20, 865.

Cosenza H. & Kohler H. (1972) Specific inhibition of plaque formation to phosphorylcholine by antibody against antibody. Science 176, 1027.

Coutinho A. (1980) The self-non-self discrimination and the nature of the acquisition of the antibodiy repertoire. Ann. Immunol. (Ins. Pasteur) 132C, 131.

Finberg R. W & Ertl H. C. J. (1986) Use of T cell-specific anti-idiotypes to immunize against viral infections. Immunol. Rev. 90, 129.

Freitas A., Rocha, B. & Coutinho, A. (1986), Lymphocyte population kinetics in the mouse. Immunol. Rev. 91, 5.

Gorczynski R. M. & Robillard M. (1986) Anomalous prolonged allograft survival avter deliberate immunization against graft specific alloantigens, in "Paradoxes in Immunology" ed. by Hofffmann G. W., Levy

J. G. & Nepom G. T. CRC Press, Inc. Boca Raton, Florida.

Hart D. A., Wang A., Pawlak L. L. & Nisonoff A. (1972) Suppression of idiotypic specificities in adult mice by administration of anti-idiotitypic antibody. J. Exp. Med. **135**, 1293.

Hiernaux, J., Bona, C. & Baker, P. J. (1981) Neonatal treatment with low doses of anti-idiotypic antibody leads to the expression of a silent clone. J. Exp. Med **142**, 106.

Hoffman G. W. (1980) On Network theory and H-2 restrictions (1980) Contemp. Top. Immunobiol. **11**, 185

Holberg D., Freitas A., Portnoi D., Jacquemart F., Avrameas S. & Coutinho A. (1986), Antibody repertoires of Normal BALB/c mice: B lymphocyte populations defined by state of activation. Immun. Rev. **93**, 147.

Hood L. E., Weissman I. L., Wood W. B. & Wilson J. H. (1984), Immunology. Benjamin, Menlo Park.

Hopfield J. J. (1982), Proc. Nat. Acad. Sci. (USA) **79**, 2554.

Jerne N. K. (1984), Idyotypic networks and other preconceived ideas. Immunol. Rev. **79**, 5.

Jerne N. K. (1974), Towards a network theory of the immune system. Ann. Immunol. (Inst. Pasteur) **125C**, 1127.

Lanzavecchia A. (1987) Antigen uptake and accumulation in antigen specific B cells. Immunol. Rev. **99**, 39.

Mezard M. & Parisi G. (1988) (in preparation)

Mezard M., Parisi G. & Virasoro M. (1987) Spin glass theory and beyond. World Scientific, Singapore.

Monroe J. G., Lowy A., Granstein R. D. & Greene M. I. (1984) Studies of Immune Responsiveness and unresponsiveness to the p-azobenzenearsonate (ABA) hapten, Immunol. Rev. **80**, 103.

Oudin J & Michel M. (1963) Une nouvelle forme d'allotypie des immunoglobulines du serum de lapin. C. R. Seances Acad. Sci. **257**, 805.

Rajesky K. & Takemori T. (1983) Genetics, expression and functions of idiotypes. Ann. Rev. Immunol **1**, 569.

Rajewsky K. (1983) Symmetry and asymmetry in idiotypic interactions. Ann. Immunol. (Inst. Pasteur) **134D**, 133.

Rubistein L. J., Goldberg B., Hiernaux J., Stein K. E. & Bona C. A. (1983) Idiotype anti-idiotype regulations V. The requirement for immunization with antigen or monoclonal anti-idiotypic antibodies for the activation of $\beta 2 \rightarrow 6$ and $\beta 2 \rightarrow 1$ polyfructosan-reactive clones in BALB/c mice treated at birth with minute amounts of anti-A-48 idiotype antibodies. J. Exp. Med. **158**, 1129.

Sack D. L., Easer K. M. & Shen A. (1982) Immunization of mice against African trypsonomiasis using anti-idiotypic antibodies. J. Exp.

Med. 155, 1108.

Slaoui M., Urbain-Vansanten G., Demeur C., Leo. O. Marvel J., Moser M., Tassignon J. Green M. I. & Urbain J. (1986) Idiotypic games within the immune network. Immunol. Rev. 90, 73.

Urbain J., Wikler M., Franssen J-d. & Collignon C. (1977) Idiotypic regulation of the immune system by the induction of antibodies against antiidiotypic antibodies. Proc. Natl. Acad. Sci. USA 74, 5126.

ON SLOW RHYTHMICAL FLUCTUATION OF NEURONAL EXCITABILITY IN CORTICAL BRAIN SLICES/AN EXPERIMENTAL STUDY

Igal Madar

  Medical biophysics Department. Hadassah medical organization and the Hebrew University - Jerusalem

ABSTRACT

In the present study the activity of single cells and cortical neurons populations in slices of rat's neocortex activated by short electrical pulses was monitored simultaneously. Cyclical fluctuations in the excitability were observed, having a cycle length from 10 to 100 seconds. Stimulation frequency and intensity did not affect total cycle length although changing the relative quiet/active phase ratio. Cycle length on the other hand, was affected by temperature.

1. INTRODUCTION

Oscillatory mechanisms constitute an essential component in fundamental processes such as digestion, respiration, circulation, homeostasis, and motor functions. A variety of neuronal components are capable of generating physiological rhythmic activity of various kinds. Some commonly examples are the cardiac pacemaker, periodicity in the smooth muscle, the bursting pacemaker in mollusc and some types of EEG waves. Neuronal oscillators cover a broad range of frequencies, from a fraction of a second up to several hours. In mammalian CNS rhytmic activity in the second-to-minute range has been largely neglected. The present study documents cyclical fluctuations in the excitability of cortical neurons having a cycle lenght of a few up to several tens of seconds.

2. RESULTS

Experiments were conducted on slices (0.3 - 0.4 mm in thickness) of the rat's neocortex. To keep the vitality of the slice preparation throughout the experiment which lasted 10 to 15 hr., the tissue was kept under a constant flow of a physiological solution (Ph. 7.3) saturated wiht a gas mixture of CO_2 5%/O_2 95%, the atmosphere surrounding the slice was saturated with a moist CO_2 5%/O_2 95% gas mixture and the slice temperature was balanced around 35°C. Under these conditions the cortical cells displayed a normal membrane resting potential and overshoot. The population of cortical neurons was activated by short (0.2 s) electrical pulses (1-10V) delivered through glass coated tungsten electrodes (50 m-tip diameter) placed in the terminating zone of the apical dendrites of the pyramidal cells-layer 1- and in the afferent/efferent axonal zone - the white matter. The electrically - evoked activity of single cells and population of cells was monitored simultaneously through two NaCl 3M-filled glass pipettes placed in two different layers.

Electrical activation evoked two major types of response: (1) A transient response composed of solitary spikes of a short and constant latency, frequently devoid of underlying EPSP's; (2) A sustained response composed of an accumulation of EPSP's that evoked a burst (8 to 15-20) of spikes with a relatively long and variable latency. The sustained response amplitude was either of all-or-nothing type or was proportional to the intensity of electrical stimulation. The transient response was concurrent with a local, rapid and monophasic field potential containing a late depolarizing wave lasting up to 100-300 msec.. Some cells displayed both types of activities, namely, a short latency unitary spike followed by a late burst of spikes.

The two response types were distinguished by their sensitivity to the electrical stimulation frequency. The intensity of the transient response was not affected by the stimulation frequency. Stimulation at rates of 0.1 up to 100Hz evoked a stable response. Whereas, the amplitude of the sustained response was stimulation rate-dependent. Electrical stimulation at rates lower than 0.1 to 0.3Hz evoked a consistent sustained response of a stable amplitude. At stimulation rates higher then 1 to 10Hz the response amplitude decreased and vanished rapidly, whereas at stimulation rates of 0.3 to 1Hz the response amplitude spontaneously rose and fell rhythmically. In cells which displayed both a transient response followed by a sustained one, only the sustained component was fluctuating while the transient one kept its stability. A comparable range of stimulation frequencies in which oscillations were induced was measured in different cells and in different cortical regions.

The oscillatory behavior was either developed spontaneously following 5 to 30 minutes of repetitive stimuli given at cyclicity-inducing frequency, or after a preceding application of high rate (10-50Hz) stimulation given for several minutes, during which the response was vanished. Returning to the proper stimulation rate the response gradually recovered and oscillation started.

The oscillatory behavior was observed in the activity of both single and populations of cells and following stimulation of layer 1 or the white matter (see Fig. 1). During the recording of the activity of single cells the decay phase could be abrupt or gradual, with a response evoked by each second and then third stimulus until the discharge frequency gradually diminished and completly disappeared. During the recording of field potentials, the decay phase was expressed by a reduction of amplitude of the late depolarizatory wave. The early period of the quiet phase was characterized by refractoriness. Activation of the other pathway, layer 1 or the white matter, using maximal electrical currents, failed to evoke any response during this period. Intracellular measurement of the membrane potential revealed an hyperpolarization of 5MV which lasted throughout the quiet phase. At the end of the quiet phase the response spontaneously resumed, the discharge frequency gradually increased and response latency decreased. In some cells during the peak of response a second burst of spikes appeared accompanied with the development of a second depolarizatory wave.

The duration of the cycle length of the rhythmic behavior in different cortical regions, measured from the beginning of the active phase up to the end of the quiet phase, ranged from 10 to 100sec. The cycle duration was very stable throughout the experiment and was only slightly affected by changes of stimulation frequency and intensity.

An increase of stimulation frequency within the cyclicity-inducing range produced a proportional decrease of the active phase duration, but hardly affected the cycle lenght (see Fig. 2). An increase of stimulation intensity up to 4-folds of threshold intensity produced a proportional increase of the active phase duration and an increase of the oscillation amplitude, but only slightly affected the cycle duration (see Fig. 3). The cycle length was substantially affected by the temperature of the slice tissue. A gradual reduction of temperature from 34°C to room temperature led to a relatively small decrease of the active phase duration and to an exponential increase of the quiet phase duration (see Fig. 4).

The oscillatory behavior appeared concurrently throughout the cortical layers. Simultaneous recording from two or more distinct sites along the radial axis, revealed that the oscillation in activity of units located in different layers was invariably phase-locked (see Fig. 1). Similar temporal coincidence was found between oscillations induced by stimulation of layer 1 and the white matter (see Fig. 1), which correspond the existence od the refractory period mentioned above.

3. DISCUSSION

In the mammalian brain, oscillatory phenomena are largely related to pathological events such as aphasia, aphaxia, stress, epilepsy or are chemically-induced. The present demonstration of periodic oscillations in evoked responses is not an isolated phenomenon in the nervous system. Thus periodic changes have been demonstrated in cortical evoked response (CHN, ALD, NOR), in Ach content in the electric organ of Torpedo (DUN) and in transmitter release at the frog nueromuscular junction (MEI).

Several findings suggest that the oscillatin activity observed in the present preparate reflects a physiological rather than pathological phenomenon. The above mentioned pathological states are charachterized by the oscillations are overwhelming the activity of the entire population of cells, whereas in the present case oscillations characterized only bursting cells and were not found in cells which were recorded in the same cortical column but dispaly a transient response. Moreover, rhythmic behavior was observed in bursting cells having physiological membrane properties - resting potential of 40 to 60mV and action potential amplitude of 70 to 90mV, which is not the case, for example, in epileptic neurons.

Recordings in the intact brain show that phasic and tonic slow waves or DC-potentials of a period from a second up to more then a minute comprise a small fraction of the total cortical activity and they are hard to be detected (GIR). The question is raised whether the rhythmic fluctuation observed in the present study reflects a physiological mechanism which underlies the slow waves that comprised the spectral extreme of the EEG in the intact brain, and if so, why is this activity is so dominant in the slice preparation.

When the functional state of local neuronal circuits is considered, the difference between isolated cortical slices and the intact brain lies in the afferent inputs they received rather than the membrane properties of individual neurons. Isolation of most of the afferent pathways in the slice preparation may lead to an increase of synchronization of local activities and, consequently, will enhance the rhythmic activity fraction. In the intact cortex, invasion of incoming inputs from different sources, which in turn are affected by variably oriented inputs, lead to an increase of desynchronization in their target areas. In fact, a clear sign, during EEG recording, for active processing of sensory information is the differentiation and diversification of cellular firing, as expressed by desynchronization of EEG waves. Whereas a reduced flow of external inputs, as occurred during sleep, is accompanied with an increase of synchronization in cortical activity which reflected by higher amplitude and longer duration EEG waves. When two important parameters of the oscillatory behavior, the phase and period, are considered, a temporal coincidence was found between activity of units located in discrete layers, between activity of a single cell and a population of cells, and between activities evoked by stimulation of layer 1 and stimulation of the white matter. Moreover, the reduced sensitivity of the postsynaptic cell to one set of inputs (activated by stimulation of layer 1 or the white matter) was generalized to other sets of inputs (activated by the other stimulus). These parametric characteristics suggest that synchronized activity of cortical cell populations coexists with the rhythmic modulations and may even be essential for generating them.

References

- Aladjalova, N.A. Slow electrical processes in the brain. Progress in Brain Res. 7:1-240. (1964).
- Chang, H.T. Cortical response to stimulation of lateral geniculate body and the potential thereof by continuous illumination of retina. J. Physiol (Lond.) 162:45p-46p. (1962).
- Dunant, Y., Jirounek, P., Israel, M., Lesbats, B. and Manaranche, R. Sustained oscillations of acetylcholine during nerve stimulation. Nature 252:485-6. (1974).
- Girton, D.G., Benson, K.L. and Kamiya, J. Observation of very slow potential oscillations in human scalp recordings. EEG 35:561-68. (1973).
- Meiri, A. and Rahamimioff, R. Clumping and oscillations in evoked transmitter release at the frog neuromuscular junction. J. Physiol. 278:513-23. (1978).
- Norton, S. an,d Jewett, R.E.. Frequencies of slow potential oscillations in the cortex of the cat. EEG. 19:337-86. (1965).

FIGURE LEGENDS

Figure 1.

Simultaneous recording of activity of population of cells in Layer-3 (A) and activity of single cell in Layer-5 (B) which evoked by electrical stimulation (4V, 50usec-pulse width, 0.5Hz-stimulation rate).

A. Upper trace: Oscilloscopic photographing of single field potentials which evoked by stimulation of Layer-1 and taken at different fractions of the cycle.

Lower trace: The amplitude of consecutive evoked field potentials during a period of 3.5 cycles.

B. Upper trace: Oscilloscopic photographing of a sustained response of single cell which evoked by an alternate stimulation of Layer-1 (upper row) and of the white matter (lower row), taken at different fractions of the cycle.

Lower trace: Response histogram of consecutive sustained responses during 3.5 cycles. The response which diminished to a greater extent was elicited by stimulation of the white matter. The other response was elicited by Layer-1 stimulation.

Figure 2.

The effect of stimulation rate on the cycle duration.

Each point represents an average of 20 measurements of an oscillating field potential which recorded in Layer-5 and evoked by electrical stimulation (6V, 50usec) of the white matter. After transition to the next stimulation frequency the response allowed to stabilize for 10 to 15 min and then the measurements were conducted.

Increasing of stimulation frequency (from 1.66 to 0.35 Hz) yielded a gradual reduction of the active phase duration (from 60 to 30 sec) and an increase of the quiet phase duration (from 20 to 55 sec) but, hardly affected the overall cycle lenght wich modified from 80 to 85 sec.

Figure 3.

The effect of stimulation intensity on the cycle duration.

Histograms were compiled from sustained responses of a single cell located in Layer-4 and activated by electrical stimulation (20 usec, 1 Hz) of Layer-1, given at different intensities. The row of the small bars in each histogram represents the stimulus artifacts.

Gradual increase of the stimulation intensity (from 3 to 9V) leads to a substantial increase of the response amplitude and prolongation of the active phase duration but, only slightly affects the overall cycle duration. Following the quiet phase the response resumed with strong bursts (higher density of histogram columns in the early part of the active phase). Further increase of the stimulation intensity (from 9 to 12V) abolished almost completely the cyclic behavior, probably due to prolongation of the active phase duration beyond the cycle length. In the middle of histogram (a), however, a short quiet phase replaced by an active one can be seen.

Figure 4.

The effect of temperature on the cycle duration.

Traces represent measurements of a sustained response of a cell in Layer-5 which evoked by electrical stimulation (3.6V 40usec, 0.5Hz) of Layer-1. The temperature of the slice was evaluated by a miniature probe placed in the perfusing medium close to the slice tissue. The temperature was gradually lowered from 34°C to room temperature. At each measurement point the slice temperature was allowed to stabilize for 20 to 30 min and then measurements of 5 cycles were taken.

A decrease of the temperature produced a relatively small decrease of the active phase duration (from 28 to 10 sec) but, produced an exponential increase of the quiet phase duration (from 50 to above 280 sec.).

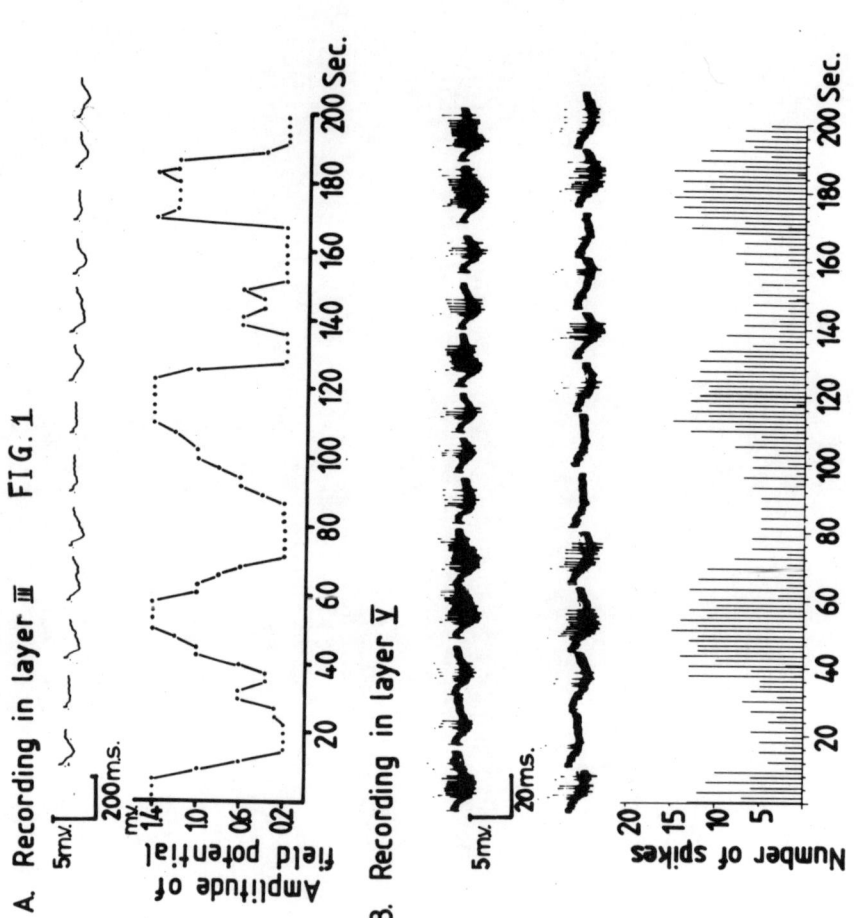

FIG. 1

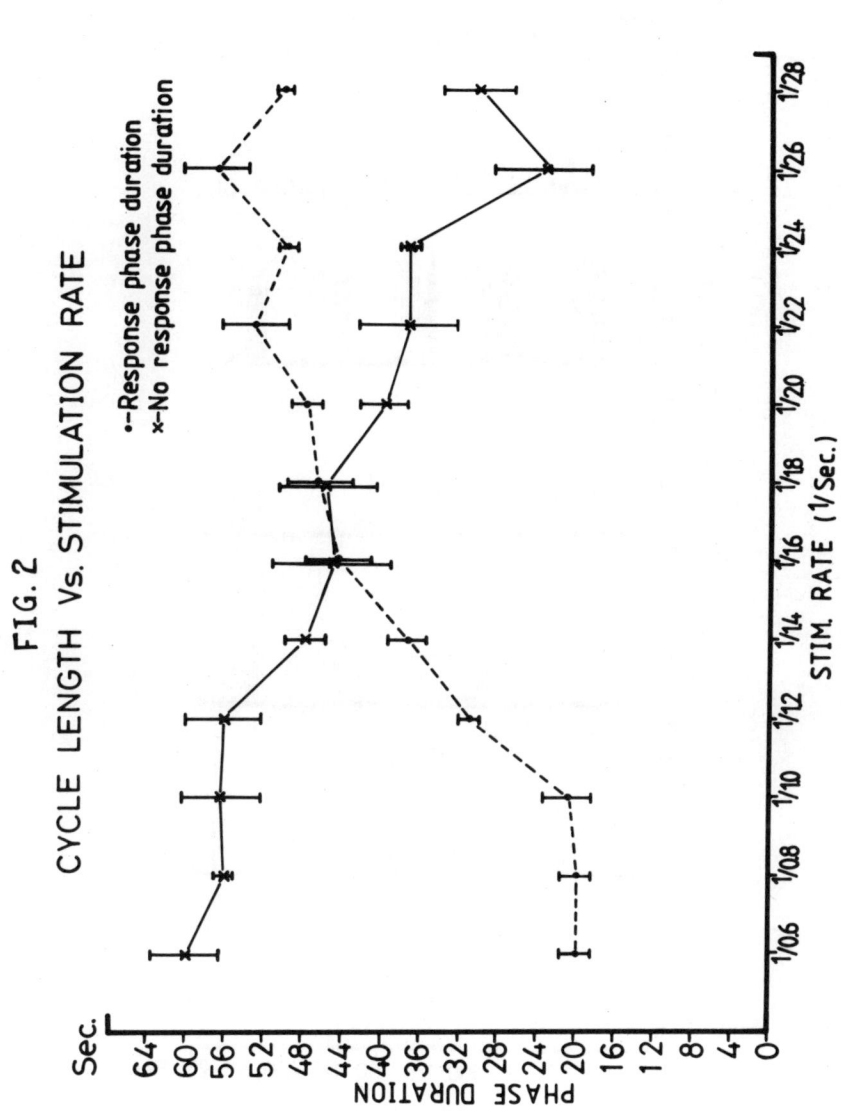

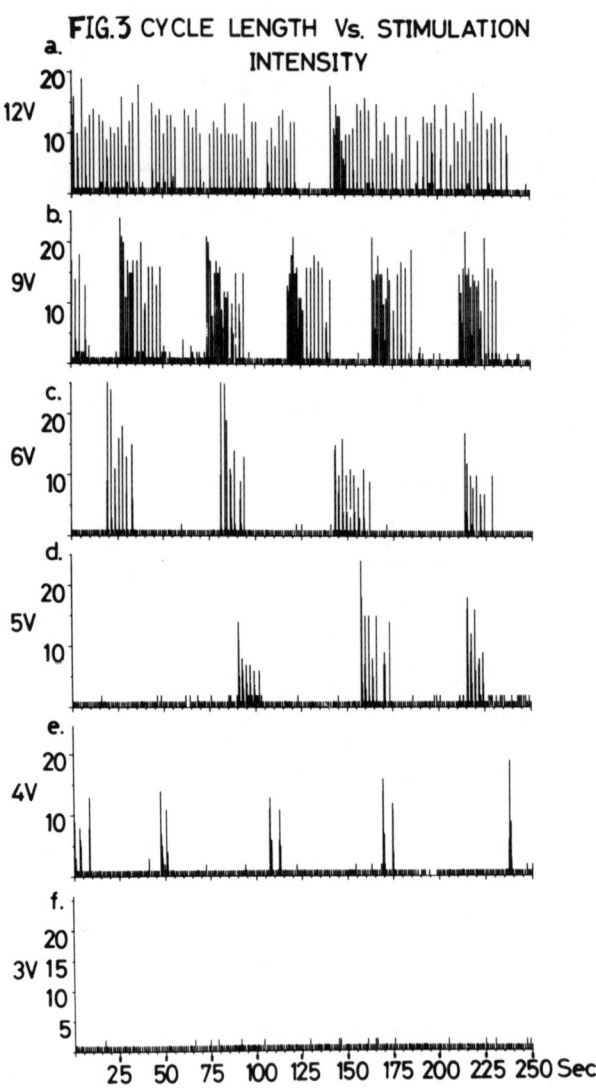

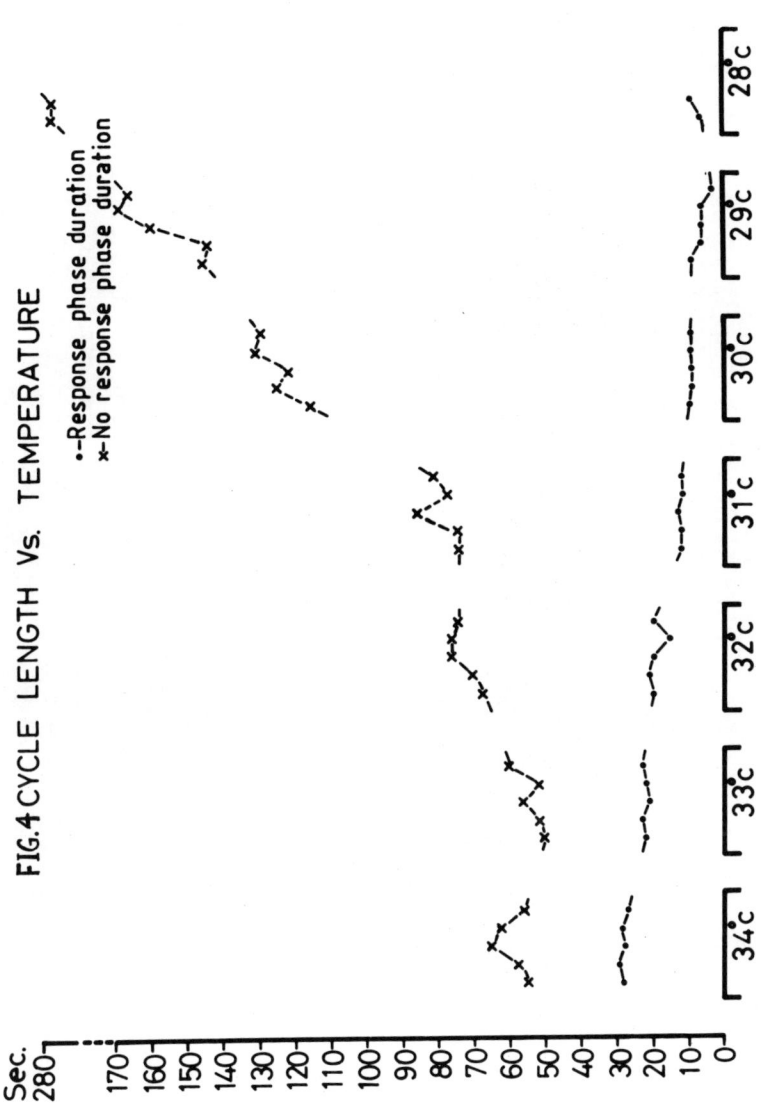

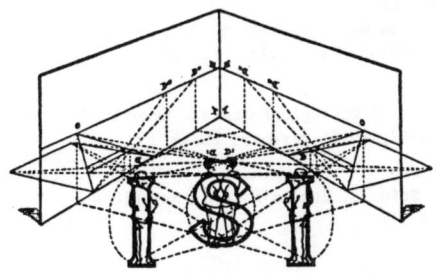

INSTITUTE FOR SCIENTIFIC INTERCHANGE

Workshop on
"CHAOS AND COMPLEXITY"
Torino, Villa Gualino, October 5-10, 1987

LIST OF PARTICIPANTS

C. ALABISO
Dipartimento di Fisica, Università di Parma
Via M. D'Azeglio 85, 43100 Parma, Italy

T. ARECCHI
Istituto Nazionale di Ottica, Università di Firenze
Largo E. Fermi 6, 50125 Arcetri, Firenze, Italy

R. BADII
Institut für Theoretische Physik
Schönberggasse 9, 8001 Zürich, Switzerland

F. BAGNOLI
Dipartimento di Fisica, Università di Firenze
Largo E. Fermi 2, 50125 Firenze, Italy

N. BELLOMO
Dipartimento di Fisica, Politecnico di Torino
Corso Duca degli Abruzzi 24, 10129 Torino, Italy

G. BENETTIN
Dipartimento di Fisica, Università di Padova
Via Marzolo 8, 35131 Padova, Italy

P. BERGE'
CEN/SACLAY
91191 Gif Sur Yvette, France

J. BERNASCONI
BBC Brown, Boveri & Company
CH-5405 Baden-Dättwil, Switzerland

E. BIENENSTOCK
Laboratoire de Neurobiologie du Development, Bât. 440, Université de
Paris-Sud, 91405 Orsay Cedex, France

M. BUIATTI
Dipartimento di Biologia Animale e Genetica, Università di Firenze
Via Romana 17, 50125 Firenze, Italy

S. CARACCIOLO
Classe di Scienze, Scuola Normale Superiore
56100 Pisa, Italy

P. CARNEVALI
IBM Italia, ECSEC
Via Giorgione 159, 00147 Roma, Italy

M. CASARTELLI
See C. Alabiso

B. CHOPARD
Departement de Physique Théorique, Université de Genève
24, quai Ernest-Ansermet, CH-1211 Genève, Switzerland

S. CILIBERTO
See T. Arecchi

G. COSENZA
Dipartimento di Fisica, Università di Napoli
Mostra d'Oltremare Pad. 19, 80125 Napoli, Italy

P. COULLET
Laboratoire de Physique Theorique, Université de Nice
Parc Valrose, Nice Cedex 06034, France

M. DROZ
See B. Chopard

M. DUBOIS
See P. Bergé

F. FOGELMAN-SOULIE
L.D.R./CISTA
1, Rue Descartes, 75005 Paris Cedex, France

A. FRANCESCATO
See F. Bagnoli

L. FRONZONI
Dipartimento di Fisica, Università di Pisa
Piazza Torricelli 2, 56100 Pisa, Italy

F.G. GASCON
Facultad de Ciencias Fisicas, Dpt. Metodos Matematicas
Universidad Complutense, Madrid 28040, Spain

A. GEORGES
Centre National de la Recherche Scientifique, Laboratoire de Physique
Theorique de l'Ecole Normale Supérieure, Université de Paris-Sud
24, Rue Lhomond, 75231 Paris Cedex 05, France

R. GIACHETTI
Istituto Nazionale di Fisica Nucleare, Sezione di Firenze
Largo E. Fermi 2, 50125 Firenze, italy

J.P. GOLLUB
Haverford College, Department of Physics
Haverford, PA 19041-1392, U.S.A.

P. GRASSBERGER
Bergische Universität Gesamthochschule Wuppertal
Gauss-Strasse 20, 5600 Wuppertal, F.R.G.

S. ISOLA
See F. Bagnoli

K.E. KURTEN
Universität zu Köln, Institut für Theoretische Physik
Zülpicher Strasse 77, D-5000 Köln 41, F.R.G.

S. KAUFFMAN
University of Pennsylvania, School of Medicine, Department of Biochemistry
and biophysics, Philadelphia, PA 19104-6059 U.S.A.

P. LALLEMAND
Département de Physique de l'Ecole Normale Supérieure
24 Rue Lhomond, 75231 Paris Cedex, France

R. LIMA
Centre National de la Recherche Scientifique, Centre de Physique Théorique,
Section 1 LUMINY, Case 907, F-13288 Marseille Cedex, France

R. LIVI
See F. Bagnoli

Y. MADAR
Hadassah Medical Organization
PO Box 12 000, IL-91 120 Jerusalem, Israel

A. MARITAN
Dipartimento di Fisica "G. Galilei", Università di Padova
Via F. Marzolo 8, 35131 Padova, Italy

J.P. NADAL
Groupe de Physique des Solides de l'Ecole Normale Supérieure
24 Rue Lhomond, 75231 Paris Cedex, France

R. NOBILI
See G. Benettin

M. OPPER
Institut für Festkorperforschung, KFA
Julich, F.R.G.

G. PARISI
Dipartimento di Fisica, Università di Roma "Tor Vergata"
Via orazio Raimondo, 00173 Roma, Italy

S. PATARNELLO
See P. Carnevali

L. PELITI
See G. Cosenza

A. PELISSETTO
See S. Caracciolo

C. PEREZ-GARCIA
Departamento de Fisica, Università Autonoma de Barcelona
08193 Bellaterra (Barcelona), Spain

A. POLITI
See T. Arecchi

R. RECHTMAN
Departamento de Fisica, Facultad de Ciencias UNAM
04510 Mexico D.F.

M.A. RUBIO
Departamento de Fisica E.T.S., Arquitectura-UPM
Avenida Jan de Herrera, 28040 Madrid, Spain

J. RUBNER
Physik Department der Technischen Universität München, Theoretische Physik
James-Franck-Strasse, 8046 Garching, F.R.G.

S. RUFFO
See F. Bagnoli

R. SERRA
Tema
Viale Aldo Moro 38, 40127 BOlogna, Italy

D. STAUFFER
Institut für Theoretische Physik, Universität zu Köln
Zülpicher Strasse.77, 5000 Köln 41, F.R.G.

R.H. SWENDSEN
Department of Physics, Carnegie-Mellon University
Schenley Park, Pittsburgh, PA 15213, U.S.A.

G. VICHNIAC
Plasma fusion Center, M.I.T.
Cambridge, MA 02139, U.S.A. and
BBN Communications Corporation
50 Moulton Street, Cambridge, MA 02238, U.S.A.

A. VULPIANI
Dipartimento di Fisica, Università di Roma "La Sapienza"
Piazzale A. Moro 2, 00185 Roma, Italy

G. WEISBUCH
Laboratoire de Physique Theorique de l'Ecole Normale Superieure
24 Rue Lhomond, 75231 Paris Cedex 05, France

M. WOLF
Institute of Theoretical Physics, University of Wroclaw
ul. Cybulskiego 36, 50-205 Wroclaw, Poland